EXPLOITING BIOTECHNOLOGY

EXPLOITING BIOTECHNOLOGY

by

Vivian Moses
Division of Life Sciences, King's College,
University of London,
London, UK

and

Sheila Moses
College of North West London,
London, UK

harwood academic publishers
Australia • Austria • Belgium • China • France • Germany • India
Japan • Malaysia • Netherlands • Russia • Singapore • Switzerland
Thailand • United Kingdom • United States

Copyright © 1995 by Harwood Academic Publishers GmbH.

All rights reserved.

No part of this book may be reproduced or utilized in any form or by any means, electronic or mechanical, including photocopying and recording, or by any information storage or retrieval system, without permission in writing from the publisher. Printed in Singapore.

Harwood Academic Publishers
Poststrasse 22
7000 Chur
Switzerland

British Library Cataloguing in Publication Data

Moses, Vivian
 Exploiting Biotechnology
 I. Title II. Moses, Sheila
 338.476606

 ISBN 3-7186-5570-5 (hardcover)
 ISBN 3-7186-5571-3 (softcover)

Contents

Preface		vii

INTRODUCTION

Chapter 1	What biotechnology is all about	1

THE TECHNOLOGY BASE

Chapter 2	The science...	7
Chapter 3	...the engineering...	49
Chapter 4	...and something about feedstocks	59

THE ORGANIZATIONAL AND COMMERCIAL FRAMEWORK

Chapter 5	Lawyers, legislators, communicators, funders — and the public	65
Chapter 6	The job of a biotechnology manager	77

THE BIOTECHNOLOGY BUSINESSES

Chapter 7	Healthcare: prophylaxis, diagnosis and therapy	103
Chapter 8	Chemicals, enzymes, fuel and new materials	139
Chapter 9	What we eat: agriculture and food	163
Chapter 10	Wealth from the earth: metals and hydrocarbons	197
Chapter 11	Keeping the place clean	219
Chapter 12	Bioelectronics: a courtship between technologies	249
Chapter 13	The future: biotechnological bonanzas?	263

APPENDICES

Appendix 1	Help! Where do I go from here?	275
Appendix 2	Glossary of technical terms	285

INDEX

	307

Preface

Biotechnology is at once intensely technical and intensely commercial: looking at either aspect without the other might do for some applied sciences but would not make much sense for this important and growing industry.

The scientific basis of biotechnology powerfully excites the attention of biologists, chemists, engineers and indeed all manner of technically minded professional and lay people. Its commercial relevance inevitably attracts the curiosity not only of people who just wish to understand what is going on around them but also of those whose business it is to take stock of the world of industry, and to perceive patterns of development and the nature and sizes of present and future markets, in order to decide where investment should or should not be made. Understanding in any depth the impact that biotechnology will increasingly have on human affairs, or making commercial decisions about technical investments, must be more than a little difficult for anyone whose education and training does not include the science and engineering on which the technology is based.

It is for those readers, students of biotechnology as well as people without much direct experience of science, that the authors planned which topics to include, how to present them and the detail each one would merit. While mainly scientific and technical in its approach and in the material it covers, the book hopes to satisfy both scientific and business interests. It outlines the commercial and business contexts in which biotechnology contributes products and services to society; it also reviews the most relevant national and international regulatory and legal issues, and touches briefly upon some of the wider economic and ethical concerns. Primarily, however, the book seeks to explain to people with no special knowledge of chemistry, and with no more than a layman's appreciation of biology, both the promise and the limitations of using living organisms and their products in industry.

Mostly, but not entirely, taking that technical point of view, we look to future developments which might take place if the economic climate were right and the appropriate business opportunities perceived. Recognizing that people with scientific interests often have little understanding and limited experience of commercial activities, activities which are nevertheless critical for the success of biotechnology in the marketplace, we have tried to describe what commercialization in biotechnology is like, how it is managed, what has to be done and what avoided. But we do not attempt to predict, company by company or industry by industry, what is going to happen; commercial organizations are reticent about their future plans and the authors do not attempt to divine their intentions.

Rather than undertaking in-depth analyses of specific sectors — matters for the relevant industrial and commercial experts — the purpose of this book is to make accessible a general understanding of the technical base on which biotechnology rests. It also offers a broad view of the commercial and industrial applications which have already been made or are likely to be developed before too long. Having read it, those who do want to understand the significance of future biotechnological advances as they are reported, or who have to make decisions involving biotechnology, may do so with as great an understanding of the technology as a busy person can reasonably expect to acquire.

Preface

Inevitably, there are omissions. In no sense is this book a comprehensive catalogue of all that is going on. To satisfy that objective would have required a volume many times longer and demanded of its readers an extensive knowledge of the underlying sciences. Moreover, the subject is moving so fast that, however hard the authors might have tried to include every last bit of news, publication delays would inevitably have led to the omission of many interesting items. We have tried instead to offer a balanced presentation of current activity and where it might lead in the short- and medium-term. Long-term guessing is necessarily very risky and only rarely have we dared to do it. Those guesses should be taken in the spirit in which they were written: to provoke but by no means to predict.

1 What biotechnology is all about

The Word "Biotechnology"

Two or three decades ago the word would have meant little to biological scientists — something about biology, obviously, but what exactly? Had it been uttered at all it would have conveyed a sense of doing things, presumably in a useful way, perhaps involving machinery. But what connection had biology with machines? By the mid-1970s, however, *biotechnology* as the name for a new industry was clearly gaining popularity and in many people's minds was associated with some of the spectacular advances in what was then the new subject of *molecular genetics*.

Genetics is the study of inheritance. It has been used empirically throughout human history for breeding better strains of domesticated animals and crop plants. People knew how to achieve beneficial results but why their efforts were successful did not begin to become clear until the late 19th century and real understanding in terms of chemistry had to wait a further hundred years. About two decades ago geneticists discovered a series of biochemical activities, universal and normal in living systems, which showed them how to manipulate genetics to produce totally new effects; these permitted industrially convenient microorganisms to be used as living factories to make products until then available only from animals or people. That was the birth of what became known as *genetic engineering*.

Those discoveries had startling effects both on the scientific world and on sections of the industrial, commercial and financial communities. Excitement mounted as speculation grew about the possible impact of the new genetics. There were clearly going to be ethical considerations and effects on public policy. Parallels were drawn with the growth of computing, another area of novel development profoundly affecting human activity. And thus appeared that new word which at times, and for some people, became virtually synonymous with genetic engineering.

What Biotechnology Is Really About

Biotechnology is not just genetics. It is a technology, a set of techniques for doing practical things, all of them with implications for the commercial and/or the public sectors: making products and providing services which can be sold for a price in the marketplace, or paid for from the public purse. In that way it differs fundamentally from the underlying sciences which are neutral with respect to industry and commerce. The biological sciences are dedicated to the exploration and comprehension of natural phenomena; biotechnology is about using those

Exploiting biotechnology

sciences as a basis for industry and commerce. Although some definitions are rather elaborate and perhaps a bit woolly,* biotechnology is fundamentally about making money with biology.

Logically it must include all the commercial ways of using biology, the old and traditional as well as the new, agriculture and brewing as well as manipulating genes to produce new drugs or designing biological procedures for mining minerals from the earth. While some people do take that broad view, for others biotechnology retains a strong genetic engineering connotation. It really does not much matter: what is important is an appreciation of the commercial opportunities which have developed and are developing in the current invigorating climate of biological innovation.

Recognizing its promise for the existing activities in agriculture, pharmaceutics, sewage treatment and fermentation, the broad view takes modern biology into areas where it has never had more than a tenuous presence: waste management, electronics, new materials, and the recovery and processing of minerals are among the more prominent. Even civil engineering is beginning to throw up opportunities for a biotech. dimension. Biotechnology, to the extent that it has something valuable to offer these activities, is becoming an integral part of modern practice and it might be more realistic in some of those contexts to use the term *biological engineering* rather than biotechnology — some people are beginning to do just that.

BIOTECHNOLOGY AND SCIENCE

One of the more remarkable things about biotechnology is that fundamental scientific advances proceed hand-in-hand with industrial development and commercial exploitation. Change has been and continues to be very rapid. The immediate scientific discoveries which led to genetic engineering were made in the late 1960s and early 1970s; the article which first brought the new possibilities clearly into public focus was published in 1975 and frontier research continues to be an essential part of technological advance.

By the end of the 1970s a number of commercial companies were already working directly to generate saleable products based on the new knowledge. Genentech and Cetus, among the first of them, became public corporations in 1980 and 1981 with major and highly successful public stock offerings. The laboratories of these and similar new companies were almost indistinguishable from the best in the universities except perhaps for the greater lavishness of their capitalization. And the scientists, too, were indistinguishable, hardly surprising as most had joined directly from university laboratories and many who remained professors accepted part-time consultancies with the companies.

Now, 20 years on, biotechnology has acquired maturity without losing zest and excitement. The earlier dedicated companies have long since recognized that they cannot do everything and have identified the areas in which they will concentrate their main efforts. While some of the original companies have merged with or been taken over by others, much of the pharmaceutical

---

* This one, for example, is from a document published by the UK Department of Trade and Industry: *Biotechnology, defined as the application of advancing understanding of living organisms and their components to create new and improved industrial products and processes.....*

What biotechnology is all about

and some of the chemical industry, once a little slow in perceiving what the advent of the new mood in the biological sciences might mean in industrial terms, made up for lost time and established their own biotechnological activities. Gradually, but gathering strength with each successive year, came the more obviously engineering-related developments: in mining, enhanced oil recovery, electronics and elsewhere. The incubation period to the market place has generally proved to be longer and the capital investment greater than many expected. But new products and services are now being sold profitably and many who keep a close watch on developments predict a rapidly rising revenue curve for the decades to come.

PROSPECTS, SHORT- AND LONGER-TERM

A number of near-and longer-term biotechnological products and services can tentatively be identified with different degrees of clarity in the various industries. Science and technology permeate these categories, each of which primarily represents a set of marketing opportunities. Most of them will be discussed in greater detail in succeeding chapters and are listed here as a convenient summary.

Healthcare

Diagnostics and therapeutics are convenient to group together although their markets are rather different:

- Vaccines, including ones effective against AIDS and malaria;
- Anti-cancer drugs;
- Antibiotics;
- Gene therapy to treat inherited disorders;
- New diagnostic methods for improved recognition of specific disease;
- Genetic fingerprinting for medical and other purposes.

Industrial Products

Including new materials, they number:

- Polymers;
- Surfactants;
- Steroids;
- Intermediates in the production of chemicals;
- Bioplastics;

Exploiting biotechnology

- Alcohol from sugar;
- Bone and tissue replacement materials.

Agriculture and Food

In an enormous area of activity with profound economic, social and political implications, important developments are likely to include:

- Genetically engineered plants via tissue culture and clonal propagation;
- Better resistance of crop plants to herbicides, insecticides and disease;
- On the other hand: better, especially "natural", pesticides;
- New plant strains for better quality, storage properties and processing;
- Reducing and avoiding the need for fertilizers;
- Animal healthcare;
- Use of hormones to give higher product yields;
- Genetically improved animals;
- "Biopharming" — using plants and animals to make foreign proteins;
- Novel foods and food ingredients, including single-cell protein;
- Modification and upgrading of existing raw materials;
- High-intensity sweeteners;
- New and improved flavours and fragrances.

Waste Management and Pollution Control

A cleaner environment is now everybody's dream. Biotechnology can help with:

- Oil spill clean-up;
- Removal of toxic metals, noxious organic materials and radioactive elements from discharge effluents;
- Landfill management and methane emission control;
- Bioremediation — the cleanup of contaminated sites;
- Improvement of natural waters.

Mineral Resource Recovery and Processing

As the best mineral ores and oil reservoirs become progressively exhausted, new methods will be needed to work both the remaining residues and the poorer deposits which will gradually become the main sources. Biotechnology is already involved with:

- Microbial mining: the leaching of metal from *in situ* ore bodies;
- Metal leaching from low grade ore dumps;
- Desulphurization of coal;
- Crude oil production: microbial methods for the enhancement of oil recovery.

Electronics

Interactions between biotechnology and electronics are already established. Some products are now on the market and others must be expected to follow in due course:

- Biosensors for detecting and measuring specific chemicals in medical diagnostics and manufacturing processes;
- "Biochips" — a new concept of biotechnology-based computers.

These are the real and prospective end products based on the discoveries of the basic sciences coupled to the practices of business, management and finance. Together they make up the present reality of biotechnology. The future is naturally more difficult to define but in a final chapter we will try our best to look ahead to see if it is possible to discern the shape of things to come.

Supplies for Biotechnology

The wide range of resources, many of them commercially available, needed for research, development, manufacture and the marketing of biotechnological services represents another major category well worth a mention. Because of their technical complexity and diversity, they are difficult in this general book to discuss in detail and are therefore presented in outline only. Biotechnological supplies include:

- Equipment for laboratory investigations, product testing and manufacturing;
- Chemicals, biochemicals, radioactive compounds and other consumable materials for the same purposes;
- Analyses, custom-syntheses and other specialist services;
- National and international collections of microorganisms. Payment of a small administrative fee will usually make a microbial culture available for academic research; their

Exploiting biotechnology

use for industrial (profit-making) purposes is likely to involve a more extensive payment, perhaps via an up-front or royalty fee, or both;

- Information sources. Scientific and technical information is available in national, university and professional society libraries. Access may be entirely open, free to members of the relevant society or permitted for a small fee. Photocopies of individual articles can be obtained from central sources, again for modest fees. Many on-line data bases may be accessed on a commercial basis; searching can be expensive if a wide trawl is undertaken.

The supplies sector is currently the most mature in biotechnology and certainly the most profitable: more than half the companies involved report profits. In therapeutics, by contrast, no more than a quarter of the existing companies have sufficient resources to survive for three years while, with their present level of business activity and profitability, 40% will not be able to last more than a year without additional capital.

2 The science...

Biotechnology is essentially about commercializing biology; every aspect of this very broad subject may sooner or later provide opportunities for business development and as biotechnology grows so does the range of its biological involvement. The word "biology", however, conveys a sense of uniformity and cohesion that active biologists proclaim philosophically but do not necessarily practise. Biology is just too big. Individual practitioners usually affiliate to one or another of the biological sciences, of which there are many: a botanist, a biochemist or a zoologist are all biologists but all specialists.

The branches of biological science most intimately related to biotechnology at the present time are:

- *biochemistry*, which addresses the chemical structure and behaviour of all types of living beings; hence it is directly concerned with employing biological processes to provide useful products and services;
- *genetics*, the study of inheritance and the relationships between individuals and populations; in recent times it has become increasingly concerned both practically and theoretically with the biochemical mechanisms of inheritance and development, and the ways they can be used to do and make new things in biotechnology;
- *microbiology*, a field closely integrated with both biochemistry and genetics, which explores and manipulates microbes of all sorts, the main category of living organisms used industrially.

Each of them warrants some discussion here.

BIOCHEMISTRY

In a sense, biology is a special sort of chemistry and, as biotechnology is a way of using biology, it too is based on chemistry and things that change or are changed in a chemical way. Biotechnology uses biological mechanisms to generate products and services for sale. The products are chemical ones and the services chemical services. For the non-specialist, a short general introduction to chemistry might at this point be useful because people who have never studied it will have little familiarity with what chemistry is about or why it is so helpful for understanding biotechnology.

The Importance of Chemistry

All materials and substances constitute the province of chemistry, a subject concerned with the way that *atoms* are joined together to form *molecules*. A material made up entirely of one

Exploiting biotechnology

kind of atom is called an *element*. There are 105 of them and some, like iron, aluminium, chromium, nickel, platinum, carbon, chlorine, iodine, copper, lead, gold, silver, mercury, tin, zinc and tungsten are often encountered in a fairly pure form in everyday life. Others, including sodium, potassium, calcium, oxygen, nitrogen and neon, sound familiar enough but are rarely or never encountered in the pure elemental state outside the laboratory or factory. That leaves dozens more: phosphorus, silicon, uranium and plutonium are well-known words although the pure elements are beyond common experience, while obscure and exotic elements such as praseodymium, ytterbium and gadolinium abound.

Certain elements can be mixed together in any proportion: metallic alloys are mixtures like that. Brass, made of various proportions of copper and zinc, is a widely used material and its actual composition can readily be varied in order to fine-tune the properties of the alloy. *Compounds*, by contrast, are not mixtures whose composition can be altered at will but are collections of molecules: precisely defined groupings of certain atoms in specific proportions, joined together in a unique and invariant manner. Many common materials like wood, cloth, foods, air, petrol and most rocks are actually mixtures of chemical compounds which can more or less readily be separated from one another. Water, common salt, refined sugar, washing soda, chalk, glycerine, acetone and alcohol virtually exhaust the range of pure, or almost pure, compounds familiar to most people. Many elements which are rarely found in the pure state as elements are nevertheless major constituents of commonplace compounds. Salt is an example; it is composed of the two elements sodium and chlorine, neither of which occurs in nature as the free element although chlorine manufactured industrially is used to purify water and many people will recognize its odour.

The number and variety of atoms in each molecule of a compound, and the way they are joined together, are all critically important in determining the properties of that compound. For example, water and salt are made of entirely different atoms and are obviously very different materials. Sugar and alcohol are also very different from one another, yet they are made of the same atoms but in different proportions and joined together in different ways. Chemistry is about understanding the behaviour of the huge number of known chemical substances and exploring how they interact with one another, inventing and discovering new ones and manipulating them to make useful products.

Chemical Change

Apart from the elements, everything around us, including ourselves and all other living things, is made up of chemical compounds, many of them in a fairly continuous state of change. A good part of industrial activity is dedicated to channelling those changes along pathways useful to us as producers and consumers. We refine oil, smelt metal ores, treat sewage, ferment sugar to alcohol, make cheese from milk and convert a whole host of feedstocks into antibiotics, flavourings, industrial chemicals and other useful materials. All that work needs energy; indeed, the classic definition of energy is *a capacity for doing work*.

Energy is required for these chemical activities because the final products are more complicated and often purer when they are sold, than their raw materials. Effort is needed to put complicated chemical structures together, just as it is for building bridges or making cars. Energy is also needed for purification, that is, separating a product from the feedstock and any by-products, just as it is needed in the extraction of a metal from its mineral ore or separating the wheat from the chaff. High temperature and pressure, electric discharges, radiation and other sources provide energy for chemical processes in industry: none of these is directly useful for living systems. They have to get their energy in other ways.

Fuel, Energy and Feedstocks

Much of the energy used by human societies comes from burning fuel in boilers, motors and engines, either on site or at electricity generating stations. In chemical terms, burning a fuel usually means *oxidizing* the chemicals of which it is composed, combining it with oxygen from the atmosphere and liberating energy. In total there is more energy in the fuel and the oxygen before they interact in the process of burning than in the products of their combustion — the balance is released as heat and light or in other forms. Compared with the non-living world, biological systems need a great deal of energy. By inanimate standards, both their chemical compounds and their internal and external organizations are extremely complicated; building such highly complex structures is very energy-demanding. So living organisms need fuel for energy and many of them, including most of the microbes and all the animals, get it by oxidizing (burning) the chemicals in their food. Since those chemicals come from other living things, they are already complicated substances with a high potential energy content, entirely suitable as raw materials for animals and microbes to build their own bodies and to oxidize as fuels to supply themselves with energy.

The chemistry of biological oxidation is basically similar to any other but compared, for example, with the explosive combustion of petrol in a car engine, the rate of energy release in biological systems is low so living organisms do not become as hot as engines. But the biological process is far more efficient, with more useful energy being produced for each unit weight of fuel burned.

Green plants do it differently. Most of their energy comes not from oxidizing materials derived from other living things (although a few of them do that, too) but from the sunlight which the green pigment in their leaves is designed to trap, a process called *photosynthesis*. The sun powers the plants; the plants power the herbivorous animals; they in turn fuel the carnivores while many microbes live off the dead remains (and sometimes the living bodies as well) of all of them, including other microbes. The whole system goes round and round, driven by sunlight.

Energetics is obviously fundamental to life and much of the biochemical apparatus of living organisms is directly concerned with mobilizing and using energy, activities tightly integrated with all the rest of the organism's biochemistry. It is time to explore that biochemistry a little further because it is mostly there that biotechnology is based.

As well as needing a supply of fuel, living things have to have raw materials for building their own substance and for carrying out internal repairs. For green plants most of this need is easily met by carbon dioxide from the atmosphere. But neither animals nor most microbes

Exploiting biotechnology

(the group of organisms of greatest importance for product generation in biotechnology) possess within their bodies the capacity for using a feedstock as simple as carbon dioxide to make the range of chemicals they require for their own internal biochemistry. They therefore need more elaborate "food": other living organisms or complicated chemical products derived from them. The food each individual ingests, much of it chemically different from its own characteristic compounds, has to be converted into exactly the right building materials for its own needs. It is digested down to simple components, many of which have to be refashioned before later being assembled into the organism's own characteristic substances.

Biochemical Complexity

Biochemical compounds exist at two levels of complexity. Most of the body chemicals collectively contributing to the bulk of an individual are actually made up of large numbers of small units, or individual "building blocks". Thus, *proteins*, about which we will have a good deal more to say, are long chains made of small *amino acid* units joined together in a row. Many *carbohydrates*, like *starch* or *glycogen* ("animal starch") consist of variable length chains of individual sugar (*saccharide*) units. *Deoxyribonucleic acids (DNA)*, the compounds so important in inheritance which convey genetic information from parent to offspring, are also constructed of chains of units, in this case called *nucleotides*. Only the fats are not unitized in this way. Compounds made of long chains are called *polymers*; the word implies a series of multiples but does not specify precisely the number of units. Each individual unit (amino acid, sugar or nucleotide) is a *monomer* and other simple words describe chains of two, three or four units, etc. (*dimer, trimer, tetramer*, and so on).

There is a critical difference between *polysaccharides* (long chain carbohydrates) on the one hand, and proteins and nucleic acids on the other. The units in polysaccharides are either all the same or may comprise two or three different ones arranged in a simple alternating sequence. If the letters A, B and C represent three different monomeric sugar units, typical carbohydrates might be represented by:

$$-A-A-A-A-A-A-A-A-A-A-A-A-A-A-$$

or

$$-A-B-A-B-A-B-A-B-A-B-A-B-A-B-$$

or

$$-A-B-C-A-B-C-A-B-C-A-B-C-A-B-$$

Proteins and DNA are very different. The amino acids in protein chains are of 20 different sorts. Because the chemistry of each type of amino acid differs from that of all the others, and many different interactions are possible among the component amino acids within the protein molecule, the properties of the protein of which they form a part are governed both by the total number of amino acids in the chain and by their sequence in the linear array. It

The science . . .

would be hard to exaggerate the importance of these factors in determining the structure and behaviour of proteins.

From a simple chemical point of view, devoid of any biological or functional significance, every conceivable sequence of those 20 amino acids is possible. Biochemists use the following letters as shorthand for the amino acids: A, C, D, E, F, G, H, I, K, L, M, N, P, Q, R, S, T, V, W and Y. So a particular protein, more than a hundred amino acid units long, might start from one end:

G–L–A–S–Y–Q–E–G–G–L–D–G–P–N– – – – – – –

while another might begin:

R–F–F–E–Q–Y–D–A–H–K–S–S–S–D– – – – – – –

With no more than slight reservations, it is correct to say that each protein is unique in its properties. The above examples are only two possibilities chosen at random: the actual number of variants, all of them different from one another, is unbelievably large because each position in any sequence can in principle be occupied by any one of the 20 amino acids. Since there are 20 possibilities for the first amino acid and 20 independent possibilities for the second, there will be 20×20 alternatives for a group of two, $20 \times 20 \times 20$ for three, $20 \times 20 \times 20 \times 20$ for a sequence of four, and on and on. The number of possibilities for a chain only 30 units long is 20 multiplied by 20 twenty-nine times or, in more conventional terms:

1 073 741 182 400 000 000 000 000 000 000 000 000

And this is for a short chain! For a polymer 76 units long (the largest with which the authors' calculator will cope) the number is about 7 followed by 98 zeros, with each variant having its own set of properties; some chains are even longer than that. This dependence on correct sequence in proteins (and, as we shall see later, in DNA and its close relatives) distinguishes them from the boring repetition of polysaccharide chains and led to their being called *informational macromolecules*.

Proteins as Catalysts

Many proteins serve as *catalysts* (*enzymes*), enabling specific chemical reactions to proceed much faster than they would if left to themselves. In order to do this, a protein has to bind to the participating chemicals in just the right way to encourage them to interact. That binding requires a recognition procedure analogous to fitting a key into a lock: there are many possible keys and many possible locks but only the right key will fit any particular lock. Similarly, a particular enzyme must have exactly the right structure to permit the correct binding and catalyse the reaction between the chemicals; it is able to do so by having the right number and sequence of amino acids in its chain. Each cell has thousands of different enzymes to perform the multiplicity of catalytic functions it needs, each type distinguishable from all the others.

Exploiting biotechnology

Figure 2.1 demonstrates just how complicated proteins are. The individual atoms represented as spheres and related shapes are shown in their actual three-dimensional configuration. It is possible to get an idea from the diagram of the cleft into which the reacting chemicals fit exactly.

Different species of plants, animals and microbes share many of their biochemical details with respect to the small molecules: the sugars, amino acids and nucleotides in bacteria are identical with the ones in people. It is the proteins which differ because of their different amino acid sequences. So a protein which in bacteria catalyses the a change of chemical A to chemical B carries out the same job as the protein which does that conversion in man but will have a somewhat different structure for doing so. The amount of difference varies from one case to another: most proteins with parallel functions usually have some parts of their molecules more or less in common, while other regions may differ very considerably. Even individuals of the same species have subtle differences between their proteins, or in the proportions of their various types of protein, which is why particular frogs, or even bananas, though obviously closely related, are not identical with one another.

Metabolism

All the biochemical reactions in an organism or a cell, collectively encompassing the breakdown of food substances, the synthesis of complex chemicals to make the individual's own substances and the generation of a host of products (each catalysed by a specific enzyme), are integrated into *metabolism*, a network of activities which underlies much of biotechnology. The conversion of the sugars in grapes to the alcohol in wine is a process of metabolism. So is making an antibiotic from a degraded starch or other agricultural feedstock. And so also is the synthesis of products like polymers to aid crude oil recovery or surfactants to help clean oil spills from beaches.

An individual organism with a metabolic system that does not work properly may suffer grave disadvantages, the precise details of which vary from one case to another. Suppose the individual is unable to make one of the twenty amino acids which it needs to construct its own proteins: it will surely be unable to live unless it can acquire the missing substance directly from its food. That might not be difficult for an animal living on a mixed diet but could pose insuperable problems for a bacterium trying to exist in a stream where it had to live on simple chemicals. Even if survival were not directly threatened, the failure of a metabolic function might so incapacitate the individual as to put it at a grave disadvantage with respect to acquiring or being able properly to use the resources necessary for survival.

The Organization of Metabolism

The relationship between the pattern of metabolism and the catalytic function of the enzymes in a cell or an organism is all important. Chemicals spontaneously interact among

The science . . .

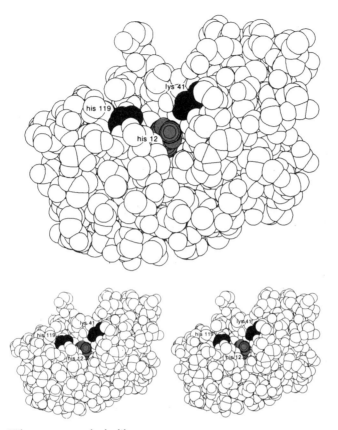

Figure 2.1 What an enzyme looks like.
This illustration is a diagrammatic representation produced by computer graphics from detailed physical analysis. The enzyme (called *ribonuclease*) catalyses the breakdown of RNA. Based on a single chain of 124 amino acids, it folds spontaneously to the compact structure shown in the figure. The reacting chemicals bind specifically to several amino acids located in the cleft (but which are not identified in the drawing). The three highlighted amino acids are histidine at position 12 on the chain (*his 12*), lysine at position 41 (*lys 41*) and another histidine at position 119 (*his 119*). Because of their chemical structures, they interact with the chemicals undergoing reaction and are particularly important for promoting the enzyme catalysis.

The smaller drawings are two slightly different perspectives of the same enzyme. You can see the enzyme in three dimensions without stereo glasses by holding the page level about 10 inches (25 cm) from your eyes and staring at the drawing with your vision relaxed as if you were looking at something far away. A double image will form and the central pair should drift together and fuse. Once you have succeeded it becomes easier next time! The page must be held so that a horizontal line is exactly parallel with your eyes. If you really cannot get the effect, try using two toilet roll tubes to channel a view of the two diagrams separately to your two eyes. Focus each eye with the other one closed; then with both open you should see the 3-D image. (*Reprinted with the permission of Macmillan College Publishing Company from Biochemistry, 2nd edition by Geoffrey Zubay. Copyright© 1988 by Macmillan College Publishing Company, Inc*) themselves: with thousands of different chemical substances in each cell, the range of interactive possibilities is prodigious. But not every potential interaction is useful and most have no metabolic purpose. So how is the cell to distinguish between valuable and useless chemical activities and ensure that it is only the former which actually take place? The answer lies in its catalysts. Almost all of the possible reactions which can occur spontaneously do so very slowly. Enzymes speed up these rates so much that reactions stimulated by catalysis might take place thousands or, in extreme cases, millions of times faster than if the specific enzymes were absent. Thus, the variety and quantities of enzymes which are actually present determine which among the many possible reactions take place rapidly and thereby set the pattern of metabolism. Uncatalysed reactions, for which no enzymes are present, are effectively of negligible significance. There can hardly be anything biochemically more important for an organism than to have its protein structures correct and its enzymes working properly. With so many conceivable structural variants, every organism needs very precise and specific instructions about which proteins to make and how to make them. That information comes from its parents via its genetic inheritance, a matter to be discussed below.

In every case metabolism actually depends totally on the available enzymes. That is why it is so critical for each cell (or organism) to have the proper complement of enzymes for its needs. If it does not, its metabolic activity will not be fully effective and properly integrated: it will be unable to operate at maximum efficiency. Life is a highly competitive matter. The price paid for inefficiency is well known to anyone familiar with human activities and it applies just as much in biology (indeed, it originated there): given half a chance, the more efficient competitors, likely always to be present in natural situations, will surpass and dominate their less efficient neighbours. Those that have the best enzymic equipment and organization survive and flourish; those that do not will be outperformed, outgrown and may eventually be eliminated entirely. Getting catalysis right for the circumstances is a matter which has been refined through the aeons of biological evolution. As we will explore later, each modern individual benefits from this historical process by inheriting from its ancestors, via its parents, the genetic information for making its own enzymes correctly — with the proper sequence of amino acids in all the protein chains and with all the proteins in the right proportions.

The organization of metabolism along these lines works extremely well but not perfectly. Mistakes can be made in the transfer of information from parent to offspring, in the further copying of that information as the new organism grows and in "reading" the genetic messages; occasionally the messages themselves become corrupted. Furthermore, the needs of the cell or organism change with time, often with bewildering frequency. If the organism is an animal developing from a fertilized egg, it will need to be doing different things at different stages of its growth: producing a head at one time, limbs later on, enlarging its body at different rates as it grows into an adult, acquiring sexual maturity, and so on. All these activities are based on biochemistry and all must be integrated and controlled according to plan or the end result may be deformed and defective. Perhaps the cell is a microbe bathed in an ever-changing bath of nutrient chemicals which are usually in short supply and competed for locally by other microbes; our cell will need the right enzymes rapidly to convert the available foodstuffs into chemicals which it can use in its own metabolism before its competitors grab everything for themselves. But enzymes are complex and energy-expensive items to make; it is obviously not

The science . . .

only wasteful to make them for foodstuffs which are for the moment not actually present but also disadvantageous not to be able to make them when the foods do appear. The microbe therefore needs mechanisms for sensing the presence or absence of a range of potential nutrients as well as "switches" to enable it to turn on and off the manufacture of specific enzymes as the need arises. These instructions, too, are all encoded in the genetic inheritance.

Metabolic By-products

The commercial interest in biotechnology lies both in metabolic by-products and directly in the enzymes or other proteins themselves. For biotechnological services, such as certain aspects of waste management or microbial mining (described in Chapters 10 and 11), the products are not offered separately packaged. The service nevertheless depends on specific compounds being generated on location in order to promote desirable chemical or physical changes, possibly in an effluent stream or in an ore body.

Although metabolism is highly optimized for the natural environment in which the host cell or organism has evolved, that optimization may not also be right in a different environment or under artificial conditions. Factors of nutrition or enforced lifestyle may exceed the capacity of the metabolic organization to adapt and so it becomes unbalanced. Efficiency may fall in this new situation, perhaps because the foodstuff can no longer be utilized as fully and extensively as before and some parts of it, or some products from it, may have to be discarded.

Oxidation and Fermentation — a Gentle Excursion into Biochemical Detail

A good example of a metabolic by-product is alcohol produced by yeast, a microbe which can grow either with oxygen or without it. If oxygen is available, some of the sugar in the yeast's diet is completely oxidized to carbon dioxide and the energy yielded by the oxidation enables the yeast to use the rest of the sugar to build its own body substance and produce more yeast cells — which, from the yeast's point of view, is what life is all about. Nothing is left over. But if the yeast is deprived of oxygen (as it is in a beer fermentation vat), it must adopt another strategy because it still has to use the sugars for energy but can no longer oxidize them fully. Circumstances force it to do what it can, taking the oxygen it must have from some of the food chemicals and then, because it can do no more with them, discarding the residues from which some or all the oxygen has been removed. One such residue is alcohol.

In the Preface to this book we promised not to get too technical and said we were writing for people with no special knowledge of chemistry. Without reneging on that commitment, it is a good idea at this point to edge just a little into chemical detail to explore briefly the difference between oxidation and fermentation.

First, we must take a quick look at a chemical "formula" which is actually a description of a compound. There are two sorts: the *structural formula* is complicated, virtually providing a picture of what the compound looks like and where all the atoms are located; we do not need that here. The simpler *empirical formula* merely states the number of atoms of each element

Exploiting biotechnology

present in the molecule. Every element is designated by a one- or two-letter abbreviation: the ones we need are C for carbon, H for hydrogen and O for oxygen. The sugar glucose can then be written $C_6H_{12}O_6$ which means that each molecule comprises six atoms of carbon, twelve of hydrogen and six of oxygen.

Next, we need to consider briefly the meaning of "chemical energy". It takes effort (i.e. energy) to construct complicated chemicals, rather like the effort needed to wind up a watch spring. Complex chemicals and wound-up watch springs are both sources of "potential" energy — under the right conditions, the energy in the chemicals can be released to be used for something else (like making other chemicals) just as the spring in a watch slowly unwinds releasing energy to drive the hands. When a complex chemical is constructed from simple components energy has to be supplied; if that chemical is broken down, the energy comes out.

That is exactly what happens when an organism metabolizes a substance like glucose (the most common sugar) to obtain energy to make complicated chemicals for its own growth and for many other purposes. It first converts each glucose molecule into two molecules of a substance called *pyruvic acid*. As this is no more than a partial breakdown, a fairly small portion only of the energy contained in the glucose is released, most of it still being locked up in the pyruvic acid:

$$C_6H_{12}O_6 \longrightarrow 2\ C_3H_4O_3 \quad (+\ \text{a little energy released})$$
$$\text{glucose} \qquad\quad \text{pyruvic acid}$$

Those two molecules of pyruvic acid each contain only four hydrogen atoms, a total of eight, while the original glucose contained twelve. To make things add up, those four hydrogen atoms have to be put into the equation:

$$C_6H_{12}O_6 \longrightarrow 2\ C_3H_4O_3 + 4\,H \quad (+\ \text{a little energy released}) \qquad (1)$$
$$\text{glucose} \qquad\quad \text{pyruvic acid}$$

Oxidation

If the yeast is living in air, and hence in the presence of oxygen, those four hydrogens combine with two oxygen atoms to make water:

$$4\,H + 2\,O \longrightarrow 2\ H_2O \quad (+\ \text{lots of energy released}) \qquad (2)$$
$$\qquad\qquad\qquad\quad \text{water}$$

Pyruvic acid itself can also be oxidized by the oxygen in air:

$$C_3H_4O_3 + 5\,O \longrightarrow 3\ CO_2 + 2\ H_2O \quad (+\ \text{lots of energy released})$$
$$\text{pyruvic acid} \qquad\qquad \text{carbon dioxide} \quad \text{water}$$

The science . . .

Two molecules of pyruvic acid simply need twice as much oxygen and generate twice as much product:

$$2\,C_3H_4O_3 + 10\,O \longrightarrow 6\,CO_2 + 4\,H_2O \quad (+\,2 \times \text{lots of energy released}) \qquad (3)$$

Thus, the complete description of glucose oxidation in air is obtained by adding up lines (1), (2) and (3) and cancelling out the pyruvic acids and hydrogens which would appear on both sides of the arrow:

$$C_6H_{12}O_6 + \cancel{2\,C_3H_4O_3} + \cancel{4\,H} + 12\,O \longrightarrow \cancel{2\,C_3H_4O_3} + \cancel{4\,H} + 6\,CO_2 + 6\,H_2O$$

(+ lots and lots of energy released)

i.e.

$$C_6H_{12}O_6 + 12\,O \longrightarrow 6\,CO_2 + 6\,H_2O$$

(+ lots and lots of energy released)

Oxidizing glucose all the way to carbon dioxide and water squeezes out almost all the energy in the sugar but it does require the presence of oxygen and, if none is available, this process cannot take place.

Fermentation

So what does yeast do without oxygen? Go back to line (1) and look again at those four hydrogens. Their chemical form is not just "H"; in reality they are joined onto another (carrier) molecule of which the cell has no more than a limited supply. Reaction (1) can therefore more accurately be written as in line (4), where "X" is the carrier:

$$\underset{\text{glucose}}{C_6H_{12}O_6} + 2\,X \longrightarrow \underset{\substack{\text{pyruvic}\\\text{acid}}}{2\,C_3H_4O_3} + 2XH_2 \quad (+\,\text{a little energy released}) \qquad (4)$$

XH_2 cannot pile up indefinitely because there is so little X available — something must be done with the hydrogens or the X carriers become choked with them and whole process will come to a halt. What happens in the yeast is that the hydrogen atoms are dumped (at no energy cost) into the two molecules of pyruvic acid (each one of which, as it happens, then also loses a molecule of carbon dioxide) and the carrier X is freed to pick up more hydrogens from the metabolism of another glucose molecule:

$$\underset{\substack{\text{pyruvic}\\\text{acid}}}{2\,C_3H_4O_3} + 2\,XH_2 \longrightarrow \underset{\substack{\text{ethyl}\\\text{alcohol}}}{2\,C_2H_6O} + \underset{\substack{\text{carbon}\\\text{dioxide}}}{2\,CO_2} + 2\,X \qquad (5)$$

Exploiting biotechnology

Adding up (1) and (5) and cancelling out as before:

$$C_6H_{12}O_6 \longrightarrow 2\ C_2H_6O + 2\ CO_2 \quad (\text{+ a little energy released})$$
$$\text{glucose} \qquad \text{ethyl} \quad\ \ \text{carbon}$$
$$\text{alcohol} \quad \text{dioxide}$$

This is the biochemical origin of the beverage alcohol in all the world's beer, wine and spirits.

Some bacteria do things a little differently: when they add the four hydrogens to the two molecules of pyruvic acid there is no loss of carbon dioxide and instead lactic acid is formed:

$$2\ C_3H_4O_3 + 2\ XH_2 \longrightarrow 2\ C_3H_6O_3 + 2\ X$$
$$\text{pyruvic} \qquad\qquad\qquad \text{lactic}$$
$$\text{acid} \qquad\qquad\qquad\quad \text{acid}$$

So for those organisms the summary is:

$$C_6H_{12}O_6 \longrightarrow 2\ C_3H_6O_3 \quad (\text{+ a little energy released})$$
$$\text{glucose} \qquad\quad \text{lactic}$$
$$\text{acid}$$

Lactic acid is the cause of milk souring and is thus part of all cheese and yoghurt production.

Both alcohol and lactic acid are half-way products of glucose breakdown and only some the energy originally present in glucose is released. Neither of these products is of any further use without oxygen and, from the organism's point of view, simply build up as wastes. As they are both poisonous in high concentration, there is a limit to how much can be accumulated, which is why the maximum alcohol concentration produced in wine by fermentation is about 12%. Fortified wines (ports, sherries, etc.) have more alcohol added to them, while spirits are obtained by distillation to concentrate the alcohol, separating it from the water. The biological lesson to be drawn from the difference between oxidation and fermentation is that the former is a much more efficient way of using food molecules because so much more energy is released. If an organism is equipped to do both, it will almost invariably oxidize rather than ferment: the energy returns are so much greater.

Metabolic Balance

In nature, microbes generate just the right amount of each compound for their needs but commercial manufacture usually benefits from excess production of particular chemicals in a balance of metabolism different from the natural one. Placed in a novel and properly designed chemical environment, the microbes are forced into a pattern of chemical behaviour in which

The science ...

doing the manufacturer's bidding now becomes the best way to optimize their own growth and reproduction. Thus, yeast in the brewer's vat does its best to make more yeast, even if it has to divert much of its food into making the alcohol which it excretes as a waste product. There are many cases of microbes being subverted into making industrial or pharmaceutical products in quantities far exceeding anything which is useful to them. Of course, given a chance, they will stop doing so and use all their resources for growth; the restraints on them must therefore be unremitting. No foreign organisms insensitive to the constraints must be allowed admission or they will sequester the feedstocks for their own growth, fail to yield the desired product and abort the manufacturing process.

It is the responsibility of the biotechnologist to design operating systems as foolproof as possible. He uses not only environmental and nutritional conditions to maximize product yield but he may directly modify the organisms so that they become even more productive: he can do this by adding additional enzymes to their complement of catalysts or by persuading them to make larger amounts of certain of their own enzymes than they would otherwise do. In other words, he changes their genetic instructions, a technique often called "genetic engineering", and one to which we will return in due course.

Protein Engineering: Making New Enzymes to Order?

Proteins, for all their enormous catalytic power, are fragile molecules. Their very complexity renders them susceptible to degradation by many sorts of adverse conditions. The way the proteins in egg white congeal when an egg is boiled or fried is an illustration of their sensitivity to change by heat, in many cases at temperatures well below the boiling point of water. But heat-resistant proteins do exist and some of them are found in microbes which live in hot springs or in central heating systems. Industry would benefit from having more of them because chemical reactions go faster at high temperatures and heat-resistant proteins are often also resistant to other damaging influences and last longer in industrial processes.

Sometimes heat-resistant enzymes with the right catalytic properties for a particular process cannot be found in nature and it would clearly be an advantage if they could be made to order. There are other reasons for wanting tailor-made enzymes. Many chemical reactions in industry have to be catalysed by high temperature or pressure; it would often be much cheaper to do so with enzymes if only the right enzymes existed. But because the reacting chemicals might be totally artificial and unknown in nature, no natural enzymes exist to catalyse their interaction. If we knew enough about enzyme structure and how it is related to catalytic properties, it might become possible to design enzymes to do certain specified jobs.

Protein chemistry is well on its way to doing so. Designing the structure is not enough: it also has to be put together. How that can be done will come up again when we consider genetics.

GENETICS

This chapter has already made a number of allusions to genetic information, the need to specify accurately the amino acid sequence of each protein and the transmission of this information

Exploiting biotechnology

from one generation to the next without incurring errors. The biochemistry underlying these activities is complex, but the principles of their operation can be fairly well understood without any reference at all to chemistry. It is worth making the effort because an appreciation of how protein structures are determined and how information is protected from corruption during transmission is all-important to knowing what biotechnology has to offer now and in the future. In particular, it opens the door to genetic engineering which, more than any other aspect of biotechnology, has received massive publicity and interest.

Why Genetics is so Important

We have already seen that the characteristics of an organism are governed by its chemistry which, in turn, depends on the particular properties, quantities and locations of its enzyme proteins. So the catalytic powers of proteins are absolutely crucial because they determine the pattern of chemical activity. For an individual, whether a bacterium or a human, suffering from a defective protein function can be very serious. *Haemophilia* (the absence of certain clotting agents in blood) is one example among many of a disability resulting from the failure of part of the chemical mechanism to perform properly.

For each protein molecule to work properly, its own structure has to be just right for the job; adapting protein structures to particular catalytic functions is what much of 3.5 billion years of biological evolution has been about. Not surprisingly, most modern proteins are very well suited to their roles.

The exact structure of every protein in all living beings is specified by information contained in the sequence of units in other long-chain compounds called *nucleic acids*; one particular variety (DNA) is by far the most common as an information store. Like proteins, DNA is built of a chain of monomeric units whose chemistry needs concern us only to the point of recognizing that it is very different from the chemistry of amino acids. It is the sequence of these units in DNA which specifies the parallel sequence of amino acids in proteins.

The structure of each individual protein is encoded in a particular section of the DNA molecule: such a section, specifying one type of protein, is called a *gene*. Most genes do just that but a proportion of them have other functions concerned with regulating the biochemical activities of the organism. They act as sensors and switches by providing the instructions for special mechanisms which recognize the chemical needs of the moment and determine whether or not each one among the many enzymes or other proteins which could be made by the organism should actually be synthesized at that time and, if so, in what quantities.

In this section of the chapter the following issues will be addressed:

- how information is encoded in DNA;
- how this information is copied with a minimum of errors;
- how it is used to make proteins;
- how it can be manipulated artificially in order to extended the range of biotechnological activities.

The science...

Encoding of Information in DNA

There are four different units (monomers) making up the chains of DNA. They have biochemical names, of course, but their initials will suffice here: A, C, G and T. It so happens that among these compounds there is a strong chemical tendency to form pairs, A with T and C with G. So each molecule of DNA is actually composed of two chains, with As and Ts, and Cs and Gs, cross-pairing between the chains; like this:

```
-C-T-T-G-A-T-G-C-A-T-G-G-C-A-T-C-T-A-C-G-
 : : : : : : : : : : : : : : : : : : : :
-G-A-A-C-T-A-C-G-T-A-C-C-G-T-A-G-A-T-G-C-
```

in which the horizontal hyphens represent firm chemical links forming the backbone of the chain while the vertical colons indicate a weaker affinity between the members of each pair. The actual chemical structures of these components force the two chains to twist round one another in a characteristic manner, the well-known *double helix*. Figure 2.2 shows what the double helix looks like.

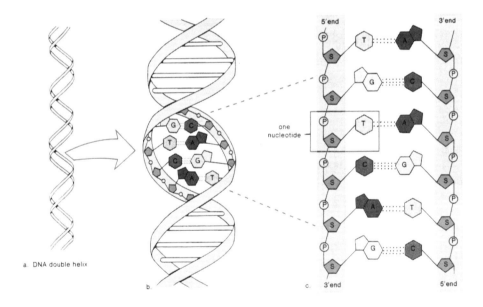

Figure 2.2 Overview of DNA structure. (a) gives a general perspective; (b) shows the complementary base ("letter") pairing, still in the context of the twisted helix; (c) shows the ladder effect of the pairing when the helix is unwound. (*From Sylvia S Mader, Biology, 3rd edition, Copyright© 1990 Wm. C. Brown Communications, Inc., Dubuque, Iowa. All rights reserved. Reprinted by permission*)

Exploiting biotechnology

If these sequences of units in DNA are to specify the sequence of units in proteins, there must be a way in which one type of sequence (DNA) with only four different units can relate directly to another (proteins) with twenty: individually four DNA units could specify precisely no more than four amino acids. This difficulty is overcome by relating more than one DNA unit to each amino acid. If two DNA units taken together were to specify each amino acid there could be any one of four units for the first member of the pair and, quite independently, any one of four for the second: in total $4 \times 4 = 16$ possibilities. That is not enough because there are 20 different amino acids in protein so the DNA units need to be "read" in threes or triplets: $4 \times 4 \times 4 = 64$. Now there is too much information but that does not matter. A few of the 64 possibilities are used for "punctuation" to indicate the beginnings and ends of genes, and all the rest are assigned to the 20 amino acids, most of which therefore have multiple representation. The groups of three DNA units are called *codons*.

In order not to have to go into chemical detail, or use the names or symbols of the 20 amino acids which will mean little or nothing to those unfamiliar with them, it will be convenient to explore these systems by using English rather than biochemistry. The simplified models to be used here will be broadly correct and easy to understand; to make it possible to form real words they will draw as necessary upon all the 26 letters used in English, always in 3-letter groupings.

A simple "message" in English encoded in a linear sequence might be:

THECATSATANDATETHERATANDRANOFF

It becomes more intelligible if a "reading frame" of three letters to each word is imposed upon the linear message:

word →	THE	CAT	SAT	AND	ATE	THE	RAT	AND	RAN	OFF
	1	2	3	4	5	1	6	4	7	8

The same basic mechanism can be used for punctuation, indicating where a particular message starts and stops. In DNA terms, the message about the cat sitting and eating the rat and then running off will be part of a much longer message, just as the actual sentence in English might be part of a book. In such a book the beginning and end of the message would be marked by a capital letter and full stop: "...off the hut. The cat sat and ate the rat and ran off. The owl saw its egg..." The capital letter and full stop could be replaced by three-letter groupings, say XXX and GGG, respectively. Then the message might read "...OFF THE HUT GGG XXX THE CAT SAT AND ATE THE RAT AND RAN OFF GGG XXX THE OWL SAW ITS EGG..." Anyone used to this sort of punctuation would have no difficulty identifying the sentences.

Replication of DNA

In order both for the parents to provide their offspring with exact copies of their genetic material, and for the developing organism to make more of its own DNA correctly as it grows

The science . . .

larger and eventually also reproduces, there needs to be a foolproof way of copying DNA as nearly perfectly as possible. There is. DNA molecules act as their own templates for making copies. Again detailed chemistry can be avoided by using letters to represent chemical entities. In the English language model the pairing between units ("letters") is represented as:

one in plain type and the other in bold. The plain and bold letters are very good at recognizing their counterparts and rarely make mistakes. Individual DNA molecules consist of hundreds of thousands or more of these "letter" pairs in sequence in long chains. Still with the cat and the rat, but using hyphens to show how the "letters" are linked in chains, and colons to illustrate the pairing, we can start with a model of double-stranded DNA like this

```
···T-H-E-C-A-T-S-A-T-A-N-D-A-T-E-T-H-E-R-A-T-A-N-D-R-A-N-O-F-F···
   : : : : : : : : : : : : : : : : : : : : : : : : : : : : :
···T-H-E-C-A-T-S-A-T-A-N-D-A-T-E-T-H-E-R-A-T-A-N-D-R-A-N-O-F-F···
```

When the DNA is going to replicate, the two strands separate:

```
···T-H-E-C-A-T-S-A-T-A-N-D-A-T-E-T-H-E-R-A-T-A-N-D-R-A-N-O-F-F···

···T-H-E-C-A-T-S-A-T-A-N-D-A-T-E-T-H-E-R-A-T-A-N-D-R-A-N-O-F-F···
```

Living cells, in addition to containing DNA carrying messages, also have their own supply of uncombined DNA "letters", both plain and bold. The individual "letters" linked into these separated strands or chains each pair up with their free counterparts, plain with bold and bold with plain:

```
···T-H-E-C-A-T-S-A-T-A-N-D-A-T-E-T-H-E-R-A-T-A-N-D-R-A-N-O-F-F···
   : : : : : : : : : : : : : : : : : : : : : : : : : : : : :
   T H E C A T S A T A N D A T E T H E R A T A N D R A N O F F
```

and

```
···T H E C A T S A T A N D A T E T H E R A T A N D R A N O F F···
   : : : : : : : : : : : : : : : : : : : : : : : : : : : : :
   T-H-E-C-A-T-S-A-T-A-N-D-A-T-E-T-H-E-R-A-T-A-N-D-R-A-N-O-F-F
```

but the new "letters" are not yet linked together in their own chain.

Exploiting biotechnology

Finally, a special enzyme links up the newly assembled "letters":

···T-H-E-C-A-T-S-A-T-A-N-D-A-T-E-T-H-E-R-A-T-A-N-D-R-A-N-O-F-F···
   :
···T-H-E-C-A-T-S-A-T-A-N-D-A-T-E-T-H-E-R-A-T-A-N-D-R-A-N-O-F-F···

and

···T-H-E-C-A-T-S-A-T-A-N-D-A-T-E-T-H-E-R-A-T-A-N-D-R-A-N-O-F-F···
   :
···T-H-E-C-A-T-S-A-T-A-N-D-A-T-E-T-H-E-R-A-T-A-N-D-R-A-N-O-F-F···

From one original double chain we now have two identical copies. Each of the new molecules has one chain derived from the original parent (the upper plain one in the top example or the bottom bold one in the lower) and one which is new (the bold lower one in the top example or the plain upper one in the lower). The system works so long as the letters always recognize one another without error. Like all biological systems, it is not quite perfect but the error rate is very low indeed.

Importance of Accuracy

How is this "code" used and why is it so important that no mistakes occur in DNA replication? Each gene specifies the order of the individual amino acids in the chain of a protein and hence its properties. Proteins, as we have already remarked, are the catalysts which determine the chemical structure and performance of every living cell in every organism. If the structure of any protein molecule is distorted or corrupted it is not likely to function normally and the chances are overwhelming that it will be worse, not better. That might not matter much if an occasional protein molecule is in poor shape because many copies of each sort are present. But if *all* the copies of any particular type are defective because the basic template had been altered, things could be very serious indeed. An organism carrying such a defect might be severely handicapped or unable to survive.

In nature, defects, especially serious ones, tend to be eliminated in the competition for resources which is why those living beings and their constituent chemicals which have actually survived to the present day are well fitted to their individual life styles. They might therefore not take kindly to any changes in their enzymes, especially those responsible for their most critical functions. Haemophilia is a case in point. It results from a genetic defect affecting a protein called *Factor VIII* which participates in blood clotting. One of the advantages of living at the present time is the possibility of benefiting from the advances in medical biotechnology which can now begin to correct human genetic faults. Factor VIII can be isolated from human blood supplied by donors but it is probably more economic to manufacture it in genetically modified microorganisms. The promise of biotechnology is that there will be many more such products in future. Even more exciting is the prospect of actually correcting such genetic defects in the patient; *gene therapy*, as this is called, will come up again in Chapter 7.

The science . . .

Mutation

There are several ways in which corruption of genetic messages can occur:

- by changes in the chemical structure of one or more of the chemical "letters" (A, C, G and T) making up the genetic message, changes which result from alteration of the structure of the letter units by agents called *mutagens* which may be physical (X-rays, ultraviolet light, radioactivity) or chemical;
- through occasional mistakes in the copying procedure, when a mispairing takes place, a letter unit is inadvertently omitted or an extra one added.

Alterations in the genetic message are called *mutations*; they occur naturally for the reasons just given and some can be produced in the laboratory. They might give rise to altered protein products which could have profound effects on an organism trying to live with enzymes or other proteins which do not work as well as the originals. Mutations are the very stuff of evolution, the genetic basis for the Darwinian concept of *natural selection*, popularly described as "the survival of the fittest". In the course of biological history chance mutations, occurring at random, have generated many variants of protein structures, some more effective than the originals and others less so. Organisms carrying mutated genes coding for "better" proteins will be able to carry out their activities more effectively and will tend to out-compete their rivals. Thus, beneficial mutations tend over time to spread through a population. Deleterious mutations, by contrast, disadvantage an organism and are progressively eliminated through diminished fecundity in the carrier. Mutation and evolution are not simply aspects of history; they continue to be important factors in determining the genetic inheritance of individuals.

The following examples illustrate mutation (note that everything in italics in these examples represents nonsense; the point will be appreciated when the message is broken into groups of three letters and the sense/nonsense implications become clear).

The "message"

THECATSATANDATETHERATANDRANOFF

can be expressed within a three-letter reading frame:

THE CAT SAT AND ATE THE RAT AND RAN OFF

If a single error (in bold) is introduced only the word containing the error is wrong:

THECATSATA**Y**DATETHERATANDRANOFF

word → THE CAT SAT A**Y**D ATE THE RAT AND RAN OFF
               1   2   3   (4)   5   1   6   4   7   8

If an extra letter (in bold) is added, the message from that point on is in the wrong reading frame. Most of the message is now nonsense:

THECA**A***TSATANDATETHERATANDRANOFF*

Exploiting biotechnology

THE *CAA TSA TAN DAT ETH ERA TAN DRA NOF F*

A similar consequence follows the deletion of a letter (indicated by '):

THECATSATANDATET'ERATANDRANOFF

THE CAT SAT AND ATE *T'ER ATA NDR ANO FF*

But a deletion can be at least partly compensated for by an insertion elsewhere. Only in the section between the two is the message nonsensical:

THECA*ATSATANDATET'*ERATANDRANOFF

THE *CAA TSA TAN DAT ET'E* RAT AND RAN OFF

and the organism may still be partly functional.

Using the Information in DNA to Make Protein

It is superfluous here to recount in biochemical detail how DNA sequences specify protein structures, but broadly it works as follows.

The information in DNA serves as a template on which to build a similar, but not identical, substance called *ribonucleic acid (RNA)*. This process, which closely resembles DNA replication itself, is called *transcription* because RNA is distinguished from DNA only by minor differences in the chemical structure of the individual units ("letters") along the chain; their original sequence is preserved. For our purpose, lower case letters will be used for RNA versus the capitals for DNA. Transcription looks like this:

The DNA chains separate as in replication:

···T-H-E-C-A-T-S-A-T-A-N-D-A-T-E-T-H-E-R-A-T-A-N-D-R-A-N-O-F-F···

···T-H-E-C-A-T-S-A-T-A-N-D-A-T-E-T-H-E-R-A-T-A-N-D-R-A-N-O-F-F···

One of the chains only is used as the template for aligning the (lower case) RNA "letters":

···T-H-E-C-A-T-S-A-T-A-N-D-A-T-E-T-H-E-R-A-T-A-N-D-R-A-N-O-F-F···
   :
···t h e c a t s a t a n d a t e t h e r a t a n d r a n o f f···

The science...

The RNA "letters" are joined together:

···T-H-E-C-A-T-S-A-T-A-N-D-A-T-E-T-H-E-R-A-T-A-N-D-R-A-N-O-F-F···
   :
···t-h-e-c-a-t-s-a-t-a-n-d-a-t-e-t-h-e-r-a-t-a-n-d-r-a-n-o-f-f···

The two chains separate:

···T-H-E-C-A-T-S-A-T-A-N-D-A-T-E-T-H-E-R-A-T-A-N-D-R-A-N-O-F-F···

···t-h-e-c-a-t-s-a-t-a-n-d-a-t-e-t-h-e-r-a-t-a-n-d-r-a-n-o-f-f···

and, while the DNA returns to pair again with its erstwhile mate, the single-stranded RNA is used as a template for ordering the sequence of amino acids along a protein chain.

Many RNA molecules are made from each DNA, and many protein molecules are in turn based on each RNA, so it matters little if an occasional error creeps in during RNA and protein manufacture because many copies are made of each and none participates in the inheritance of the next generation of organisms resulting from reproduction. Any defects are therefore of localized and short-term significance only.

Amino acids are chemically very different from the units making up RNA. A conversion mechanism is accordingly needed to translate each three-letter "word" of RNA ("the", "cat", "sat", etc.) into the correct sequence of amino acid units making up the protein chain. This mechanism incorporates the chemical equivalent of a dictionary.

Thus, in principle:

$$\text{DNA} \xrightarrow{\text{transcription}} \text{rna}^* \xrightarrow{\text{translation}} \text{БЕЛОК}^{**}$$

*   same "language" but different "letters"

**   *byelok*, Russian for "protein"; both a different "language" and different "letters".

Back to our earlier model (easier to illustrate in German than in Russian):

DNA message:	THE	CAT	SAT	AND	ATE	THE	RAT	AND	RAN	OFF
RNA message:	the	cat	sat	and	ate	the	rat	and	ran	off
Amino acids:	Die	Katze	saß	und	aß	die	Ratte	und	lief	weg

Exploiting biotechnology

Chromosomes

Living cells carry their DNA in highly organized structures called *chromosomes*. In their simplest forms in bacteria and viruses, chromosomes are no more than the very long molecules of double-stranded DNA, each folded and super-folded into a very compact body. Typically, when a bacterium is at its smallest size immediately after its formation by the division of its mother cell, it has either a single chromosome or at most two identical ones. As it grows in size towards its own division, DNA/chromosome replication takes place ensuring that the number of chromosomes per unit volume of cell remains more or less constant. By the time the cell is ready for division, the one (or two) chromosome(s) with which it was "born" has become two (or four), ready to be shared out evenly between the next pair of daughter cells. Do note, however, that while the terms *mother* and *daughter* are widely used in describing the reproduction of bacteria, they do not have exactly the same meanings as in common usage, as we will shortly see.

In the more elaborate cells of "higher organisms" (i.e. all those other than bacteria and viruses), DNA is organized in a more complicated manner:

- there may be dozens of different chromosomes in each cell (human beings have 46);
- the chromosomes, in addition to DNA, contain certain proteins which probably help to regulate the flow of genetic information;
- each chromosome comprises two identical components called *chromatids*;
- the chromosomes themselves are grouped in pairs, one member of each pair coming from the original mother and the other from the father. Mother and father now mostly have their common meanings because the majority of higher organisms undergo a real sexual stage in their reproduction, fundamentally similar in biological terms to the human: chromosomes are contributed by both parents to form the genetic endowment of the offspring;
- within each cell of a higher organism, the chromosomes are packaged into a discrete organelle called the *nucleus*; collectively organisms having such a distinct nucleus are termed *eukaryotes* (meaning "true nucleus") while the simpler bacteria, in which the chromosome(s) lies free and unpackaged in the cell, belong to the *prokaryotes* ("before the nucleus" [i.e. primitive nucleus]). There are many ways in which the cells of eukaryotes and prokaryotes differ: the nuclear membrane is one of the most important;
- nuclei are not the only bodies in eukaryotes which contain DNA. Most of their cells also contain organelles called *mitochondria*, the location for much of the biochemical mechanism for providing the cell with energy. Far back in history mitochondria are believed to have originated in bacteria which were somehow incorporated as part of the eukaryotic cell at an early stage of evolution. Mitochondria contain their own DNA separate from that in the nuclei, DNA, moreover, which has some of its organizational characteristics more in common with bacterial genetic material than with eukaryotic chromosomal DNA. Because mitochondria are not part of the cell nucleus, the genetic material they contain is not distributed between the two daughter cells in the same way: half the total number of mitochondria simply pass randomly to each daughter cell.

The science . . .

The significant aspect of mitochondrial distribution comes in the development of the *gametes*: egg cells contain mitochondria but sperm do not. Thus, all of an individual's mitochondria come from its mother, fathers making no contribution to mitochondrial DNA.

Chromosomal complexity in eukaryotes has some very important implications. We have already seen how simple double-stranded DNA (e.g. the chromosomes of bacteria) replicates by each strand acting as a template for the formation of a complementary copy. Of course, DNA replication also takes place in eukaryotes but something additional needs to happen because of the existence of two chromatids per chromosome and the pairing of the chromosomes themselves. Using the first part of the THE CAT SAT AND ATE THE RAT AND RAN OFF example, the eukaryotic situation can be illustrated like this:

```
DNA strand M1a      ···T-H-E-C-A-T-S-A-T···  )
                       : : : : : : : : :     ) chromatid M1  )
DNA strand M1b      ···T-H-E-C-A-T-S-A-T···  )               ) chromosome M:
                                                              ) inherited from
DNA strand M2a      ···T-H-E-C-A-T-S-A-T···  )               ) the mother
                       : : : : : : : : :     ) chromatid M2  )
DNA strand M2b      ···T-H-E-C-A-T-S-A-T···  )


DNA strand F1a      ···T-H-E-C-A-T-S-A-T···  )
                       : : : : : : : : :     ) chromatid F1  )
DNA strand F1b      ···T-H-E-C-A-T-S-A-T···  )               ) chromosome F:
                                                              ) inherited from
DNA strand F2a      ···T-H-E-C-A-T-S-A-T···  )               ) the father
                       : : : : : : : : :     ) chromatid F2  )
DNA strand F2b      ···T-H-E-C-A-T-S-A-T···  )
```

One consequence of having multiple copies of genes is that the inheritance from the two parents may not (in fact, will not) be identical for all genes. Instead of the above pattern, the message carried by the chromatids of one chromosome might be corrupted:

```
DNA strand M1a      ···T-H-E-C-Y-T-S-A-T···  )
                       : : : : : : : : :     ) chromatid M1  )
DNA strand M1b      ···T-H-E-C-Y-T-S-A-T···  )               ) chromosome M:
                                                              ) inherited from
DNA strand M2a      ···T-H-E-C-A-T-S-A-T···  )               ) the mother
                       : : : : : : : : :     ) chromatid M2  )
DNA strand M2b      ···T-H-E-C-A-T-S-A-T···  )


DNA strand F1a      ···T-H-E-C-A-T-S-A-T···  )
                       : : : : : : : : :     ) chromatid F1  )
DNA strand F1b      ···T-H-E-C-A-T-S-A-T···  )               ) chromosome F:
                                                              ) inherited from
DNA strand F2a      ···T-H-E-C-A-T-S-A-T···  )               ) the father
                       : : : : : : : : :     ) chromatid F2  )
DNA strand F2b      ···T-H-E-C-A-T-S-A-T···  )
```

Exploiting biotechnology

while all the others remain correct. Such a combination may have no effect on the organism carrying the corruption so long as the uncorrupted message (THE CAT SAT.....) performs its proper function and the corrupted message (THE CYT SAT.....) is neutral in its effect. But if the corrupted message were *not* neutral (say, THE BAT SAT.....) the balance of corrupted and correct information might have deleterious and unfortunate consequences for the individual.

Chromosomal inheritance

As a eukaryotic organism grows, its cells divide and the genetic material in each cell (in the form of the two chromatids of each chromosome) is shared out among the daughters, much like in bacteria. The example above shows the four chromatids of one chromosome, M1 and M2 from the original mother, and F1 and F2 from the original father. At the time of cell division these are shared out to give M1 + F1, M2 + F1, M1 + F2 or M2 + F2. Every daughter cell thus receives one of the two chromatids from each of its mother's chromosomes. The chromatids in the daughter cells replicate before they next divide, so restoring two complete chromosomes to each cell which is described as being *diploid*. For example:

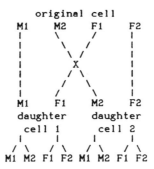

During sexual reproduction, which entails the production of reproductive cells (*gametes*), the sharing is of complete chromosomes, not chromatids. Thus, the whole of the originally maternal chromosome (comprising M1 + M2) goes to one daughter cell with the whole of the originally paternal one (F1 + F2) going to the other. Depending on the sex of the individual, each daughter cell of that division will ultimately develop into a sperm or an egg (or their equivalents in plants and such eukaryotic microorganisms as fungi) containing only half the number of chromosomes characteristic of the adult cell; the gametes are said to be *haploid*. When an egg is fertilized by a sperm the original number of chromosomes is restored, half from the father via the sperm and half from the mother via the egg, back to the diploid state.

Many of the genes donated by the father will be identical with those from the mother — that is what characterizes the species — but there are always some differences which is why individual members of a species differ within a general pattern. Thus, all humans have eyes but some are coloured blue and others brown, with varying shades in between; that goes, too, for hair colour, physique and all sorts of other properties. Sometimes the paternal genes are more strongly expressed than the maternal ("doesn't little Jennifer resemble her father?"),

The science . . .

sometimes the other way round. Quite often characteristics do not resemble either parent so little Jennifer might have blue eyes although both of her parents have brown ones (but never the other way because brownness is always dominant to blueness).

One highly significant difference between maternal and paternal inheritance refers to the sex of the offspring. In humans, for instance, there are 22 pairs of identical chromosomes but the components of the 23rd pair exist in two forms, X and Y. If an individual person has XX, she is female; if XY, he is male (no normal person has YY). Because all eggs come from an XX mother, they each have one X and do not determine the sex of the offspring. But since a man is XY, half his sperm will be X and the rest Y. Whether the offspring is XX or XY is determined by which type of sperm fertilizes a particular egg.

Some inherited diseases result from mutations of genes resident on the X or Y chromosomes and are therefore sex-linked: they show up mainly in one sex although the members of the other can be carriers. One of the best-known examples is haemophilia, a defect in the blood clotting mechanism mentioned earlier. The gene for the protein in question is located on the X chromosome; let us call the haemophilia defect X'. A female could thus be XX', with one fully functional X ensuring proper clotting, obscuring the presence of the defective X' and so not displaying the symptoms of the disease (only if she had inherited X' from both parents, and was thus X'X', would she herself be a haemophiliac). On average, half her daughters will also be carriers and the rest normal. Half her sons will be XY, and thus normal, but the other half will be X'Y, possess no functional gene and display the symptoms of the disease. All the daughters of male haemophiliacs will be carriers and all the sons normal. The following diagram shows the pattern of inheritance:

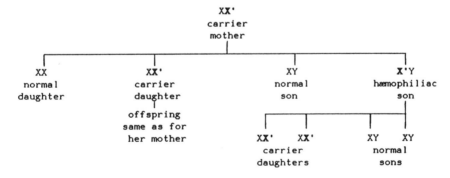

Genetic Engineering (Recombinant DNA Technology)

During the past few decades one of the most important advances in *molecular biology* (a name often applied to the interface between biochemistry and genetics) has been the development of techniques to allow the transfer of genetic information from one living organism to another; and to do so, moreover, in such a way that the protein(s) specified by the transferred gene(s) can be made (*expressed*) in the new host.

Its practical significance was recognized as soon as the scientific papers detailing these new discoveries were published in the 1970s. For example, many hormone substances which

Exploiting biotechnology

control the internal biochemical activities of animals (including man) are proteins. Although hormone function is similar over a wide range of animals, every protein hormone is unique, or nearly so, to the species which produces it. Just like enzymes, hormones are specific because each has to fit exactly with a receptor recognition site (a lock and key relationship) to which it must bind in order to achieve some defined chemical effect. It must not be possible for a hormone to attach itself to the wrong site or the whole array of interacting biochemical events would collapse in confusion. So protein hormone structure, like enzyme structure, must be absolutely correct for the job: being proteins, their structures are, of course, specified by genes.

Several medical syndromes result from a deficiency of one or other protein hormone in the patient: for example, some types of diabetes are due to a shortage of insulin. (Diabetes is a disease of sugar metabolism; untreated, some varieties may have serious consequences for the patient.) Therapy usually involves supplying the missing component which, in order to act correctly, ideally should originate from a human source although in some cases (insulin again) animal versions do function well. It is extremely difficult to obtain human material other than blood; living people can hardly be used as sources of supply and access to cadavers is a limited and very sensitive matter.

While each type of living organism has its own characteristics which differentiate it from all others, there is at the biochemical level an underlying unity embracing wide areas of metabolism, including the way proteins are made with DNA as the information source and RNA as the intermediary. Even more critical to the transfer of genes between different species is the fact that the primary "languages" of DNA (the four units A, C, G and T), RNA (another four units a little different from those of DNA) and protein (the 20 amino acids) are identical in almost all living systems. So are the three-letter codons specifying each amino acid. Thus, in principle, a gene from one bacterium could be used to make its protein in another; and different though people and bacteria are, a gene from a person ought also to work in a bacterium if a way could be found of getting it there.

Equipping a microorganism (a bacterium or a yeast) with the functional genetic information to make the human protein under production conditions could guarantee a supply of that protein without any of the problems of human sourcing — except for the gene, of course. But human genes are easy to obtain from very small samples of tissue taken from a living person. Even scrapings from inside the cheek can be sufficient. Hence the interest in moving genes from one organism to another, provided always that in their new biochemical environment they continue to function normally.

There are many technical problems attendant on these procedures; the biochemical details need not enter this discussion but some of them will be noted briefly towards the end of this chapter.

The Basic Concepts of Genetic Engineering

In nature, when individual microorganisms die their cell coverings usually dissolve and release their contents to the environment, including some or all of their DNA. Many microbes have the ability to absorb DNA from their surroundings; it is, indeed, one of the ways they

The science . . .

have for exchanging or passing genetic information encoded in DNA from one individual to another. That is a useful property because genetic improvements, when they arise, can be rapidly shared among many members of a population. It is, however, not without its risks: the DNA a microbe acquires from its environment may not come from a close relative but perhaps from something very different. The genes carried by the newly-acquired DNA and the proteins they specify would not then fit into the recipient's metabolism and might cause great damage. How can the recipient gain the potential benefits without incurring the risks? By being able to distinguish between DNA from its relatives and that from strangers, and taking steps to destroy the foreign material.

Each type of microbe therefore "marks" its own DNA in a characteristic chemical way and possesses a mechanism which continuously searches for these marks in all the DNA within the cells, including any absorbed from outside. If a bacterium finds that its mark is missing it knows the DNA is foreign and takes steps to destroy it before damage can be done.

Microbes sometimes generate internally or acquire from their environment useless short lengths of DNA which they quickly destroy. Their own genuine genetic material is untouched because that exists as very long strands. Once any foreign DNA is recognized by an absence of the marks, certain *restriction enzymes* cut the chain at very specific places to yield short lengths which are then degraded. Some enzymes cut across the two chains of DNA in a "dog-leg" fashion, with very interesting consequences.

The places where the cutting occurs have a sequence of DNA units, usually palindromic, which is absolutely characteristic for the particular enzyme doing the cutting. The following example is based as before on a hypothetical section of DNA with a continuous three-letter word sequence, but with a new message: THEOLDREDHENSAWTHEDUDEGGWAS-BAD. Let us represent a palindromic recognition site by the grouping

```
· · · E–D–U–D–E · · ·
    :  :  :  :  :
· · · E–D–U–D–E · · ·
```

Whenever this grouping is located, the cutting enzyme snips the two chains in the position shown by the arrows; the upper chain between the first E and D and the lower chain between the second D and E:

```
                    ↓
· · ·?-?-?-?-?-?-?-E-D-U-D-E-?-?-?-?-?-?-?· · ·
    :  :  :  :  :  :  :  :  :  :  :  :  :  :  :
· · ·?-?-?-?-?-?-?-E-D-U-D-E-?-?-?-?-?-?-?· · ·
                          ↑
```

This gives a structure in which the two strong chemical links (between the upper first E and D, and the lower second D and E) have been broken and the left- and right-hand ends of the whole structure are held together only by the weak links (colons) between -D-U-D- letter pairs:

Exploiting biotechnology

```
···?-?-?-?-?-?-?-E D-U-D-E-?-?-?-?-?-?-?···
   : : : : : : : : : : : : : : : : :
···?-?-?-?-?-?-?-E-D-U-D E-?-?-?-?-?-?-?···
```

Those links are not strong enough to hold the two long sections firmly together and they tend to come apart:

```
···?-?-?-?-?-?-?-E          D-U-D-E-?-?-?-?-?-?···
   : : : : : : :                : : : : : : :
···?-?-?-?-?-?-?-E-D-U-D        E-?-?-?-?-?-?···
```

Suppose that THEOLDREDHENSAWTHEDUDEGGWASBAD is the message in question:

```
THE OLD RED HEN SAW THE DUD EGG WAS BAD
::: ::: ::: ::: ::: ::: ::: ::: ::: :::
THE OLD RED HEN SAW THE DUD EGG WAS BAD
```

The site is recognized:

```
                              ↓
···T-H-E-O-L-D-R-E-D-H-E-N-S-A-W-T-H-E-D-U-D-E-G-G-W-A-S-B-A-D···
   : : : : : : : : : : : : : : : : : : : : : : : : : : : : : :
···T-H-E-O-L-D-R-E-D-H-E-N-S-A-W-T-H-E-D-U-D-E-G-G-W-A-S-B-A-D···
                                    ↑
```

The chains are snipped:

```
···T-H-E-O-L-D-R-E-D-H-E-N-S-A-W-T-H-E D-U-D-E-G-G-W-A-S-B-A-D···
   : : : : : : : : : : : : : : : : : : : : : : : : : : : : : :
···T-H-E-O-L-D-R-E-D-H-E-N-S-A-W-T-H-E-D-U-D E-G-G-W-A-S-B-A-D···
```

and the left- and right-hand sections come apart:

```
··T-H-E-O-L-D-R-E-D-H-E-N-S-A-W-T-H-E         D-U-D-E-G-G-W-A-S-B-A-D·
  : : : : : : : : : : : : : : : : : :             : : : : : : : : :
··T-H-E-O-L-D-R-E-D-H-E-N-S-A-W-T-H-E-D-U-D       E-G-G-W-A-S-B-A-D·
```

The same process can take place with any message so long as the correct recognition site is present:

The science...

```
···JIM FOX HAD ONE DUD EYE TOO···
   ::: ::: ::: ::: ::: ::: :::
···JIM FOX HAD ONE DUD EYE TOO···
                |
                ▼
```

```
                ↓
···J-I-M-F-O-X-H-A-D-O-N-E-D-U-D-E-Y-E-T-O-O···
   : : : : : : : : : : : : : : : : : : : : :
···J-I-M-F-O-X-H-A-D-O-N-E-D-U-D-E-Y-E-T-O-O···
                                ↑
                |
                ▼
```

```
···J-I-M-F-O-X-H-A-D-O-N-E  D-U-D-E-Y-E-T-O-O···
   : : : : : : : : : : : :  : : : : : : : : :
···J-I-M-F-O-X-H-A-D-O-N-E-D-U-D  E-Y-E-T-O-O···
```

The propensity of the two exposed trailing D-U-D groups to pair together means that left-hand and right-hand sections tend to stick to one another under appropriate chemical circumstances: the jargon for this phenomenon graphically describes them as being *sticky ends*. So two (or more) messages may be *recombined*. In the first case, the separate left-hand end of the message THEOLDREDHENSAWTHEDUDEGGWASBAD and the right-hand end of JIMFOXHADONEDUDEYETOO meet each other:

```
···T-H-E-O-L-D-R-E-D-H-E-N-S-A-W-T-H-E           D-U-D-E-Y-E-T-O-O···
   : : : : : : : : : : : : : : : : : :           : : : : : : :
···T-H-E-O-L-D-R-E-D-H-E-N-S-A-W-T-H-E-D-U-D     E-Y-E-T-O-O···
```

The two sections find one another and are held weakly in place by their sticky ends:

```
···T-H-E-O-L-D-R-E-D-H-E-N-S-A-W-T-H-E D-U-D-E-Y-E-T-O-O···
   : : : : : : : : : : : : : : : : : : : : : : : : : : :
···T-H-E-O-L-D-R-E-D-H-E-N-S-A-W-T-H-E-D-U-D E-Y-E-T-O-O···
```

and the missing bonds (arrowed) are reformed:

```
                                   ↓
···T-H-E-O-L-D-R-E-D-H-E-N-S-A-W-T-H-E-D-U-D-E-Y-E-T-O-O···
   : : : : : : : : : : : : : : : : : : : : : : : : : : :
···T-H-E-O-L-D-R-E-D-H-E-N-S-A-W-T-H-E-D-U-D-E-Y-E-T-O-O···
                                        ↑
```

Exploiting biotechnology

The new double-stranded molecule with the reading frame restored is thus:

···THE OLD RED HEN SAW THE DUD EYE TOO···
::: ::: ::: ::: ::: ::: ::: ::: :::
···**THE OLD RED HEN SAW THE DUD EYE TOO**···

The two messages could just as easily be recombined the other way round:

···J-I-M-F-O-X-H-A-D-O-N-E   D-U-D-E-G-G-W-A-S-B-A-D···
: : : : : : : : : : :   : : : : : : : : : : :
···**J-I-M-F-O-X-H-A-D-O-N-E-D-U-D**   **E-G-G-W-A-S-B-A-D**···

|
▼

···J-I-M-F-O-X-H-A-D-O-N-E D-U-D-E-G-G-W-A-S-B-A-D···
: :
···**J-I-M-F-O-X-H-A-D-O-N-E-D-U-D E-G-G-W-A-S-B-A-D**···

|
▼

↓
···J-I-M-F-O-X-H-A-D-O-N-E-D-U-D-E-G-G-W-A-S-B-A-D···
: :
···**J-I-M-F-O-X-H-A-D-O-N-E-D-U-D-E-G-G-W-A-S-B-A-D**···
↑

and restoring the reading frame:

···JIM FOX HAD ONE DUD EGG WAS BAD···
::: ::: ::: ::: ::: ::: ::: :::
···**JIM FOX HAD ONE DUD EGG WAS BAD**···

The palindromic sequences in different sections of DNA do not have to be in register in the same reading frame. Unfortunately the analogy with English words breaks down here and the authors could not find a way of manipulating three-letter word combinations to illustrate a change of reading frame while simultaneously retaining the implications it would have in DNA. In English, changing the reading frame by inserting or deleting one letter produces three-letter nonsense "words" after the insertion or deletion. In DNA, however, insertions and deletions produce a different kind of nonsense: the amino acid sequence becomes nonsense but the codons themselves do not. For instance, the DNA sequence ···ACG-ATA-GTA-TGC-CGA-CCT··· specifies the amino acids ···C-Y-H-T-A-G··· Deleting one "letter" of DNA (') to give ···ACG-ATA-G'AT-GCC-GAC-CT··· changes the amino acid sequence to ···C-Y-L-R-L··· Thus, the altered DNA sequence does not cease to issue interpretable instructions for adding amino acids to a growing protein chain but the resulting chain is aberrant and hence "nonsense".

The science . . .

By DNA standards these messages are very short; in real genetics they might be 100-200 three-letter groups long for each gene, thousands of genes being carried on a strand of DNA. Every so often, perhaps thousands or tens of thousands of triplet groups apart, the configuration which we have called

$$\cdots \text{E-D-U-D-E} \cdots$$
$$: \quad : \quad : \quad :$$
$$\cdots \text{E-D-U-D-E} \cdots$$

will recur; being susceptible to attack by the enzyme, whole sections, perhaps containing many genes, will be excised and liberated as long as they are not marked for protection against the particular cutting enzyme being used. Suppose that DNA, a section of which is to be inserted into a bacterium in order to make a particular protein, is isolated from a source organism and, being appropriately unmarked, is in the test tube artificially cut into sections by the enzyme; care has to be taken, of course, that there is no susceptible site within the gene coding for the protein of interest. A recipient bacterium is chosen which does not produce that cutting enzyme and which possesses DNA with only one unmarked and hence susceptible site for the enzyme to cut. Its DNA, too, is isolated and nicked by the enzyme which, because there is only one site for attack, will open a gap in the bacterial DNA without fragmenting it into many pieces. One side of the gap will have the characteristic left-hand sticky end, the other the right-hand version. Since DNA is a flexible molecule which can twist and turn to widen the physical gap between the sticky ends, enough room is made for the sticky-ended piece excised from the other source to fit in place as an extra section in the recipient bacterium's own DNA; the following diagram shows what happens.

If the bacterial DNA is:

$$\text{X- - - - - - - -E-D-U-D-E- - - - - - -Y}$$
$$\textbf{X- - - - - - - -E-D-U-D-E- - - - - - -Y}$$

and the foreign DNA (in italics) after excision is:

$$\textit{D-U-D-E- - - - - - - - - -E}$$
$$\textit{\textbf{E- - - - - - - - - -D-U-D-E}}$$

fitting the two together will give:

$$\text{X- - - - - - -E-}\textit{D-U-D-E}\text{- - - - - - - - -E-D-U-D-E- - - - - -Y}$$
$$\textbf{X- - - - - - - -E-D-U-D-}\textit{\textbf{E- - - - - - - - -E-D-U-D-E}}\text{- - - - - -Y}$$

The missing bonds can be reformed with the result that a foreign piece of genetic material has been "stitched" into the bacterium's own DNA. With the proper chemical stimulus this new piece of DNA can be absorbed into the bacterial cell where it serves as an additional new template for making protein.

Exploiting biotechnology

Genetic Engineering with Animals and Plants

Manipulating the genes of plants and animals presents more technical problems than those in bacteria. Not only is their genetic organization very much more complex but the cells themselves are part of a large structure, the body of the animal or plant. In order to insert foreign genetic material into the cells they have to be cultivated outside their parent bodies as single-cell preparations, rather like bacteria. But animal and plant cells are very sensitive to their local environment. They normally live under constant and stable conditions within the body, a conducive environment which must be reproduced in the laboratory if the cells are to survive and flourish. It can be done but the techniques tend to be elaborate.

There are several ways of inserting foreign DNA into such isolated cells:

- in certain chemical environments, animal cells will take up DNA from their environment, just as bacteria will. The process can be helped by applying an electric current which helps to enlarge the natural pores in the membrane surrounding the cells, a method called *electroporation*;
- because of their large size it is possible by *microinjection* to inject a DNA-solution directly into animal cells using a minute syringe and micro-manipulator;
- another way is to incorporate the DNA into certain viruses which are then allowed to infect the animal cells;
- using other viruses, it is sometimes possible to persuade cells from different animals to fuse together, resulting in cell hybrids containing the DNA from both;
- in the technique of *particle acceleration*, the DNA to be transferred is coated onto minute gold particles which are fired at the target cells in an electric discharge apparatus. Perhaps surprisingly, many of the target cells survive this shotgun approach and incorporate the foreign DNA.

Because of their rigid walls, cells from plants are more difficult to deal with. Some of the animal techniques will work if first the wall is stripped off (the naked plant cell is called a *protoplast*), although the cells are very delicate until the wall is regrown; protoplast fusion is one of the methods of generating new plants from two species which do not cross naturally. There is also a special method which works in plants susceptible to the bacterium causing crown gall disease. Because this bacterium transfers part of its own DNA into the plant cells, any genes which it is desired to introduce to the plant are first incorporated into the bacterial DNA.

Genetic engineering of plants presents yet more complexity. Photosynthesis in plant cells is sited in organelles called *chloroplasts* which contain a third genome in addition to the ones in the nuclei and mitochondria. Most attempts at gene manipulation have sought to integrate foreign genes into the nuclear DNA but there is a fear that they could be spread from the transgenic plants to wild relatives and weeds as a result of cross-pollination. Since pollen has no chloroplasts, it might be possible to avoid such spreading by inserting the foreign genes into chloroplast DNA.

The science . . .

Turning Genetic Engineering into a Working Technology

Putting foreign genes into bacterial DNA is no more than the beginning of the process to get the bacteria to make the foreign protein. Problems which have to be overcome include actually persuading the bacterium to read the information from its new DNA. Since bacteria already have mechanisms, some well understood, for indicating to them when to read specific parts of their own DNA in response to outside chemical signals, it is often possible to modify the foreign DNA so that it will respond to the same signals.

Another difficulty is that the foreign protein, because it is foreign to the bacterium, has no part to play in the bacterium's normal biochemistry: the bacterium does not "know" what to do with it. The foreign protein might simply accumulate inside the cell and choke off vital functions. But bacteria do have mechanisms for exporting certain proteins which they make in order to attack and break down food molecules too large for the cells to absorb in their original forms: more genetic manipulation of the foreign gene may be undertaken to arrange for its protein also to be exported into the liquid in which the bacterium is growing. That would not only allow the bacterium to continue indefinitely to make the product; it would also greatly simplify its subsequent purification.

Although the fundamental system for encoding genetic information is similar in all living organisms, a problem arises from an organizational difference between bacteria and all the others. In bacteria, the genetic messages are encoded, as it were, in plain language:

THECATSATANDATETHERATANDRANOFF

In all other living organisms the message contains extraneous information (called *introns*) which corrupts the message and has to be removed by "editing". The following is the same message with the addition of two introns (underlined):

THECATS<u>QQWYPHYGBNRZZ</u>ATANDATE<u>QQHNDGOFCHZZ</u>THERATANDRANOFF

Putting in the reading frame makes the effect of introns on the plain language message more obvious (letters in italics are nonsense as the message stands):

THE CATS *QQ WYP HYG BNR ZZA* TAN DAT *EQQ HND GOF CHZ ZTH* ERA TAN DRA NO

(In this model introns start with QQ and finish with ZZ. There will be no confusion with real three-letter words because there are none with those letter combinations.)
"Editing out" the two introns gives:

THECATSATANDATETHERATANDRANOFF

and with the reading frame in place:

THE CAT SAT AND ATE THE RAT AND RAN OFF

Exploiting biotechnology

Before such a genetic message can be used to make protein the introns must be recognized and removed by editing. Higher organisms have the mechanism for doing so; in the model used here this implies locating QQ and ZZ, then excising both the markers and the portion of DNA between them. Bacteria cannot edit out introns so if genes from higher organisms, including man, are to be usefully inserted into bacterial DNA the introns have first to be removed. There are techniques for doing this and unless it is done bacteria cannot produce human and other animal proteins, although other microbes such as yeasts, which are eukaryotic fungi — not prokaryotic bacteria — and do possess the editing mechanism, can do so. This is one of the reasons why yeasts are tending to find favour for the production of animal proteins.

This view of genetic engineering is of necessity very simplified. Many other manipulations and rearrangements are possible but even the examples used here may give some sense of the potential and versatility of recombinant DNA technology and the promise it holds for the future of biotechnology.

A Word About Terminology

Not every genetic potentiality of an organism is used all the time. In the human case, we obviously change as we grow older: body proportions and shape alter, sexual characteristics develop and may in time decline, hair changes colour and may cease growing, and so on. Those properties which actually take physical form (sex, height, physique, colour of eyes and hair, shape of nose, maybe a tendency to get duodenal ulcers, etc.) are said to be *expressed*. The totality of an organism's genetic information is called its genome and we can distinguish between the *genotype*, the sum of all its genetic potentialities, and the *phenotype*, that collection of properties which at any one time and in any one individual is actually expressed.

Genetic engineering, like most new technologies, has generated its own jargon. Most of it need not concern us but one word does need clarification: *clone*. It means "a population of identical individuals derived from a single parent". The population can be a genetic one: thus a large number of identical genes produced by replication from a single original are referred to as being a "clone". Transferring a gene from its original source to another (recipient) organism is, by extension, called *cloning*. The extension sometimes goes further: a protein expressed by such a gene is also said to be "cloned". Thus, a protein cloned from organism A to organism B is one whose gene has been so manipulated that it now forms part of the genetic complement of organism B; the latter might also be able actually to use that new genetic information to make A's protein itself.

Microbiology

Just like plants and animals, almost all microorganisms are living beings complete in themselves but most are so small that they can be seen properly only with a microscope. They are found everywhere on earth from the equator to the poles, in the air, in the waters, in soil and apparently also deep underground. They are the causes of many diseases of humans, plants, animals and also of other microbes; they degrade plant and animal carcasses and facilitate the recycling of their chemical components; and some of them play a vital role in fixing nitrogen from the air into a form in the soil which plants can employ for their growth. There is an enormous variety of microorganisms and some are of immense value to biotechnology and to manufacturing industry for the following reasons:

- between them they produce a wide range of chemical products, many of direct industrial value;
- they are usually cheap and quick to grow and easy to employ;
- they are comparatively readily manipulated genetically which means that they can be redesigned and reconstructed to produce both their own by-products in greater yield and chemicals originating from other organisms, including humans.

Microbiology deals with five categories of organisms, of which three (*bacteria*, *viruses* and *fungi*) are of major biotechnological importance and will be discussed here most extensively. *Protozoa*, tiny animals, have little involvement in biotechnology while the *algae*, equally tiny and the most primitive sorts of plants, are gradually growing in importance but have not yet quite enough significance to merit more than a brief occasional mention in this book. The *viruses*, the smallest of all microorganisms, lie on the borderline between true living things and the inanimate, chemical world. They are responsible for many human, animal and plant diseases (influenza, mumps, polio and perhaps even some cancers in man; foot-and-mouth in cattle and rabies in mammals generally) and their significance for biotechnology is twofold: in the production of vaccines for the prevention and treatment of certain diseases and as tools for some types of genetic manipulation. But as this book is not a technical exposition and viruses are not production organisms they, like the algae, will appear no more than occasionally.

Bacteria

These minute organisms are the very epitome of efficient biological organization. Most of them are shaped like spheres or short rods, so small that a thousand spheres or 300–500 rods laid end-to-end would span no more than a millimetre. If a teaspoon were packed solid with them it might contain as many as 5, 000 000, 000 000 individuals. Within the extremely small volume occupied by each cell, of which about 80% is water, there exist up to 5,000 different types of protein with perhaps a total of 1 million protein molecules in all. In itself each bacterial cell is a complete viable entity and, placed in a compatible environment well supplied with the raw materials it needs, each will grow and produce progeny. Their organizational efficiency is so great that under the most favourable circumstances some bacterial cells multiply every ten minutes. A collection of bacteria colonizing a surface is shown in Figure 2.3; the individual rod-like cells are each about 1/500th of a millimetre in length.

The common pattern of increase is for each cell to grow and, when it reaches a certain size, to divide into two identical daughters. Each daughter in turn enlarges and divides, each

Exploiting biotechnology

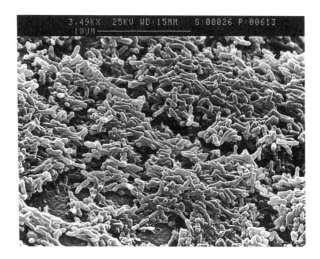

Figure 2.3 Bacterial cells: This electron microscope picture of bacteria growing on a surface shows how densely they pile one on top of another. (*From J Jass and H M Lappin-Scott*)

granddaughter..., each great granddaughter... and so on until the food runs out. Under a prescribed set of nutritional and environmental conditions, the time between divisions is approximately constant but for different organisms, or under different conditions, the times vary greatly — from minutes for the most rapid to months or even longer for the most laggardly.

This pattern of growth means that a burgeoning population undergoes a very characteristic course of development and decline. Imagine a population of bacteria which had been deprived of nutrient for some days and had therefore had no recent opportunity for growth; they would be rather moribund and small sized in their comparatively starved state. Suppose a million of them are suddenly placed in a flask of warm water containing the foodstuffs they needed and the flask is shaken to make sure that plenty of air was dissolved in the water and there is no shortage of oxygen. The bacteria immediately begin to change their chemistry in response to their new-found good fortune: this stage of growth is called the *lag phase*. Each cell grows larger. They become chemically more active but for a while their numbers do not increase. Once they reach a critical size, each cell divides into two. The offspring begin immediately to increase in size and they, too, divide when they become big enough; this happens over and over again at some constant interval, say one hour. Thus, an hour after the lag phase is over, the original million cells have become 2 million, an hour later 4 million, an hour after that 8 million, then 16 million, 32 million, 64 million... faster and faster, doubling in numbers every hour through the *exponential phase* of growth, until a shortage of resources develops and everything has to slow down and stop. The *stationary phase* has begun. No significant further growth is possible because the foodstuffs have all gone. Some cells die and dissolve, thereby providing a little food for a flicker of growth activity among the survivors, but the population gradually declines and in the end all are dead. Unless, that is, they belong to a type which

The science...

can wall themselves off into tiny resistant *spores* able to weather the catastrophic shortage of nutrition which has befallen them until the next opportunity for a growth boom arises.

Bacteria in Biotechnology

Bacteria are very versatile chemically. Many different products can be made. They are simple and rapid to grow and often their food requirements are inexpensive and easy to satisfy. Some of the products they make appear during the exponential phase of growth while others are released actually during the stationary phase, when the cells have stopped growing, or in between the two phases when growth is slowing down. Their store of genetic information is rather limited, and its organization relatively simple, so that by using the genetic methods described above it is fairly easy to introduce into them the instructions to make a human protein like insulin, for example, which they then make as if it were one of their own.

But there are problems with making human proteins in bacteria. One is that some human (and other animal and plant) proteins have sugars attached to them and, because they are absent from their own proteins, bacteria do not naturally have the biochemical apparatus for putting such sugars in place. Another problem is that most of the protein molecules made by a bacterium are retained inside the cell. That complicates manufacturing processes because the cell has to be broken open before the protein can be recovered, and breaking such tiny particles presents a number of technical problems. But, as bacteria often excrete some of their proteins into the environment, it may be possible to use this property to arrange for them to export a foreign (perhaps human) protein, the instructions for which have been grafted into their genetic information store.

Mutation and Selection

As we noted in the discussion of genetics, replacement, deletion and insertion mutations are all ways in which DNA messages may be altered in an inheritable way. Unless the mutations are *lethal*, i.e. so damaging to one or more essential enzymes or other proteins that the cells are unable to survive and reproduce, they will be transmitted to the offspring.

Mutations occur naturally in all species of living organisms. One reason is that the biochemical mechanisms governing the recognition and pairing of bases (letters in our examples) during DNA replication are not quite free from error. Very rarely mistakes are made and the wrong base/letter occasionally inserted. In a genetic sense the cell may not "know" that an error has been made. Only when a protein is subsequently synthesized using the corrupted information may the error show up; however, some errors are actually neutral in their effect and can be detected only by detailed chemical analysis of the DNA itself. The reason why not all DNA changes are reflected in altered protein structures is because 64 DNA triplet codons are in use, three for punctuation and 61 to code for the 20 different amino acids of proteins. The code is therefore "redundant", with most amino acids being specified by more than one codon — three amino acids have 6 codons, five have 4, one has 3, nine have 2 and only two have a single codon each. This means that not all mutations necessarily change protein structure, as the following example with the amino acid called *serine* will show. This amino acid is one of those having 6 codons: AGA, AGG, AGT, AGC, TCA and TCG. If a mutation were to occur in the third base of AGA to give AGG, AGT or ACC, the codon would still specify serine. Only if it affected the first or second bases (perhaps to change AGA to CGA [specifying alanine] or to ATA [tyrosine]) would the consequences be serious.

Mutations may be promoted artificially by chemicals, X-rays, ultraviolet light, radioactivity and other agents which change the chemical structure of one or other of the four bases A,C,G and T. Some agents cause one type of mutation rather than another. All can be used deliberately to mutagenize organisms, say bacteria intended for an industrial process, in the hope of generating in the population one or more mutants having more desirable properties: the mutants might metabolize faster, produce a higher yield of the desired product, or accept a simpler feedstock. The chance of such a mutation is likely to be low, perhaps no more than one for every 10 or 100 million treated bacterial cells. The question is then how to find the mutant among all the other cells: the needle in the haystack seems simple by comparison and looking for a needle by sifting manually though the mass of hay may not be so very different from growing each one of those 10 million bacteria separately in flasks and testing for the new property. Fortunately there are short-cuts: finding a steel needle is much easier if you have a magnet and something analogous is done with bacteria.

Consider a bacterium able to grow only slowly unless it is supplied in its growth medium with a small quantity of some essential chemical (a bacterial vitamin). For a valuable industrial process, and if the vitamin were expensive, it might be cost-effective to use genetic methods to eliminate the requirement. Suppose the bacterium is not totally incapable of making the vitamin for itself but just cannot make it fast enough to sustain rapid growth: perhaps one of its control mechanisms is defective and might be repaired by mutation. But even if mutagenesis does by chance repair a few cells, how is one to find them among so many of the originals? By spreading all the 10 or 100 million mutagenized cells onto the surface of a nutrient jelly containing everything they need for growth except for the vitamin, the original cells will fail to grow while any vitamin-independent mutants will rapidly develop to form visible colonies on the jelly (see Figure 2.4). The colonies can be picked off with sterile needles and each of the mutant colonies purified from any unchanged cells adhering to them by spreading them onto a fresh jelly minus vitamin. This time round, most of the cells on the jelly will be the mutants, with few of the unchanged originals still present. Do it once or twice more and the mutants will be entirely freed from any originals. Selection procedures like this for finding needles in microbial haystacks are the very essence of experimental microbiology.

Using Bacteria to Make Products

Growing microbes in a flask, the way it was described a few pages back, is called a *batch culture*. Its merit lies both in its simplicity and its usefulness for accumulating those products which are generated in the twilight period as growth slows down. However, its value is more

The science . . .

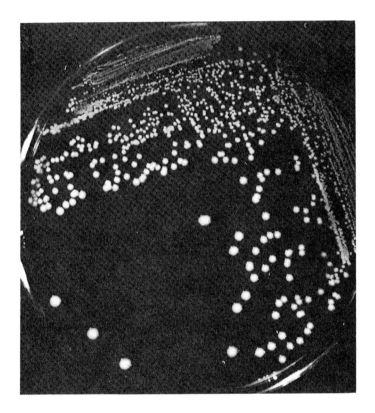

Figure 2.4 Bacterial colonies on a jelly (agar) surface. The dish is about 4 inches (10 cm) in diameter. A tiny drop of culture in a liquid medium was placed near the top left-hand rim and repeatedly smeared with a small wire loop further and further over the jelly surface. The dish was covered and kept warm for a day or two to allow the colonies to develop. Each white blob represents a colony containing millions of individual bacterial cells. Where the smearing started, the cells were deposited so densely that they grew into colonies which merged together. Further across the dish individual cells were left much further apart and each gave rise to a discrete colony. (*Reproduced from General Microbiology by R F Boyd, 2nd Edition, Mosby, Missouri, USA*)

limited for products made during the exponential phase which simply does not last long enough in batch culture: once growth has produced a population large enough for the rate of product formation to have become significant, only a few more doublings of the population will be possible before resource depletion becomes limiting and exponential growth stops. An alternative design is a *continuous culture* in which fresh growth solution is fed into the vessel continuously as fast as the bacteria exhaust their supply, the excess volume, after removal of the bacteria, being processed to recover the product. In the vessel itself growth is always exponential at its maximum rate, thus ensuring the continuous and rapid generation of products which are made only in that growth phase.

Exploiting biotechnology

Many bacteria need oxygen for growth. Some can use it when it is available but can otherwise do without. Yet others cannot grow at all in oxygen, and some are rapidly killed by its presence. Bacteria in all these categories have their value for production processes but each demands its own proper and correct handling procedures.

Maintaining Efficient Production

Because bacteria are so small, and many are resistant to adverse conditions, they are found everywhere, carried by wind and water, on the skins and in the intestines of animals and on the surfaces of objects. A bacterial production system, carefully optimized for a particular product, would fail disastrously if unwanted organisms were able to invade the production vessel and capture the feedstocks for their own growth without generating the desired products. Keeping them out can be an elaborate and expensive business. All equipment must be sterilized before use, often with live steam, to exterminate adventitious microbes. All nutrient medium for the production process is commonly also sterilized: heating under high pressure in an autoclave (a large pressure cooker) will eliminate all contaminants but may inactivate some of the chemical feedstocks. Solutions of heat-sensitive chemicals might have to be filter-sterilized, a very mild but expensive procedure.

Air injected to maintain the oxygen supply also has to be sterilized, usually by filtration, but that is easier and cheaper for a gas than for a liquid. During operations, all manipulations of the process vessel which involve opening it to add or remove anything must be conducted with proper precautions to prevent extraneous contaminants from gaining access. Seals through which stirrer drive shafts enter the vessel are particularly vulnerable points for microbial entry and must be protected. Important though all these precautions are with batch processes, they are even more critical with continuous cultures which might be designed and intended to be run for months without being stripped down, cleaned and resterilized. But continuous cultures have an important economic advantage in not requiring frequent down time to clean out, recharge and sterilize the equipment. Further consideration of production is reserved until the next chapter.

Fungi (Moulds)

While the higher forms of fungi (the mushrooms and their relatives) are beyond our scope here, much of what has been said about handling bacteria applies equally to lower fungi, including yeasts. Like bacteria, they are biochemically versatile, many are fast growing, they generate large numbers of valuable products and the techniques for handling them in production are similar to those for bacteria. From a practical point of view the distinctions reside primarily in the different styles of growth and the more complicated biochemistry and genetics of the fungi.

Most lower fungi grow as filaments: most people are familiar with the extremely fine hair-like structure of the white, pink or blue-green mould colonies found on decaying bread

The science . . .

or fruit. The colonies grow by the filaments elongating and branching, sometimes throwing up special reproductive structures for producing resistant spores analogous in purpose to those of bacteria, though different in structure. Exponential growth obviously differs from the simple division of most bacteria but it has its similarities in the growth pattern of the filaments. Since the latter do not fragment in the way that daughter bacterial cells separate from one another but may give rise to large, matted clumps, there are problems in some cases with achieving continuous culture regimes for fungi. The yeasts are a special case among fungi; they do not produce filaments and are much more like bacteria in the way in which cells divide regularly to form daughters. Most fungi, but not all, need oxygen for growth: yeasts employed for alcoholic fermentation are among the ones that do not.

The Matter of Viruses

Earlier in this chapter viruses were dismissed in a few lines. They deserve a little more attention.

Unlike all the other types of microbes, viruses are incomplete in themselves and cannot live and breed independently; they are entirely parasitic and can develop only within the context of other cells. The form of a virus is simple, usually consisting of no more than a protein coat which encloses the DNA constituting its genetic information.

Infection takes place when a virus attaches itself to a particular protein recognition site on the host cell's surface. Both the host's and the virus's interacting proteins must match, so susceptibility to virus infection, like so much else in protein-related biology, is a specific process; viruses attacking one species are unlikely to attack another unless the relationship between them is very close. Having attached itself, the virus injects its own DNA into the host cell; some viruses possess extra structures, almost like molecular syringes, which facilitate this injection.

Having thus offered itself in a recognizable manner to the host cell's own mechanism for making RNA and protein, the viral DNA ensures that certain proteins coded by itself are then made by the host. These viral proteins promptly subvert the host cell's activities generally, forcing it to concentrate instead on making more *viral* DNA and *viral* proteins instead of its own products. The newly-synthesized viral components spontaneously assemble into large numbers of fully-fledged viral particles (complete viruses); eventually the host cell is either greatly modified or totally destroyed, liberating the viruses which may then infect other cells.

In a few viruses, RNA substitutes for DNA. Host cells, however, cannot replicate RNA molecules as they can DNA: they make their RNA on DNA templates as described in the section on genetics. What happens in the case of *RNA viruses* is that the viral RNA is used as a template for making a certain type of enzyme which allows the DNA $\longrightarrow$ RNA process to flow backwards, making DNA *from* RNA; for this reason such viruses are called *retroviruses*. The DNA made from the viral RNA is used by the host's protein-synthesizing machinery, acting in its normal way, to make viral protein, while fresh viral RNA is made directly from this copied DNA. We will encounter retroviruses again in discussing AIDS.

Viruses have another and very dangerous property. In some cases, viral DNA can actually become integrated with the host cell's own DNA where it might lie dormant, perhaps for

Exploiting biotechnology

decades, before becoming activated by some stimulus and behaving again as a virulent agent. There is reason to believe that some human disease conditions of viral origin are like this, with very long incubation periods lasting until a trigger of some sort sets the disease going.

All forms of life, from bacteria, through plants and animals, to man are susceptible to viral infection. Not all such infections are dangerous and life threatening, but many are: people generally survive common colds but not AIDS. As mentioned earlier, viruses might also be the causal agents for some forms of cancer. In the past methods for dealing with them have been limited, but biotechnology has begun to offer a whole new armoury for coping with viruses and viral diseases, an armoury whose extent, variety and scope will undoubtedly increase rapidly in the next few years.

3 ...the engineering...

Products And Services

Biotechnology provides its benefits both as products and as services. The products are usually either pure materials or characteristic mixtures of substances (beer, for example) with certain more-or-less well defined and recognized properties. Exactly how such products are manufactured is not usually of particular interest to the purchaser so long as he is assured that they meet the proper purity and quality standards set by the appropriate regulatory and supervisory agencies. Within that remit the manufacturer may usually use whichever technology he considers most cost effective.

Since many biotechnological products result from microbial activity, the equipment for their manufacture is derived from years of laboratory experience in microbiology, genetics and biochemistry. The relatively small size of laboratory operations has, of course, to be increased for manufacturing on a commercial scale and this in itself introduces many problems and demands appropriate equipment modifications. A whole technology has developed to meet these needs. But, in principle, the way things are done in microbiological industry clearly follows from what happens in the microbiology lab.

The services of biotechnology are offered in rather different circumstances. A service provided for an industrial operation implies integrating a biotechnological process into the ongoing and wider activities of that industry. Later chapters in this book will consider such opportunities in some detail. They include: the clean-up of polluted sites on land and in water; the management of effluent streams to eliminate or minimize the discharge of undesirable substances; the employment of microbes to remove sulphur from coal before it is burned; and the use of microbial systems to improve the recovery of metals from low-grade ores and of crude oil from reservoirs.

In all these cases and others like them, biotechnology is not the primary operating technology and the industry concerned is therefore unlikely to overturn its existing well-tried practices to accommodate a biotechnological contribution. On the contrary: to be useful and acceptable, the biotechnologists must so design their own systems that they are maximally compatible with what is already taking place. Only if the installation of a biological technology incompatible with existing procedures were to mean the difference between clear economic success and total failure is an industrial activity likely to agree to major changes in basic procedure.

Later chapters will discuss the integration of novel biotechnology with contemporary industry; this one is dedicated to a more general consideration of manufacturing technology on the assumption that the product will be offered in a bottle or packet for the consumer to use as he thinks fit.

Exploiting biotechnology

MANUFACTURING MICROBIAL PRODUCTS

The productive versatility of microorganisms makes them favoured candidates for generating a host of chemical products. But versatility and a wide range of choices are no more than starting points for the development of specific manufacturing systems: for each particular product a defined microbe (or collection of microbes) must be chosen, its performance optimized, the most cost-effective production protocols developed and the requisite equipment made available and ready for use. Each of the following considerations is important:

- choice of the production organism;
- nutrient requirements and growth conditions for the organism;
- choice of the most cost-effective nutrient feedstock;
- ensuring optimum growth conditions for the organism;
- equipment design;
- maximizing product yield;
- monitoring the production process;
- extraction and purification of the product;
- packaging and distribution.

Choice of Organism

Hundreds of thousands of different microorganisms are already known to scientists. Stock cultures (i.e. samples of the living organisms) of many are held in public and private collections from which they may be obtained for a fee. Others are in the possession of individual researchers, derived from their own work or received from colleagues; details of their behaviour and other properties may or may not have been published in the open literature. Their owners are sometimes prepared to release their microbes to those wanting to use them, but patent protection might be in force and commercial negotiation will be necessary. Corporations involved in microbiological activities frequently develop and maintain their own culture collections.

In addition to tapping those sources, there is always the option of looking for microbes in natural or industrial environments. Microbiologists are skilled in recognizing and isolating microbes with desirable properties from complex, mixed natural populations. For example, if a search is to be made for new surfactant-producing strains, samples might be taken from oil-soaked environments, since microbes using oils as nutrients for growth often make surfactants as part of their chemical mechanism for attacking and incorporating the oils into their own metabolism.

An organism with the right collection of properties for making a naturally-occurring product will probably be identified from one or other of these sources. But, if the intended product is "unnatural" in the sense that it is a chemical from another organism (as in the example in Chapter 2 of a human protein being produced by a bacterium), the production organism

...the engineering...

will need to be manipulated genetically in order to equip it with the proper information to make the foreign product. The ease with which genetic manipulation can be achieved with a particular organism might become an overriding factor in deciding whether or not to use it.

The choice of microbe is likely also to be influenced by such factors as the equipment and other facilities already in place, by the experience the scientists and technologists conducting the programme have of the relevant microbiological parameters, and by the separation and purification procedures needed to extract the product and ready it for sale.

Growth of the Production Organism

Vessels for the growth of microorganisms are usually called *fermenters* from their early use in the fermentation of food grains to alcohol. Essentially, they are enclosed tanks. Depending on the scale of production they might be anything from tens of litres to hundreds of tonnes in capacity (Figure 3.1). The largest one ever built could accommodate about 1,300 tonnes of growth medium; we will meet it again in Chapter 9 (Figure 9.3). The volume of comparable laboratory vessels, which work on exactly the same principles as production models, can be as small as a fraction of a litre.

Figure 3.1 A 1,500 litre fermenter at the former Cetus Corporation in California, used for the production of pharmaceuticals involving recombinant DNA technology. (*Copyright Bio/Technology*)

Exploiting biotechnology

Many different designs are available (Figure 3.2) and the choice between them depends on their intended purpose and cost. Production vessels are usually made of stainless steel to minimize the undesirable consequences of corrosion. If the microbial production system requires oxygen, provision will need to be made for injecting either air or oxygen into the aqueous medium containing the cells. Aerating a large bulk of aqueous solution requires energy and is expensive. If the fermentation is to be run in the strict absence of oxygen, provision must be made for excluding air — probably also blanketing the contents with nitrogen.

The liquid in the fermenter tank needs to be stirred or otherwise agitated. If it is allowed to become stagnant there will be no way of introducing and distributing fresh supplies of oxygen as the organisms use it up, the microbes will in time settle out to the bottom and local deficiencies of nutrients and excess of excreted products will develop. In some cases the sought-after product is actually poisonous to the producing organism: alcohol production by yeast is a case in point. It is then important to prevent a local accumulation of product so as not to poison the microbes in that part of the fermenter. Many designs for agitation exist such as air bubbled through the liquid or rotating propeller blades. Alternatively, large amounts of air can be injected into the vessel to circulate the liquid by driving it upwards through a column; it then overflows at the top and falls back to the base while excess air escapes and carries with it the carbon dioxide and other waste gases produced in metabolism. Agitation might result in foaming which can be controlled either mechanically or with anti-foaming chemicals.

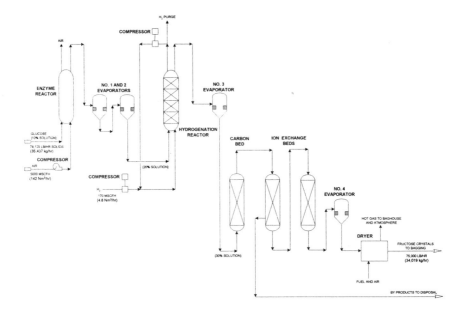

Figure 3.2 A typical flow diagram for a biotechnological production procedure. In this case the system is a speculative design for making the sugar fructose. (*Published by permission of N Hariatos*)

...the engineering...

The fermenter vessel will need entry ports for the introduction of the microbes and their feedstock as well as for probes to monitor the course of the fermentation. If running in the continuous culture mode, it will also need ports for removing the overflow culture while, if the batch mode is used, a drain will be fitted for emptying the vessel at the end of the run. There must also be provision for temperature control in the form of pipes circulating warm or cold water within or around the tank to cool or heat the reaction fluid as necessary.

Fermenters have to be cleaned and sterilized between runs. Sterilizing the empty vessel before filling is done with toxic gases or with live steam. All nutrient feedstock and other solutions must also be sterilized before introduction into the fermenter. The microbial seed culture used to start the run will have been carefully prepared to ensure that it contains only the right organisms for the job, with no unwanted contaminants. There is a slight risk that the seed culture will itself develop variants which, under some circumstances, will ruin the production run but, with proper precautions and a well-characterized microbial strain, the risk is small. If the fermenter is well designed and constructed, thoroughly sterilized before use and accepted procedures are followed during the run, the risk of extraneous infection can also be made very small.

Optimizing Product Generation

The yield of product in a microbial system is sensitive to a number of factors in the microbial environment. They include the concentrations of the various nutrients, the temperature, oxygen availability or exclusion, acidity and other variables.

Temperature control is normally effected by heating or cooling coils while oxygen is supplied by blowing in air, both as noted above. Where necessary, acidity is controlled by the measured addition of acid or alkali, often administered automatically in response to an acidity sensor immersed in the liquid. Additional supplies of some or all components of the nutrient feedstock might have to be added during the course of the fermentation.

All these parameters are extensively explored in the laboratory before large-scale production begins. Small laboratory fermenters are used to check microbial behaviour and the yield of product in response to variations in the microbes' physical and chemical environment. Scale-up takes place in intermediate sized equipment before transferring the technology to the large production vessels. It is important to realize that a large vessel is not simply a larger version of a small one of the same shape because the surface area-to-volume relationship changes dramatically as the volume increases — with profound consequences for heating/cooling, stirring and aeration. Large production vessels have to be designed with these considerations in mind.

Monitoring the Production Process

Reliable though biotechnological processes usually are, it is always important to monitor the progress of a production run. Unforeseen events do sometimes occur and reactions do not

Exploiting biotechnology

always run exactly as planned; if caught in time, it might be possible to rescue an aberrant fermentation before it is entirely wasted.

Environmental conditions within the tank, such as temperature, oxygen concentration, acidity and the amounts of some of the nutrient and product chemicals can often be monitored continuously with probes, and corrections made automatically to ensure constant conditions. Other measurements require the withdrawal of samples from the fermenter followed by analysis in the laboratory or elsewhere. Speed is often a requirement so that the production staff can coordinate operations to maximize product yield.

Extraction and Purification of the Product

Making chemical products with microorganisms is a water-based activity. The previous chapter noted that the microbes themselves are some 80% water. Microbes employed in manufacturing processes live and metabolize either submerged in water or floating on its surface. Aside from a few cases (such as yeast for the baking industry) in which the microbe itself is the product, chemical substances generated by microbial activity are dissolved in water — either the water within the cell or the water in which they are living. A few products are not soluble in water and either precipitate as solids or escape as gases, but in most cases product purification involves concentration and dewatering as an essential part of processing. It also requires separation from other chemicals.

If the product is released into the producing cells' aqueous environment it will find itself in the presence of other chemicals from which it will probably need to be separated and purified before sale. These might be:
- substances provided for microbial nutrition;
- chemical components in the nutrient feedstocks which are not necessarily essential for microbial growth but which it is too expensive to remove from bulk nutrients;
- living cells;
- all manner of compounds which have leaked from the cells or been excreted by them;
- dead and partially decomposed cells;
- soluble chemicals and particulate matter released by decomposing cells.

Some products are not excreted into the surrounding medium and must be extracted directly from the cells which have either to be broken physically to release their contents or have their outer coverings altered chemically in order to allow the desired materials to escape. Of course, other compounds will also escape and the resulting mixture of substances in the medium might become be very complicated.

Extraction will probably begin with the separation of the microbial cells (the *biomass*) from the liquid in which they grew and which, at the end of the production run, contains the chemical products. Filtration and centrifugation are the most common methods. Filtration is used for the coarser microbial accumulations like the mats of tangled fungal filaments. The much smaller bacterial and yeast cells must be separated in the centrifuge, that is, made to

...the engineering...

settle to the bottom of a tube by increasing the force of gravity acting upon them. Although bacterial and yeast cells are very small, given time they will eventually settle out of liquid by themselves. Spontaneous settling, however, is very slow and the mass of settled material so friable that it is likely to be disturbed and resuspended as soon as the liquid above it is moved. Centrifugation accelerates the process. Tubes containing the mature fermentation liquid are placed in cups mounted on trunnions distributed around a vertical axis. As the axis is rotated at high speed the tubes are swung out into horizonal positions and the small bacterial or yeast cells forced by the enhanced gravitational field into the tips of the tubes. There they are packed into solid masses which, after the tubes are again brought to rest, allows the liquid to be poured off or decanted without disturbing the sediment.

The subsequent purification of the products from the fermentation broth depends on their chemistry. They can, for example, be made to precipitate out from solution as solids (which can then be collected by centrifugation or filtration), washed out of the original aqueous liquid with various solvents or attached to certain solid matrices which are highly selective in their binding.

Selective attachment is very useful for purifying proteins. In Chapter 2 the natural recognition and binding properties between specific chemicals on the one hand, and certain proteins (including enzymes and some hormones) on the other, were likened to a lock and key relationship. In purification by the selective binding procedure the binding chemical ("key") is attached to a solid support. Its affinity for "lock", the protein to be purified, enables it to hold that protein onto the solid matrix. Other chemicals will not bind and can be washed away, leaving just "lock" protein attached to the solid support via its linkage with "key". The "lock"/"key" binding is specific and tight but not rock-solid; now and again, on a random basis, the pairs come apart. When that happens a molecule of "lock" can bind to any molecule of "key", not just the one it recently left. So if a solution containing a high concentration of "key" is flowed past the solid support, "lock" will transfer from bound "key" to free "key" and enter solution because any unattached "lock" molecules are more likely to reattach to one of the many "keys" in solution than to one of the few attached to the solid.

First the free "lock" protein is flowed past the attached "key" chemical:

```
////|
////|——key
////|                                    lock
////|——key
////|                     lock
////|——key
////|                                             lock
////|——key                                                   lock
////|
```

Exploiting biotechnology

Because of its affinity, the "lock" protein links to the "key" chemical:

```
////|
////|——key·····lock
////|
////|——key·····lock
////|
////|——key·····lock
////|
////|——key·····lock
////|
```

while other proteins and chemicals unable to bind to "key" are washed away. This is followed by adding a solution rich in *unbound* "key" so that the "lock" protein molecules now tend to detach from the bound "key" and attach to the much larger quantities of unbound "key":

```
////|                    key         key        key
////|——key·····lock            key         key        key
////|                    key key             key
////|——key·····lock            key         key        key
////|                    key             key key
////|——key·····lock            key         key        key
////|                    key key                   key
////|——key·····lock            key    key      key
////|                    key         key        key
                                ↓

////|                    key         key        key
////|——key                     key         key   key
////|                    key key·····lock        key
////|——key                     key         key   key
////|                    key             key key·····lock
////|——key                     key         key   key
////|                    key key·····lock        key
////|——key                     key    key   key·····lock
////|                    key         key        key
```

Further processing of this last preparation involves separating "lock" from "key" by one of a number of simple methods, dewatering and drying. Because they are so sensitive to drying, the removal of water from solutions of proteins and most other complex biological chemicals has to be carried out very gently, often by *freeze-drying (lyophilization)* — a process in which the solution is first frozen and, while still in that state, subjected to high vacuum. The ice vaporizes directly at the low temperature with minimum damage to the chemicals originally dissolved in the water; they remain behind as solids.

Packaging and Distribution

Finally the product is prepared for dispatch. The form in which it is shipped depends largely on its stability and how it is to be used. For commodity chemicals in the form of dry powders, packages, sacks and drums might suffice but pharmaceutical preparations will often be dispensed in accurately measured doses into sterile bottles, capsules or vials. Liquids and solutions will be sent out in sealed bottles, also probably sterile. Materials very sensitive to heat degradation require storage and shipping under cold conditions in a deep freeze or surrounded by dry ice (solid carbon dioxide which has a temperature of nearly $-80°C$). For others less sensitive refrigerator temperatures ($2-4°C$) will serve.

Production Costs

While the distribution of costs varies greatly with particular procedures, the following approximate breakdown for antibiotic production offers some indications of where they arise. The percentages refer to direct costs to which must be added overheads for management, administration, research and development, and depreciation of capital invested in buildings and equipment:

	Percentage of direct production costs
• feedstock	30–40
• other fermentation costs	20–35
• filtration to remove the microorganisms	3–5
• recovery of product	15–30
• purification and drying	10–20

4 ...and something about feedstocks

THE ROLE OF FEEDSTOCKS IN BIOTECHNOLOGY

The manufacture of biotechnological products depends on the conversion of feedstocks to useful products, a type of production technology with very special requirements because much of it also requires the cultivation of living organisms. In this book the organisms of greatest interest are mostly microbes and accordingly it is their nutritional requirements which are our primary concerns.

Aside from their uses as foods, the industrial value of microbes lies in the chemical products resulting from their metabolic and other activities. By virtue of their enzymes, microbes in industry act as catalysts for changing feedstocks into products. In some way and at some stage in the manufacturing process this involves the growth of a microorganism either because:

- product synthesis is an integral part of the growth process; or,
- even when product generation can be divorced from active growth, enough microbial biomass must be present to allow the catalysis to proceed at a satisfactory rate. That biomass will earlier have been grown for the purpose.

In practice, most well-designed fermentation processes, as well as many microbial activities which are used in such virtually uncontrollable pre-existing environments as oil reservoirs, fouled beaches and contaminated soil, seek to restrict the microbial population as far as possible either to a single type of microbe or to a mixture comprising a small number of well-characterized species. In only a few cases, of which sewage treatment is perhaps the best example, are mixed populations preferred because of the very complex and variable task to be performed.

The reasons for seeking simplicity are those of control and predictability. Like all living systems, microbial populations tend always to maximize their opportunities for growth and reproduction. Simple systems, in which the microbial components, their nutrient support and environmental conditions can all be precisely defined, are the ones offering the greatest certainty with respect to the quality and quantity of the product. The brewing of beer is a good example: contamination of the fermentation yeast with other microbes might not greatly influence the yield of the bulk product (alcohol), but may so alter the quantities and balance of those secondary products which give each beer its special flavour as to render the production batch unacceptable and unmarketable. Mixed populations, made up of different microbial species, will exhibit many complex biochemical interactions; they are correspondingly difficult to control and are avoided where possible.

Nevertheless, biotechnological manufacture does not depend only on microbial growth. Some processes are independent of growth taking place simultaneously with product synthesis; the two processes may be separable and the optimal feedstock requirements specifically for product manufacture might be different from the best ones for growth. Indeed, it may be

Exploiting biotechnology

important, once growth has provided enough biomass for the job in hand, to prevent further growth and confine as much biochemical activity as possible solely to the conversion of feedstock to product.

Each industrial process dependent on microbial activity will therefore be configured to maximize the effectiveness of the operating system through its various stages; the nature and the price of alternative feedstocks are often important factors influencing process design.

MICROBIAL FEEDSTOCKS

The nutrient substances required for microbial growth comprise:
- water — remember that this is by far the most common substance in all living systems;
- oxygen (usually as air), where necessary;
- a mixture of various mineral salts, each in comparatively small amounts. Such mixtures are rather similar to the nitrogen and phosphorus fertilizers used in agriculture;
- a source of energy (fuel);
- one or more carbon compounds for growth.

The first three categories are easy to supply and low in cost; the last two are more complicated.

Nutrients for Energy

The need for a fuel to supply energy was discussed as some length in Chapter 2. In green plants and a few exotic types of bacteria, energy demands are met by the absorption of sunlight. Some other bacteria can satisfy their needs by oxidizing certain minerals found in rocks. (This has considerable industrial importance and will be encountered again in Chapter 10.) All other living organisms, including all the other microbes of industrial significance, are dependent on the oxidation of complex chemicals which originate from green plants and those few light- and rock-using bacteria; most of these chemicals come from "current production" (consumption of living or recently deceased sources). Yet others, which can be used by a limited number of microbes, were made millions of years ago and have spent the intervening period deep underground being converted into oil.

Nutrients as Food and for Product Synthesis

All those complex fuel chemicals are compounds based on the element *carbon*, familiar as soot, graphite or diamond — although none of those forms of elementary carbon remotely resembles the carbon in chemical compounds. Once the water is removed, every familiar food material, all living organisms, and most of the products — like wood, paper, leather and wool — which are made from them are about 50% carbon. Carbon-containing compounds, therefore, have a dual role in biotechnological production: as fuel to meet the energy requirements of the living and growing microbes, and as a source of material from which the microbes build their own substance. Green plants and the rock-using bacteria are again different from the rest: they can get their carbon from an *inorganic* (that is, non-biological) source, the carbon dioxide in the atmosphere. Everything else needs the *organic* carbon of the complex chemicals. Growth and product synthesis, whether or not directly associated, usually have similar but not necessarily identical feedstock requirements.

For many microorganisms, one or more of these complex chemicals can serve both as bulk food for building and as fuel for energy. But, for their proper functioning, some microbes additionally require small quantities of a limited number of specific chemicals which they are unable to make by their own biochemical mechanisms and must therefore absorb with their food. Such chemicals are called *vitamins*; animals and people have analogous requirements. Vitamins are expensive compared with bulk nutrients and are used in industrial fermentations in the minimum quantities necessary to satisfy needs. For some purposes an unrefined source of vitamins, such as yeast or meat extract, is acceptable. For certain types of vaccine production and other fastidious pharmaceutical procedures very high-cost nutrient materials derived from animal sources are necessary, their expense being justified by the high added value of the manufacturing process.

OPTIONS FOR FEEDSTOCKS

Apart from the carbon dioxide used by plants and some bacteria, and the crude oil and its refined derivatives which, together with natural gas, can support the growth of a limited number of microbes, all carbon compounds used in biotechnological industry are agricultural products.

Views about the relative economic merits of agricultural versus petrochemical feedstocks for microbial fermentation have undergone radical revision over the past several decades. In the 1960s the widespread fear of a world food shortage coincided with a perception of a seemingly unlimited supply of cheap crude oil and petrochemical products. Consequently, industrial fermentations declined in favour of oil-derived processes and emphasis was placed on reserving agricultural production for human nutrition (see Chapter 9) and minimizing its use for industrial feedstocks: those would be supplied by the petrochemical industry. Such views changed abruptly with the steep oil price rises in the 1970s. Suddenly petrochemicals were seen as the shortage feedstocks with agricultural products more readily available, at least in many countries. Instead of converting oil and gas into food, attention turned towards using such agricultural products as sugar and starch to make fuel, especially fuel for vehicles.

Sugars

Carbohydrates are among the best general nutrient substrates for microorganisms. They are good fuels, soluble in water, fit well into the metabolism of most microbes and so are easily

Exploiting biotechnology

utilized by them. Solubility is an important property: because the microbial production of chemicals takes place in an aqueous environment, the more readily the microbes living in the water gain access to their nutrients in a water-soluble state the more easily they can use them. Sugars are very soluble indeed: think of how much sugar is dissolved when making jam!

In the discussion of biochemistry in Chapter 2 it was observed that carbohydrates often occur in long chains of repeating units; free single units (*monosaccharides*) or double units (*disaccharides*) are also found in nature. Among the latter, sucrose (table sugar from cane and beet) is one of the world's major agricultural products, with an annual production of 90-100 million tonnes. Sucrose is sold as refined and raw sugar, molasses, intermediate process streams and cane or beet juice. A disaccharide composed of two single (monosaccharide) units, sucrose can fairly easily be broken apart to its constituents, *glucose* and *fructose*. Sucrose is an excellent growth substrate for many microbes although some can make use of it only after it has already been broken down into glucose and fructose.

The use of sucrose as a feedstock is politically and economically complex. While the world price has been volatile it has also fallen to very low levels on several occasions during recent years. In the European Community, protective mechanisms at one time raised the price so much that some biotechnological production became unsustainable within the Community and moved to adjacent EFTA countries.

Certain biotechnological procedures (microbial enhanced oil recovery, for example — see Chapter 10) use molasses as a feedstock because of its low cost. Large quantities would be required if the procedure became widely established. Molasses, however, is not a primary commodity but a by-product in sugar manufacture. Sugar production in general would no doubt increase in the producing countries if the demand for it as an industrial feedstock were to grow but, because it is a secondary product, molasses production might not respond in the same way. A substantial increase in the demand for molasses could raise its price and undermine its use as a low-cost microbial feedstock for large-scale *in situ* crude oil recovery processes.

Gases

A possible microbial feedstock is methane, mainly from natural gas but also from landfill sites. Available in large quantities at low cost and in a relatively pure form, it can serve as a growth and energy substrate for a restricted range of microorganisms. Its handling, however, presents serious drawbacks. Because microbes can utilize methane only if they are able to oxidize it, a process dependent on the presence of oxygen, methane has to be supplied as a methane-oxygen mixture which is potentially explosive and therefore avoided. Furthermore, as methane is poorly soluble in water, it is comparatively difficult and therefore expensive to keep an actively growing microbial population adequately supplied. Other feedstocks are preferred.

Alcohols

A simple modification of methane, already a large-scale industrial activity, changes it to the very water-soluble compound *methanol* (*methyl* or *wood alcohol*). This is technically a much better feedstock than methane for microbiological use and has served as the basis for producing single-cell protein as an animal feed supplement (Chapter 9). It does not, however, appear to be in widespread use for other production procedures.

Solids

Two solid agricultural products have enormous potential value as biotechnological feedstocks but both need further, and perhaps extensive, development before they are likely to be deployed as widely as the sugars.

One of them is starch, produced on a huge scale agriculturally by wheat, maize, potatoes, rice, sorghum and cassava. It can easily be broken down to its constituent monosaccharide sugar (glucose) by treatment with acid or by enzymes. While glucose is one of the best general feedstocks for microorganisms, some of them produce their own starch-degrading enzymes and can therefore use starch as a feedstock without pretreatment.

The other is *lignocellulose*, the bulk material of most plants, including trees. It is thus potentially available in prodigious quantities either by direct cropping or in the form of waste products from the timber, paper and other wood-based industries. The main disadvantage is the difficulty which microorganisms have in attacking and utilizing wood and wood-containing plant materials. Compost heaps degrade plant waste over long periods; wood left in the open air under moist, warm conditions takes a long time to decay from attack by natural microbial populations which colonize dead trees and branches. In time ways will surely be found of pretreating woody plant material to render it more susceptible to microbial degradation and therefore more valuable as a production feedstock.

Cheese Whey and Other Wastes

Some existing microbial processes, like cheese production, generate waste streams which still contain fermentable chemicals and which might therefore become valuable as a feedstock for another process downstream. As wastes they might initially have negative value and be available simply for the cost of collection and transportation. However, once their use as an additional feedstock is perceived there may be a price attached to them which would reduce their economic attractiveness.

Exploiting biotechnology

ECONOMIC CONSIDERATIONS

The cost-effectiveness of a feedstock will vary according to the nature and unit value of the product; in some processes the cost of the feedstock is a major factor in production but in others it is relatively insignificant. The economics of making low volume, high value pharmaceutical and other health-care preparations is likely to be less dependent on the feedstock price than are large-scale processes with lower added value per unit of feedstock. While the procedures for making high added-value products might need and be able to afford expensive, high-quality nutrients others — like soil bioremediation or waste management — are obviously more restricted in their range of options.

5 Lawyers, legislators, communicators, funders — and the public

Just as in other businesses, management, marketing, accounting and advertising in biotechnology occupy their familiar places; we will explore them in some detail in Chapter 6. And also like the others, biotechnology naturally has its own special features. They embrace the legal status of its intellectual property, official regulation and licensing of its products and services, and the styles of communication between biotechnologists and about biotechnology. Sources of money are important too, not just commercial financing for business start-ups and growth *per se* (see Chapter 6 for those), but also for funding the basic research which mainly takes place in universities and public sector research institutions. Resourcing is needed as well for the later development work which leads directly to production and marketing. Furthermore, biotechnology has also a unique relationship with the public who, rarely if ever direct customers or clients, are nevertheless concerned and intrigued by many of its activities.

INTELLECTUAL PROPERTY

The protection of new inventions, as important in biotechnology as in any innovative business, may be achieved either by patenting or by retaining commercial confidentiality.

Protection by Patent

An analysis of patent procedure and a detailed review of patent law is beyond the scope of this chapter, not least because of the complexity of these matters and the variations between different legal jurisdictions. Nevertheless, a brief general discussion of the common pattern of practice applicable to most western industrial countries might be useful; those who wish to know more should refer to specialist publications.

Patent protection cannot be obtained for naturally-occurring living organisms or naturally-occurring chemicals but it is available for new organisms and novel, artificially synthesized chemicals. However, it might be possible to protect a new way of producing an existing chemical substance through a process patent, even though the material itself is already known and in the public domain. "Novel chemical" is obvious enough, but "new organism": what does that mean? It signifies an organism, at any level of complexity from a microbe through to a higher plant or animal, which has been produced artificially by genetic manipulation (see Chapter 2); in some cases it may also be applicable to plants and animals generated by conventional breeding procedures.

Exploiting biotechnology

Consider a biochemical substance of human origin which has therapeutic value for the treatment of a disease; for example, the use of insulin in cases of diabetes. As a natural chemical material, insulin cannot be patented. But it has in some way to be prepared from its source and that preparatory procedure could be a candidate for patent protection. Because human beings themselves are not available as sources of insulin most of it, since its discovery in the 1920s, has been prepared from pig pancreas obtained from slaughterhouses. Porcine insulin is not identical with the human variety, though it is very similar in structure and works satisfactorily except in a few patients who display allergic reactions. Human insulin should be a better drug for treating human diabetes; the advent of genetic engineering opened the possibility of transferring the human genetic "instructions" for making insulin into microorganisms which are then able to make insulin identical in all senses with the human product except that it will actually have been manufactured by a different route; that route is patentable. The microorganisms now carrying the instructions to make human insulin are novel: they exist nowhere in nature and will have been constructed artificially for a specific purpose. They, too, may be protected by patent.

An application for patent protection within the prevailing rules for a new "invention" is filed with the appropriate office, often but not necessarily in the area where the inventor works. Filing a patent application anywhere confers priority for the whole world — it establishes the claim to be first for that particular invention. The invention must actually exist physically and its properties and performance must be described in detail; protection cannot be sought for ideas which have not yet been put into practice, at least as an experimental system or model. The invention must be novel, not obvious to "someone skilled in the art" and no application can be made for the protection of an invention which is either already in the public domain or for which a patent has been issued or an application already made by someone else in any country of the world. As we shall see, evidence of priority in particular cases can be so important that some research and development organisations likely to be concerned about patent protection for their products and processes require witnessed dating of their employees' laboratory records on a daily basis as proof, if need be, of the date when an invention actually occurred.

Priority is tied up with prior disclosure, another critical and very sensitive issue. In Europe as well as in many other countries, no application can be made for the protection of any invention which has already been disclosed to the public. International conventions permit filing the same application in other subscribing countries within one year of the original filing, the priority date remaining that in the home jurisdiction. During that intervening year more work can be done and a modified application submitted; the original material retains the priority of the first filing date, the new material that of the second. In the US, by contrast, a one-year grace period is permitted between public disclosure and the latest possible filing date. Disclosure is defined widely. A statement printed in a published document is not the only form: a remark made, or a question asked at a public conference — or perhaps even at a seminar or meeting not confined to affiliates of the company — may be construed as publication. There is another important difference between the US and most other countries: if a dispute over the priority date arises between inventors, the date of invention can supersede the date of filing. Evidence is needed in any litigation and it is against such eventualities that some companies store their dated and witnessed laboratory notebooks under secure conditions.

Lawyers, legislators, communicators, funders and the public

A patent application, in addition to describing the details of the invention which will emphasize its unique character and hence distinguish it from all other inventions and knowledge in the public domain, will make claims. Those claims are written to secure for the inventors as much protection for their invention as possible, coupled with the maximum exclusion of other overlapping or superseding inventions which might come later. An example might help: suppose a new genetic procedure has been invented which enables a particular human chemical to be manufactured in a certain microbe. The narrowest claim will be just for that procedure. But, assuming that the invention warrants it, wider claims will be made to include other human chemicals, chemicals from other animals, manufacture in other microbes, and so on. In this way, the inventors will seek to prevent trivial alterations of their invention being used to circumvent their claims, and hence their own protection, while simultaneously attempting to place themselves in the most favourable position for prolonged commercial exploitation of their endeavours by excluding competitors.

Before a patent is granted, the issuing body will conduct a search to ensure the novelty of the invention and will scrutinize the claims and other details. During the search period the details of the patent application are not disclosed to the public but once the patent is granted the application is published in full for anyone to read.

Patent protection lasts for a specified number of years; during that time nobody, without permission of the patent holders, may make commercial use of the invention within the geographical area of jurisdiction of the issuing office nor, of course, can they apply for a patent for the same invention. The long time it takes for regulatory approval to be obtained, particularly for drugs for human application, and the need to file for patent protection before submission to the regulatory authorities, greatly reduce the protected period within which exclusivity allows the greatest commercial profitability. However, recent legislation in some countries has extended the protection for an additional period from the date of regulatory approval.

During the protection period the patent holders may exploit their protected invention in any legal way they wish. If they are not in a position to make full commercial use of it themselves (insufficient funds, not the right people, do not wish to become involved in a business), they may decide to sell their rights or license their use against payment of royalties. It is up to them: the patent is their property and they can do what they like with it. However, once the patent has expired others may freely make use of its ideas, although the originators hope by that time to have established an unassailable market position for which they no longer depend on the protection afforded by the patent.

Protection by Confidentiality

One difficulty with patents, aside from the costs of application and search and the delays before they are issued, is that once awarded the details of the invention are fully disclosed to potential competitors. The invention and the claims exactly *as written in the application* are indeed protected, but publication of the ideas might prompt others to seeks different solutions to the problem or to make use of an implied commercial opportunity which had

Exploiting biotechnology

not previously occurred to the inventor. In some cases, ways of circumventing the patent by design changes might not be difficult to effect and, if competitors are able to obtain protection for their own new (and perhaps better) invention, the original inventors might find themselves entirely excluded from what had once been a valuable place in the market.

For the following reasons, some inventions are better protected by commercial confidentiality rather than by patent:

- the invention might not be sufficiently novel to qualify for patent protection;
- the commercial prospects might not be worth the costs;
- publication of the patent may encourage competition or otherwise undermine the commercial position of the inventors.

Thus, it might in such cases be better not to patent but to retain the details of the invention as a trade secret.

Obviously, there are also risks in that course of action. Can the secret be retained in house: are the staff reliable and will they remain in their jobs? It may not be possible to prevent access by others to the product or process: will that in itself help another operator or supplier to start up in competition? But the risks of confidentiality may be ameliorated because the potential competitor, even if he does gain access to the new invention (perhaps a sample of the chemical or the microbe), may not properly know how to use it. In the meantime, the original inventors will have gained in experience, consolidated their market position and continually upgraded their invention so that the competitor will in any case have no more than last year's model in his possession. But if competition does arise successfully, it may be very difficult to combat. There is always uncertainty about which course to follow: patent and disclose or try to keep the secret and take the risk.

REGULATION OF BIOTECHNOLOGICAL PRODUCTS AND ACTIVITIES

As with the protection of intellectual property, details of regulatory legislation vary between jurisdictions, although the overall patterns are broadly similar in industrial countries.

The main impact of official regulation affects:

- the licensing of biotechnological products, particularly for food and healthcare applications;
- the control of certain genetic manipulations, especially when higher plants and animals are concerned. Any involvement of human genes attracts the most stringent regulation;
- the release of genetically engineered organisms into the environment (this mainly affects microbes, but crop plants and others are also important and there may be special requirements to prevent the spread of their pollen during field tests). Regulations may or may not cover such activities as the insertion of pest-resistant genes into food plants.

Table 5.1 lists regulatory bodies governing genetic manipulations and the release of genetically-modified organisms.

Lawyers, legislators, communicators, funders and the public

Table 5.1
Arrangements in a number of countries for regulating recombinant DNA activities

Name of body or legislation	Areas of jurisdiction
Denmark	
Environmental & Gene Technology Act (1986)	Use of recombinant DNA and cell hybridization techniques
Ministry of Environment	Approval for planned release
France	
Commission Nationale de Classement	Risk assessment
Biomolecular Engineering Commission	Planned releases
Germany	
Bundesministerium fur Forschung und Technologie	Code of practice, mandatory for federally funded work, voluntary for others; no planned release
Ireland	
Recombinant DNA Committee (under National Board for Science & Technology)	Non-mandatory notification of laboratory work
Netherlands	
Ad Hoc Recombinant DNA Advisory Committee	Focus for regulatory oversight
United Kingdom	
Advisory Committee on Genetic Manipulation (ACGM)	Notification and review
Planned Release Subcommittee	Review of field trials in the open environment
United States	
Department of Agriculture (USDA)	Animal biologics and plant pests
Environmental Protection Agency (EPA)	Chemicals and pesticides
Food and Drug Administration (FDA)	Human biologics and pharmaceuticals; animal and human food and additives; medical devices; cosmetics
National Institutes of Health (NIH)	Recombinant DNA research sponsored by the Federal Government
Occupational Safety and Health Administration (OSHA)	Occupational safety and health
Commission of the European Community	

Various proposals for Council Directives are under consideration:
1. Contained use of genetically modified microorganisms;
2. Deliberate release of genetically modified organisms to the environment;
3. Protection of workers from the risk related to exposure to biological agents at work.

Exploiting biotechnology

Products for Food and Healthcare

All developed countries enact legislation to control drugs and biologic preparations for medical, other healthcare and cosmetic applications as well as the use of additives in food. In general, any new product must undergo a more-or-less rigorous series of tests to satisfy the regulatory authorities of its safety and efficacy for the purpose for which it is intended. Not surprisingly, the most closely controlled are those materials used in human medicine and intended for direct human consumption. Others, like pesticide residues, which might inadvertently find their way into human food, are equally stringently monitored.

As appropriate for the product, evaluation will include biological testing in the laboratory for beneficial and toxic effects. In some cases this starts as far down the biological scale as microorganisms and proceeds to higher animals as models for human beings. If all goes well, the drug is shown in all the animal tests to have real therapeutic value and the side effects are acceptable (though with drugs for the treatment of very serious and intractable illnesses side effects may nevertheless be severe and undesirable), investigations will proceed to human subjects. Phase I testing may be carried out on human volunteers but in some cases, for example that of anti-cancer drugs, the first trials will be on terminally ill people for whom all other treatments have been exhausted, a very stringent form of testing since the chances of a beneficial outcome in such patients must be smaller than for the less severely afflicted. Testing for human application is extensive and detailed, usually involving double-blind comparisons between, say, a drug and a placebo: until all the results have been correlated, the doctors, the patients and even the organizers of the tests have no way of knowing who is receiving the drug and who the placebo.

Testing new products is time-consuming and expensive. It is estimated that a new drug for human application may take ten years to go through the whole series of tests at a cost of $100-200 million or more. The rate of attrition is high; only a very small proportion of potential new products survive the process to emerge successfully onto the market.

Genetic Manipulations

When the new genetic technologies were first being developed there was justifiable concern that their consequences might be unpredictable and difficult or impossible to control. It was already known that in nature some microbes are able to acquire genetic material from other organisms in their environment; it was feared that, if new and unnatural genetic combinations were made in the laboratory in "safe" organisms which were later released into the environment, such genetic information might find its way into microbes which are pathogenic to man, animals or plants. Would expression of that new genetic information have a harmful, even a disastrous, effect? Nobody knew for sure and it was clearly prudent to take precautions. The issues were debated hotly not only in the scientific community but also as a public issue: some cities discussed ordinances prohibiting the performance of such experimental manipulations in laboratories within their city limits.

There were also ethical problems. Some people felt that scientists were "playing God". Was it right artificially to alter the genetic makeup of living organisms even if such events did to some degree occur in nature although not, as far as anyone knew, to the extent of human characteristics being found in bacteria? Might it result ultimately in the manipulation of human genetics in order, perhaps, to influence human physique, personality or behaviour? Not surprisingly, there was widespread revulsion at any such possibility and governments generally responded to a desire for caution.

A series of guidelines and rules of procedure were drawn up in various countries. Official bodies and agencies were vested with the authority to control or advise on these genetic manipulations; Table 5.1 lists them for a number of countries. As time passed, without catastrophes and with experience showing that many novel genetic constructs were in fact biologically so feeble as to present no hazard, the guidelines, still generally conservative, were relaxed somewhat from their original rigour. In most countries, research workers wishing to conduct genetic engineering experiments and developments in categories conceivably hazardous, are obliged to seek approval from safety review committees responsible for scrutinizing all such proposals. Before signalling their agreement, the regulatory committee must be satisfied that proper precautions will be taken against the release of the manipulated organisms or the unacceptable transmission of genetic information to recipients.

Release of Genetically Manipulated Organisms into the Environment

Some genetic engineering is intended specifically for an objective which requires release of modified organisms into the environment. One of the better known examples took place in California in the late 1980s. Frost damage to strawberry and potato plants has serious commercial consequences. At low temperatures water on the leaves freezes round certain bacteria which act as nucleation centres for the growth of ice crystals. Those bacteria can be modified genetically so that they no longer aid the freezing process and are then called *ice-minus bacteria*. A proposal was made to spray the susceptible plants with ice-minus in order to displace the "ice-plus" version which allows ice to form. Some people objected on the grounds that the consequence of release was unpredictable. Injunctions were sought. Those wishing to spray were eventually allowed to do so but their first attempt was sabotaged, presumably by objectors. Eventually successful spraying took place; no untoward consequences have been reported and the treatment was efficacious.

COMMUNICATION WITHIN AND ABOUT BIOTECHNOLOGY

Biotechnologists, of course, exchange information among themselves about scientific, technical and business matters. Between practising biotechnologists (those in the laboratory as well as those in management) and the world outside information is disseminated:

- by researchers about developments in fundamental science;

Exploiting biotechnology

- by equipment suppliers about hardware needs and developments;
- by the financial community about investment for start-ups and growth;
- by the various markets to explore the opportunities for new products and services as well as to monitor the progress and competitive status of existing ones;
- by clients about specific contractual arrangements.

Communication about the Technology

As with all technical subjects, interaction between scientists and engineers working at the bench takes place via technical papers and reviews published in special interest journals, in monographs and other books, at conferences and by personal contact. All are well developed in biotechnology.

Experimental scientists regularly scan the basic scientific literature for developments of general or specific interest. They probably also keep an eye on journals devoted to the application of their work in medicine, agriculture, petroleum engineering and so on. Because there are so many published sources of potentially interesting information, extensive abstracting and indexing services are available, as well as databases online for computer access. Conferences, seminars and exhibitions offer many kinds of opportunity to learn about new concepts and results, and to exchange information. Those dealing with basic science openly present the design and results of investigations, with free discussion of their implications. Conferences on commercialization, however, tend to be more circumspect, relying largely on general presentations and case histories: companies are normally unwilling to divulge their good ideas before they judge the time to be ripe.

Communication about Business Matters

Companies with products and services to offer use the normal commercial techniques for publicizing them. Publishers and specialist consultancy companies have produced hundreds of market surveys addressing various aspects of biotechnology, and varying in price from less than a thousand to tens of thousands of dollars. There is wide variation in their quality and they are commonly sold on their prospectuses alone.

Commercial information is also available in the form of newsletters directed to various areas of the field. These usually contain items of company and financial news, patents issued, some scientific and technical advances and reports of important personnel changes. A number of biotechnology directories are available, some from private sector publishers and others from official bodies: they include company listings, research grants and contracts awarded, and summaries of research and development work both in universities and in private companies. Patent information and details of regulatory and other relevant legislation are issued through the usual official channels.

Lawyers, legislators, communicators, funders and the public

FUNDING R & D IN BIOTECHNOLOGY

There is very considerable variety in the funding of laboratory and related development work, and this account can do no more than sketch briefly the range of potential sources.

In the private sector, research and development is probably funded mainly from in-house resources or via financial vehicles, including limited and corporate partnerships, set up for the purpose. Some developments may be partly or wholly paid for by the clients and end-users. However, support might also be available from national or international agencies; while this is usually confined to "pre-competitive" work still at a comparatively early stage of development, some programmes provide funding for projects more closely approaching the market. The European-wide EUREKA programme is an example of one in which the proportion of the public contribution diminishes progressively as the development proceeds successfully towards commercialization and the associated technical risks correspondingly decline.

Fundamental scientific investigations in the (mainly public-sector) universities and research institutes are funded largely from public sources although, in areas of interest to them, significant contributions may also be made by commercial companies in the pharmaceutical, chemical, oil, food, agriculture and other industries. Cooperative funding by both public and private sectors may be available.

Publicly-supported research is very well developed in industrial countries. One or two examples of funding practices from both sides of the Atlantic might be appropriate here. In the UK, much of the support for university research in biotechnology originates from the Medical (MRC), Agricultural and Food (AFRC) and Science and Engineering (SERC) Research Councils (note that some research councils, including the AFRC and SERC but not the MRC, are being reorganized — future research funding in biotechnology will come from the Biology and Biotechnology Research Council [BBRC]). Funds are provided on a peer-reviewed competitive basis for the direct support of research projects. Some funding is programmatic: the Biotechnology Directorate of the SERC, for example, favours a number of defined general areas which are considered of potential value for biotechnological development as a whole rather than for direct market applications. The research councils also pay for the training of large numbers of research students partly, via the programme of Cooperative Awards in Science and Engineering (CASE studentships), in association with industrial concerns in which the student works during part of his training period.

In the US, the major national funding agencies, both for the support of research students and for defined projects, are the National Institutes of Health and the National Science Foundation but there also exist a host of other public and private sources of support. In Canada, the National Research Council and Medical Research Council play similar roles, shared with other federal and provincial agencies. Supranational funding bodies include the Commission of the European Community which from time to time announces the establishment of R & D programmes, including specific topics or areas of biotechnology. The common practice with EC funding is to require a partnership of researchers from at least two Community countries. On a wider international basis, biotechnological developments are funded, especially in the less developed countries, by the United Nations Industrial Development Organization which is also a major sponsor of an initiative specifically intended to aid third-world countries, and by its associated organization the International Centre for Genetic Engineering and Biotechnology, with branches in Trieste and New Delhi.

THE PUBLIC AND BIOTECHNOLOGY

There are two ways of looking at this relationship: public policy on biotechnology and the views and perceptions of individual members of the public.

Public Policy

As well as assuming responsibility for the regulatory functions discussed above, governments and other official bodies have generally taken an encouraging view. Biotechnology is seen as a sign of being up-to-date in attitude: part of the industry of the future, "high-tec.", like computers. After an initial period of uncertainty about whether monsters were being created and the likelihood of their escaping from the confines of the laboratory, governments and cities alike began to be encouraging along the lines of: "do set up your company in our science park"; "we offer tax benefits to new high-technology companies"; "yours could be the flagship company in our development"; "there may be investment capital and development grants available"; "we like the project and will match you dollar for dollar/pound for pound/(perhaps soon, rouble for rouble)".

Biotechnology companies often occupy reasonably pleasant buildings in reasonably pleasant areas; they have been and still are seen as a source not only of prestige but of employment (not necessarily for much of the local population, however) and ultimately, presumably, of tax revenues. Together with electronics and aerospace it is almost a "post-industrial" industry: better that the new biotechnology company is located in our town, in our country, or next door, and better that it should be a biotechnology company than something unsightly, smelly and less prestigious (the obverse of NIMBY!).

Education, too, offers biotechnology, usually meaning a good dose of the techniques of genetic engineering plus fermentation microbiology, together with an occasional additional speciality of the house. Universities mount degree courses, while students just entering university have learned all about it at high school. Never mind that what they learn might be many times more "bio" than "technology" and that few of their teachers have more than the vaguest notion of management, marketing or a profit-and-loss statement. Never mind either that biotechnology employers recruiting for their companies above all require of their laboratory scientists that they are well trained in their science; the management skills they will learn on the job. Biotechnology draws the students and provides enjoyment, employment and satisfaction for the teachers: it is a popular subject.

Lawyers, legislators, communicators, funders and the public

The Individual's View

Like other technical matters, biotechnology is rather an esoteric subject. Most people in the population at large have a limited understanding of any sort of science and technology, and the bio-version similarly remains poorly understood. One cynic, asked his view of the public perception of biotechnology, answered "The public has no perception of biotechnology".* Scare stories and promises of untold wealth for the successful practitioners at times alternate in the newspapers and magazines, while comic books recount again the antics of the gene zappers or the bug-eyed monsters that got loose. The recent popularity of "Jurassic Park" is a good example of the very strong messages the public receives about some fanciful implications of biotechnology.

Among more thoughtful people there is an understanding that huge benefits might ultimately flow, benefits as much social as commercial: of intractable and tragic diseases conquered, of hunger abolished, of better and safer ways of making things, and of eliminating the eyesores of industrial pollution. At the beginning there was some fear of an unknown and often luridly presented technology, but now people seem more to be intrigued and on the whole optimistic that the benefits will outweigh any disadvantages. There is also a growing realization that those benefits will take time to materialize, that more work and more complexity is involved than initially was anticipated, and that until now there really have been no disasters. Biotechnology is probably slowly becoming accepted as just one of those activities that people have come to expect in this very rapidly changing world.

---

* The cynic may have been inaccurate as well as cynical. An American survey in 1987 concluded that well over half the adults questioned could provide a comprehensible, if not always strictly accurate, explanation of biotechnology and that a third had read something about it. A European Commission survey in 1991 found that understanding was greatest in Germany, followed by Denmark, the Netherlands and the UK.

6 The job of a biotechnology manager

Biotechnology has to be managed at two levels. The first level, usually in the public sector, is the research and perhaps some development undertaken even before a product or service is clearly identified and certainly before it becomes a distinct commercial reality. This is early stage R & D to establish and define the underlying biological and engineering parameters, work sometimes called "pre-competitive". The second level is the management of the subsequent commercial activities to make and sell products or perform fee-generating services, with all their attendant organizational, financial and marketing implications. In free-market economies these are to be found almost exclusively in the private sector, one which comprises well-established major corporations (pharmaceutical, agricultural, food, chemical and equipment firms, as well as some oil companies and others) and a host of much smaller companies specializing in one or other of the many market niches.

PRE-COMPETITIVE RESEARCH

In addition to exploring fundamental biological mechanisms, pre-competitive work in biotechnology may be directed to devising new general methods and improving existing ones (methods like the isolation, purification and analysis of products), and developing basic manufacturing techniques potentially applicable to a range of products. Its concern with basics often makes pre-competitive R & D indistinguishable from biochemistry, genetics and the other biological sciences. In recent years the word has been regarded as so evocative that many biochemists, geneticists and others label their work "biotechnology" in the hope that it will be easier to find financial support even though the prospect of real products emerging may be very tenuous.

Work like this is carried out in university laboratories, government institutes and in the research departments of some large corporations. In university laboratories it is driven almost entirely by the scientists' curiosity and their wish to understand more of the world in which they live; corporations do it to keep up with the advance of scientific frontiers while naturally steering such "curiosity research" towards their own areas of business. Government institutes are somewhere in the middle, usually more mission-oriented than universities but less so than companies. Small companies very often find basic research too expensive, obliged as they are to align their R & D closely to the readying of defined products and services for the marketplace. It leaves them little if any opportunity for in-house fundamental exploration — most have to rely on public information for new knowledge, or get it by setting up collaborative programmes with university departments.

It is difficult to generalize about the management of basic research because in the universities, and also to a large degree in the research institutes, it is a very individualized activity: scientists work more or less on what they choose (or, more to the point, on what they can get funded). There is, of course, management at the level of the research groups themselves and, indeed, of the institution as a whole, but neither is necessarily closely related to the particular problems of biotechnology. Even in the larger companies, management of laboratory activities may resemble the university and institute pattern, with wider managerial decisions combining science, technology and commerce evident more in setting objectives for a research department as a whole than in defining detailed experimental targets.

MANAGEMENT OF BIOTECHNOLOGY COMPANIES

A recent annual report of a well-known biotechnology company in the healthcare area puts it rather well: "The key to competitive success is the effective management of research and development." How is it done? What are the issues?

A Business Like Any Other

In business terms, running a biotechnology company differs little from managing businesses in any other area. There are the same imperatives of ensuring profitability for the investors (There will be no investment if the investors fail to perceive profits.) by:

- controlling costs;
- selling the product(s) at a good price;
- keeping an eye on the future, deciding what next year's products will be and making sure that their development gets started in good time;
- keeping another eye on the competitors;
- ensuring job satisfaction and competitive salaries for the employees so as to recruit the best and make sure that the ones already there do not drift away to something better.

But there are also plenty of differences: keeping up with advances in the science is just one of them (although, as we will see later on in this book, not all biotechnology operates at the very frontiers of knowledge). Then there is the question of possible culture clashes which some observers see looming. Sound business management, as important in biotechnology as in any other enterprise, will also call for a delicate touch on the part of the managers because the culture of scientists is markedly different from their own. Scientists are brought up in a climate which encourages individual thought and initiative, and the satisfaction of curiosity. If the managers are too dictatorial and insist that their scientists follow specified and preordained paths without any opportunity for personal exploration, their best scientists may become discouraged and find more conducive employment elsewhere, leaving the company deprived of essential scientific talent. But if the management is too lax, sight may be lost of the commercial objectives as the scientists seek to pursue scientific and technical tangents because, as scientists, they want to know the answers. A proper balance demands skilful management.

The problems faced by the smaller biotechnology companies, many of them new start-ups,

The job of a biotechnology manager

illustrate as clearly as any how biotech. has to be run as a way of earning one's keep in the private sector. The whole of this business, as indeed any other, is dominated by decision-making: recognizing which important decisions have to be made, identifying the factors inside and outside the organization which must be taken into account, defining the mechanisms and identifying the people whose responsibility it is actually to do the deciding.

Management Structure

Organizationally, a biotechnology company is, in most respects, likely to resemble other small companies. At the top, and with ultimate authority, is a Board of Directors of some six to ten people. Their job is primarily to safeguard the interests of the investors (shareholders), those who have invested money in the business — in other words, the owners: to make sure their money is well deployed and that the company is properly managed. Some will be Executive Directors and employees of the company: each will have responsibility for a departmental function such as R & D, manufacturing, business development, marketing, finance and, as the company grows, for personnel and no doubt other specialist functions. Preeminent among them is the Managing Director (or Chief Executive) — the leader, with whom final decision-making below Board level rests. We will return later to what sort of person he is and to his functions. (In this unequal world, the overwhelming proportion of Managing Directors are still men, so perhaps we may be forgiven for simplifying the text and writing "he".) It is usual for some Board members not to be employees of the company: they are the Non-executive Directors and, because they are not closely involved in the daily running of the company, they are expected stand back and take a longer view. The Board will elect a Chairman, who is also often non-executive and will work closely with the Managing Director to help resolve major policy issues.

Answerable to the Executive Directors will be the scientists working in the laboratories, sales staff out contacting potential clients and the people who run the manufacturing facilities. Group and team leaders will be charged with ensuring that various tasks are carried out and objectives met; because these people, the leaders and the led, are all highly skilled and experienced professionals, they will be consulted about how things should be done and their views will be taken very seriously into consideration by the people making decisions for the whole company.

Biotechnology is influenced so strongly by contemporary scientific research, and the company scientists are themselves so busy developing products for the marketplace, that it is usual for biotechnology companies to maintain close links with the academic world in order to keep up with what is happening at the frontiers of research and to help their in-house scientists to sort out the wood from the trees. Companies may meet this need by engaging professors and others as specialist consultants; they will interact both with scientists at the bench about the minutiae of their experiments and advise the Board, and the Managing Director and his colleagues, on technical matters of relevance and concern to the company. Some companies assemble panels of distinguished scientists, offer them shares and get them all together periodically for in-depth technical discussions and reviews of the company's plans and proposals.

Exploiting biotechnology

Human Resources

A company is only as good as its workforce. Biotechnology companies employ:

- experimental scientists (mainly biochemists, chemists, geneticists and microbiologists);
- scientific managers (senior experimentalists now acting as supervisors);
- engineers (many biotechnology companies employ chemical engineers but, depending on the company's activities, there may also be others specializing in agriculture, the environment, mining, oil production and so on);
- accountants;
- managers;
- marketing and sales staff;
- public relations experts;
- perhaps lawyers and doctors;
- support staff: lab. technicians, secretaries, store keepers, cleaners, bookkeepers and others.

Each staff member has to be recruited and retained by a sufficiently attractive package of salary, promotion prospects, fringe benefits and congenial working conditions. Salaries need to be competitive, while fringe benefits may include holidays, paid sick leave, pension contributions, medical insurance, subsidized dining facilities, company cars and stock options. Working conditions are very important, with a sympathetic blend of direction and independence for scientists and managers. Work loads have to be controlled carefully enough for people to feel neither idle nor rushed off their feet for long periods at a time.

The scientists are very often the people who need the most careful handling, particularly when they have fairly recently joined a company from university. They greatly appreciate opportunities to keep abreast of their subject by having access to libraries, a little time to attend lectures and seminars at the local universities and occasionally to attend conferences at more distant locations. A sensible company policy allows employees to seek approval to spend some small fraction (say 10%) of their time, and the use of company chemicals and equipment, on investigations of their own choosing; new ideas of great subsequent value to the company can be tried out in this way at low cost.

In start-up companies, some of the most exciting biotechnology ventures of the past two decades, the scientists joining the new companies were mostly academics transferring from university jobs. People working in academic posts on problems of pure research became aware of the possible practical and commercial significance of their discoveries but traditionally they rarely have the experience or the inclination to do much about it. However, starting in the early- to mid-1970s, the chances of obtaining public funding for research diminished: excellent research proposals went unfunded and more and more academic scientists became gloomy about their futures. Some of them thought they might try to raise money from commercial sources but most had no good notion of how to go about it or what it might do for them if they were to succeed. Somehow they needed to liaise with the financial and/or industrial community; but, if their projects actually did go forward with industrial support or, if they were recruited into a company, they would have to cooperate with commercial management and perhaps themselves be managed. An academic scientist coming out of a university environment might have found this a difficult adjustment to make.

Thus the relationship between scientists and managers can be happy or uneasy; it depends very much on the individuals. In the university, the scientist was used to working largely autonomously and pursuing his own research inclinations. In the company he may have to narrow his sights and concentrate on a limited number of commercial objectives: if he and his fellows do not do that the company may fail. He will, of course, keep up with his subject and the firm may offer him an opportunity to run a small personal activity as we noted above, but the commercial objectives of the company are all-important. The compensations may be many-fold: a literally exciting environment, the stimulus of a multi-disciplinary team all working towards the same objectives, the discipline of the financial "bottom line", good facilities and no doubt a higher salary with the promise of better rewards in the future than were likely in university employment. The compensations may be more than adequate but, if the erstwhile academic concludes that he does not like it in industry, he may decide to return to academe, a lot wiser and perhaps a trifle richer.

The managers, too, are very special. A person recruited as Managing Director for a biotechnology start-up must have experience, competence and energy. He is likely to be someone in his mid-40s, mid-way through his career, with existing stable and attractive prospects, and bearing a host of domestic responsibilities in the form of a teenage family and a substantial mortgage. Why should he take risks in joining the new company, perhaps with a lower salary and an uncertain future? The reasons might include:

- he does not like his existing job and wants a change;
- the prospect of joining a newly founded company which, with luck and (his) good management may turn out to be very successful, making him personally rich;
- the drive to run his own company rather than being a cog in the wheel of somebody else's. For many people this factor is probably more important than either of the others.

Just as the scientists might feel they have problems with the managers, so the reverse may also be true. The scientific and managerial cultures can be so different that it takes time for the two species to learn to live with one another. But learn they must — the company is not going to do very well if its two major branches of activity cannot work together.

Making Decisions

Decision-making involves choices between alternatives. It is often a difficult matter and the various people who contribute have different motives for promoting their points of view. The strong science base underpinning biotechnology means that the professional scientists will have an important voice; not surprisingly they see things in scientific terms and will usually try (they may well not always succeed) to push for more money for science and more resources for experimentation. They need to be quite sure of their facts, strengthen the science base of the company and explore more possibilities. All these objectives may be intellectually desirable

Exploiting biotechnology

but not necessarily commercially essential in the short-term — if strictly speaking they do not have to be done, management is inevitably going to ask whether they should be done and, if so, why? After all, management is responsible for using the resources of the company wisely. One (partial) answer may be: "to keep the scientists happy and stop them looking for a job somewhere else".

The commercial/sales people may push for more market research, a greater understanding of the commercial markets, and the identification and development of products and services for defined market niches. They might not entirely get their way, either. The engineers, observing this conflict of emphasis between the scientists and the commercial team, may in turn stress the need for more resources for scale-up and manufacturing: "it's no good", they will say, "having a bright scientific idea plus a market in which to sell if you cannot actually design and make a functional product". They, too, will have reason on their side and have to be listened to. One of the most important of the Managing Director's functions is to balance these inputs in the best interests of the company. Some of the elements of general decision making in a business environment are illustrated in Figures 6.1 to 6.6; these are, of course, adaptable to the specific requirements of biotechnology.

The sociology of decision-making is complex and compounded in some organizations by an implicit discouragement of risk-taking. The way the decision maker perceives the rewards of making the right decision and the penalty of making the wrong one is critical — does he benefit more from success than he suffers from failure? Everybody knows about a failure at the time it happens but who remembers an idea that was killed ten or fifteen years ago? Is it therefore not better, and safer for his career, if he takes no action rather than chance an

DECISION CONSIDERATIONS I

IS THE MARKET REAL?

IS THERE A NEED/WANT?	CAN THE CUSTOMER BUY?	WILL THE CUSTOMER BUY?
KIND OF NEED/WANT	STRUCTURE OF MARKET	PRIORITY OF NEED/WANT
TIMING OF NEED/WANT	MARKET SIZE AND POTENTIAL	PRODUCT AWARENESS
COMPETING WAYS TO SATISFY NEED/WANT	AVAILABILITY OF FUNDS	PERCEIVED BENEFITS OR RISKS
		FUTURE EXPECTATIONS
		PRICE VS DESIGN/ PERFORMANCE FEATURES

Figure 6.1 Decision considerations I: is the market real? (*Reproduced by permission from The Forum Corporation*)

The job of a biotechnology manager

DECISION CONSIDERATIONS II

IS THE PRODUCT REAL?

IS THERE A PRODUCT IDEA?	*CAN IT BE MADE?*	*WILL IT SATISFY THE MARKET?*
WAYS TO SATISFY IDENTIFIED MARKET	ENGINEERED DESIGNED DEVELOPED TESTED VERIFIED	DESIGN/PERFORMANCE FEATURES
FEASIBILITY	PRODUCED PROCESSED INSPECTED QUALITY CONTROLLED	COST
ACCEPTABILITY	PACKAGED HANDLED DISTRIBUTED SHIPPED STORED INSTALLED SERVICED RETURNED	COST-VOLUME RELATIONSHIP AVAILABILITY

Figure 6.2 Decision considerations II: is the product real? (*Reproduced by permission from The Forum Corporation*)

DECISION CONSIDERATIONS III

CAN OUR PRODUCT BE COMPETITIVE?

ON DESIGN/ PERFORMANCE FEATURES?	*ON PROMOTION?*	*IS THE PRICE RIGHT?*	*IS THE TIMING RIGHT?*
QUALITY	CONSUMER ADVERTISING	COST	INTRODUCTION
UTILITY	TRADE ADVERTISING	PRICING POLICIES	DESIGN/PERFORMANCE CHANGES
CONVENIENCE			
VERSATILITY	PACKAGING	TERMS AND CONDITIONS	SALES PROMOTION CAMPAIGNS
RELIABILITY	TECHNICAL SERVICES	COMPETITION	PRICE CHANGES
DURABILITY			
SERVICEABILITY	OTHER SALES PROMOTION CONSIDERATIONS	OTHER PRICING CONSIDERATIONS	OTHER COMPETITIVE TIMING CONSIDERATIONS
SENSORY FEATURES			
SAFETY			
UNIQUENESS			
OTHER AREAS OF PRODUCT DIFFERENTIATION			

Figure 6.3 Decision considerations III: can our product be competitive? (*Reproduced by permission from The Forum Corporation*)

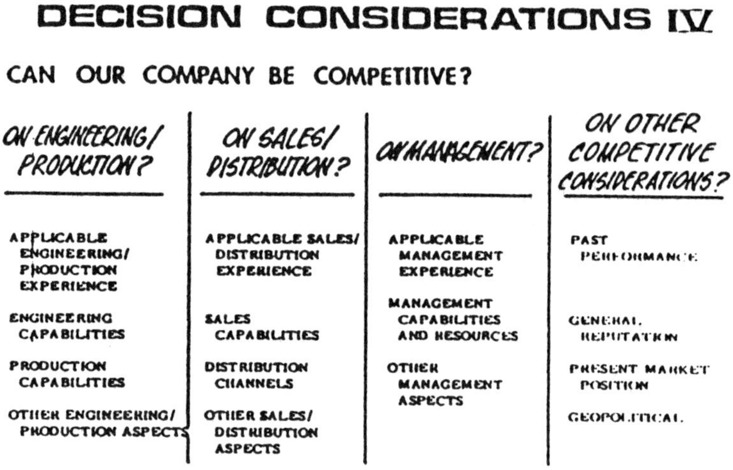

Exploiting biotechnology

DECISION CONSIDERATIONS IV

CAN OUR COMPANY BE COMPETITIVE?

ON ENGINEERING/ PRODUCTION?	ON SALES/ DISTRIBUTION?	ON MANAGEMENT?	ON OTHER COMPETITIVE CONSIDERATIONS?
APPLICABLE ENGINEERING/ PRODUCTION EXPERIENCE	APPLICABLE SALES/ DISTRIBUTION EXPERIENCE	APPLICABLE MANAGEMENT EXPERIENCE	PAST PERFORMANCE
ENGINEERING CAPABILITIES	SALES CAPABILITIES	MANAGEMENT CAPABILITIES AND RESOURCES	GENERAL REPUTATION
PRODUCTION CAPABILITIES	DISTRIBUTION CHANNELS	OTHER MANAGEMENT ASPECTS	PRESENT MARKET POSITION
OTHER ENGINEERING/ PRODUCTION ASPECTS	OTHER SALES/ DISTRIBUTION ASPECTS		GEOPOLITICAL

Figure 6.4 Decision considerations IV: can our company be competitive? (*Reproduced by permission from The Forum Corporation*)

DECISION CONSIDERATIONS V

WILL IT BE PROFITABLE?

CAN WE AFFORD IT?	IS THE RETURN ADEQUATE?	IS THE RISK ACCEPTABLE?
CASH OUTFLOW WITH TIME	ABSOLUTE PROFIT	WHAT CAN GO WRONG?
CASH INFLOW WITH TIME	RELATIVE RETURN	HOW LIKELY IS IT?
NET CASH FLOW	COMPARED TO OTHER INVESTMENTS	HOW SERIOUS IF IT HAPPENS?
		WHAT CAN BE DONE ABOUT IT?
		UNKNOWN UNKNOWNS

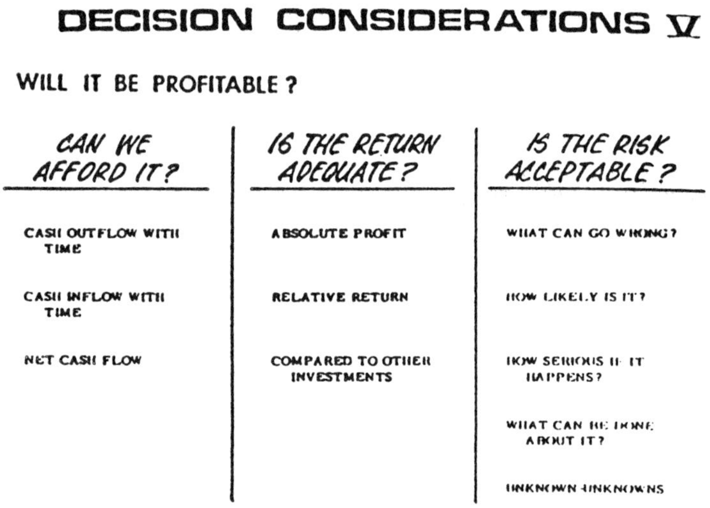

Figure 6.5 Decision considerations V: will it be profitable? (*Reproduced by permission from The Forum Corporation*)

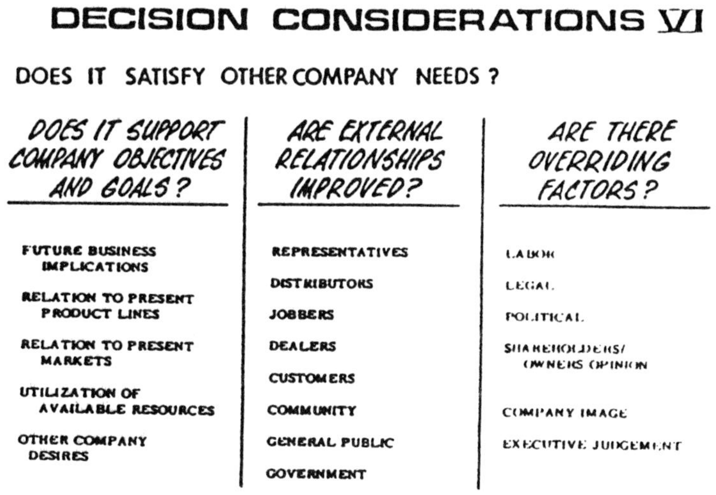

The job of a biotechnology manager

DECISION CONSIDERATIONS VI

DOES IT SATISFY OTHER COMPANY NEEDS?

DOES IT SUPPORT COMPANY OBJECTIVES AND GOALS?	ARE EXTERNAL RELATIONSHIPS IMPROVED?	ARE THERE OVERRIDING FACTORS?
FUTURE BUSINESS IMPLICATIONS	REPRESENTATIVES	LABOR
RELATION TO PRESENT PRODUCT LINES	DISTRIBUTORS	LEGAL
	JOBBERS	POLITICAL
RELATION TO PRESENT MARKETS	DEALERS	SHAREHOLDERS/ OWNERS OPINION
UTILIZATION OF AVAILABLE RESOURCES	CUSTOMERS	
	COMMUNITY	COMPANY IMAGE
OTHER COMPANY DESIRES	GENERAL PUBLIC	EXECUTIVE JUDGEMENT
	GOVERNMENT	

Figure 6.6 Decision considerations VI: does it satisfy other company needs? (*Reproduced by permission from The Forum Corporation*)

unfortunate outcome? Who will take a long-term decision which might bring success only after his own retirement or risk a failure which may come to light just before an anticipated promotion? Particularly in large organizations (private as well as public) there may be an aversion to risk-taking, commercial as well as technical; innovation is expensive and the loss of invested assets an embarrassment. There is an understandable temptation to be cautious: "let somebody else (a university, say, or a small company) do the basic science and bear the early technical risks: we can always buy in later". That is one of the reasons why some observers feel that small organizations are more inclined than large ones to be innovative and prepared to take risks with early-stage technology — their internal communications are better, risk-taking is more collective, there are likely to be fewer alternative ways of keeping the organization going and, when they reach decision-making status, the staff may be younger and less conservative.

Dealing with Innovation

A constant flow of new ideas is essential for modern science-based industries and biotechnology is no exception; there must be agreed mechanisms for choosing between alternatives and deciding how to respond. New ideas form the basis for next year's products, so it is obviously important that they should continue to flow and receive a sympathetic hearing

Exploiting biotechnology

from management. However, no organization has unlimited resources and choices therefore have to be made: to proceed with some ideas because, not only are they good, but they fit in most closely with the company's existing or planned activities and offer the best chance of a profitable outcome. Because they do not fit so well, other equally good ideas have to be jettisoned.

New ideas come from everywhere (Figure 6.7): from the company's own staff as they think about problems and share their thoughts with colleagues, from contacts with clients, suppliers and customers, from statements by governments and competitors and, of course, from reading and hearing what the world outside is doing. The problem is to deal with them all. In a lively company, most will have to be rejected because there is simply not enough time and resource. But rejection must not convey a signal that new ideas are unwelcome; they are the very life-blood of a technology company and each group of people must work out for themselves how best to keep the ideas coming without being overwhelmed by them.

Product Development

Most ideas, of course, never see the light of day as products. Rough estimates suggest that of every 100 novel suggestions, 75 are rejected pretty quickly when they are analysed in detail. Of the remainder, about ten get as far as product development but, for one reason

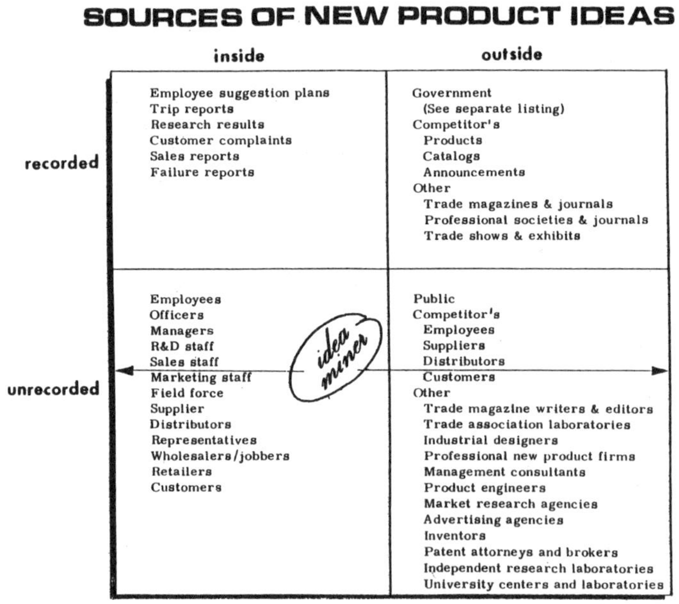

SOURCES OF NEW PRODUCT IDEAS

	inside	outside
recorded	Employee suggestion plans Trip reports Research results Customer complaints Sales reports Failure reports	Government (See separate listing) Competitor's Products Catalogs Announcements Other Trade magazines & journals Professional societies & journals Trade shows & exhibits
unrecorded	Employees Officers Managers R&D staff Sales staff Marketing staff Field force Supplier Distributors Representatives Wholesalers/jobbers Retailers Customers	Public Competitor's Employees Suppliers Distributors Customers Other Trade magazine writers & editors Trade association laboratories Industrial designers Professional new product firms Management consultants Product engineers Market research agencies Advertising agencies Inventors Patent attorneys and brokers Independent research laboratories University centers and laboratories

(idea miner)

Figure 6.7 Sources of new product ideas. (*Reproduced by permission from The Forum Corporation*)

The job of a biotechnology manager

or another, most are dropped or recycled for further thought so only about four progress to product testing. Half of those four fail the tests, leaving perhaps two out of the original hundred actually to find their way to the marketplace — and of those that do, about half are withdrawn within a year or two (Figure 6.8). One out of a hundred as a successful product seems rather a low outcome but, as that is what actually happens, the flow of original ideas needs to be kept going at a rapid rate or the chances are that no new products at all will be generated for the future prosperity of the company.

Costs go up dramatically as prospective new products proceed through the stages of development and manufacturing. More rough guesses indicate that, in typical cases, product development through to readiness for manufacturing costs ten times more than the original laboratory work, while manufacturing itself is ten times more expensive still; it is just as well that the attrition rate in product development is so high! Thus, every new product starts off costing money (Figure 6.9) — money that will not be recovered until the product reaches the market place (in biotechnology commonly a period of at least several years, longer than early investing enthusiasts had hoped) and in fact may never be recovered if some of the many things which can go wrong actually are encountered. For example, the product may fail technically during development, it may not win regulatory approval, a competitor may come up with something better and/or cheaper, and the market may have changed during the period of development so that sales are less buoyant than anticipated. A successful product, however, can be very rewarding: manufacturing and marketing costs per unit item are likely to go down as the

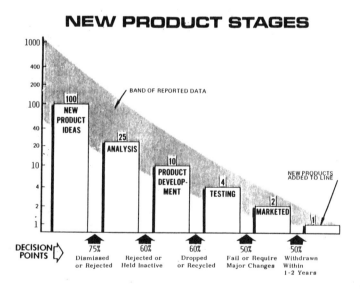

Figure 6.8 New product stages. (*Reproduced by permission from The Forum Corporation*)

Exploiting biotechnology

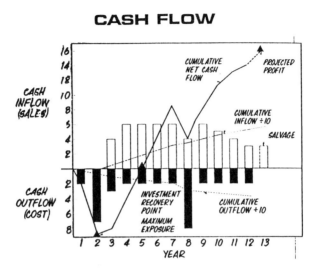

Figure 6.9 Cash flow for a new product. (*Reproduced by permission from The Forum Corporation*)

product volume grows. Eventually all the development costs are amortized resulting, hopefully, in high profits for a prolonged period before eventually the product is superseded by something better (Figures 6.10 and 6.11). Those profits, however, have to be set against the losses incurred by all the unsuccessful prospects as they failed somewhere in the research or development stages.

Every biotechnology company hopes to generate blockbusters but, of course, few do so; most are no doubt satisfied if they achieve more modest levels of success.

Keeping up with the Pack

A major problem in rapidly growing areas is whether and how to make use of the latest developments in the field. Thousands of scientific and technical journals continuously pour forth new data, new ideas and new information. Not only has the biotechnology community to be aware of what is going on, they have to decide the extent to which they should integrate new knowledge into their own activities.

Relatively minor developments will inevitably be incorporated into the daily activities of the laboratory scientists. But what of more major developments? Would it be a good idea to train existing staff in new skills, perhaps sending them off for a learning period in another laboratory? Should new specialists be appointed, with all the obvious cost implications — how quickly would such additional investment pay for itself? Might it be better to fund work on a trial basis in a university where the skill levels are likely to be very high and the overhead

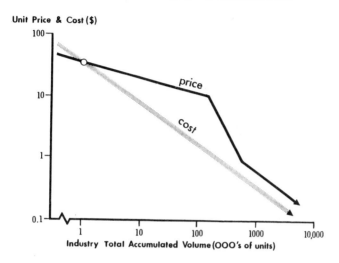

Figure 6.10 Product price/cost/volume. (*Reproduced by permission from The Forum Corporation*)

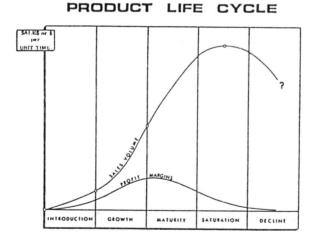

Figure 6.11 Product life cycle. (*Reproduced by permission from The Forum Corporation*)

costs lower than in the company's own laboratories? Are there government-sponsored schemes which will pay part of the costs? Is it worthwhile considering buying, or buying into, such an activity already extant in another firm, perhaps by raising more share capital to provide the cash? All these choices involve time, effort and probably money and all, therefore, involve judgement and decision-making about the best way to use a company's financial and human resources; there are clear implications for the budget, something to which we will return later.

Budgets and Financial Management

One of the main managerial tasks is preparing the annual budget and business plan for presentation to the Board. In these documents, up for discussion towards the end of the financial year, the current year's performance is reviewed against the budget agreed twelve months earlier while the proposed income and expenditure for the coming year are spelled out in detail.

The business plan will look forward to all the activities of the company in the year ahead and beyond, with products, markets and the perceived competition set in the context of the company's own position with respect to available staff skills, physical facilities, and financial resources and obligations. The budget interprets in financial terms the proposed business activity for the year to come and compares it with what happened in the year about to end.

The revenue side will show the anticipated sources of income: product sales, contracts for services, fees from consultancy, grants from government, interest from bank deposits and, eventually, royalties from licences. These will be the best and most realistic forecasts that management can make, but forecasts are what they are — inevitably there will be changes as some prospects fail to mature and (hopefully) others take their place.

Under expenditure will appear all the outgoings, separated into "direct costs" and "overhead costs". The former include salaries and fringe benefits for those directly involved in manufacturing, sales and contract research: the scientists, manufacturing staff and perhaps some of the sales force, their operating expenses (chemicals, minor equipment, travel directly connected with performing their duties, etc.) and capital expenditure for equipment of direct and immediate importance. Overheads encompass all the other expenses of the company: buildings (rent, repairs, maintenance and insurances), utilities, postage, telephone and fax, legal and patent fees, interest on loans, general travel, books and journals, and attendance at conferences. Most importantly, they encompass the costs of employment of all the staff who are not themselves direct contributors to manufacturing, sales and contract research: the salaries and benefits of the managers (including the Managing Director himself), financial officers, some of the sales team, and of the support staff: secretaries, store keepers, cleaners and others whose efforts cannot readily be assigned to specific revenue-earning activities. They will also include fees to directors and outside consultants.

The difference between revenue and expenditure is profit (or loss). It is one of the items of budgetary information to which the eye naturally gravitates. There is another: cash flow. Profits by year end are fine but meantime the company has to pay its bills and needs the cash to do so. The managers therefore have to run their affairs in such a way as to make sure that the cash is available, either from the company's own accounts or by borrowing (at a price) from the bank. Cash flow crises are events to be avoided.

At frequent and regular internal management sessions, and at less frequent Board meetings, the company's financial and business performances will be analysed in detail and compared with budgetary forecasts. Within reason, the Board and the management would like revenue to be as high as possible and expenditure as low, thus maximizing profit — remember that biotechnology is about making money and it is profit that is sought by the investors whose money supports the company. If profits appear never to be forthcoming, or are not high enough to warrant continued support, there will be a tendency for investors to withdraw. A private company will thereby be undermined, perhaps fatally, while a public one will see its share price fall, neither a happy situation. The shareholders may be prepared to live through a period when profits are not as high as everyone would wish but there is a distinct limit as to how long losses can be sustained.

It is the job of the Managing Director and his colleagues to secure profitability and avoid loss. Particularly in a young company, the Chief Executive is likely to spend a considerable portion of his time keeping the major investors informed and consulting them as appropriate. After all, they are the proprietors and, although they may be most reluctant to use it, they have the power and if necessary the duty to close the company down should they become convinced it is not going to be successful. If and when the company "goes public" and its shares are traded openly for anybody to buy and sell, the Managing Director will obviously have a different relationship with the shareholders. There are likely to be far too many for him to know personally. He will communicate in other ways, by the statements he and the Chairman make at the annual general meeting, by quarterly or annual reports plus letters on important issues, by public announcements of the company's achievements and by keeping contact with professional stock market analysts who write in newspapers or provide company evaluations and assessments for their clients together with advice about buying and selling stock.

Strategy

Every business must have a clear idea of where it is headed and what its hopes are for the future; biotechnology is no exception. Without a sense of direction, the management cannot plan effectively for the future. Because so much changes so fast and so often, forward planning may be more difficult for a growing industry than for a mature one.

The short-term deals with immediate matters: what is going on this week and this month, and what will be happening next week and next month. Experiments will already be under way in the laboratory, the manufacturing division is busy making products while the sales team is out clinching deals.

Looking at products for next year and the year after is a medium-term activity. Early-stage scientific work will be needed to ascertain the facts and understand the systems. The engineers need to think about how to turn into products the scientific information expected to emerge from the labs.; they will be planning for more equipment and perhaps new buildings.

Exploiting biotechnology

Meanwhile, the advance sales force will be informing prospective clients and perhaps taking early orders; they will liaise between the clients and their own colleagues within the company to make sure the products are the best ones for the customers' needs.

Planning for the long-term looks further ahead. How far ahead — five years; ten; twenty? The science is perhaps still a gleam in somebody's eye or at a fundamental stage in a university laboratory. The production division, however, is starting to have some very preliminary thoughts about what manufacturing techniques might be required to make such new products; they may decide to do some R & D on their own account, recognizing that novel processes and new equipment designs may be needed. The marketing group are getting ready to do some market surveys, working with the scientists and engineers to clarify the nature of the prospective new products, agreeing plausible times to the marketplace and likely selling prices; they will explore possible markets and probe for the level of potential interest in the new products.

Senior managers and the Board (all usually located on the top floor), will be weighing up the implications of all these ideas and propositions, assessing their impacts on the company in general and on budgets in particular. New initiatives are probably most successful if each is promoted inside the company by a "champion"; the Board will like to know who the champions are. They will want to be sure that the company has the appropriate scientific, technical and managerial skills, and in sufficient quantity; if not, can they be recruited in time? Will essential personnel be relocated and what effect might that have on existing activities? Are long-term plans compatible with short- and medium-term activities already further along in terms of maturity? Will all the various projects reinforce one another or are they going to be pulling resources in different directions? Ought some initiatives to be hived off into separate, subsidiary companies? Where will the up-front funding come from for the new programmes — are such programmes the best ways for the company to spend its money? If the company decides not to go ahead now, can it do so later or will the opportunity have been lost permanently? Is somebody else likely to pick it up if we don't (and if we do!). Questions, questions, questions, but without sensible and convincing answers the company will become lost in the biotechnological wilderness.

The Company and the World Outside

Quite apart from running the internal organization, the Managing Director also carries ultimate responsibility for relations between the company and all the external people and organizations with whom its personnel might interact. A review of some of the different categories illustrates the complexity:

- customers and clients;
- other biotechnology companies — each has something to learn from the others;
- collaborators;
- competitors — keep a close watch on them, they could be dangerous;

- suppliers of chemicals, equipment and utilities;
- job applicants and recruitment agencies;
- universities, research institutes and hospitals — those are the people doing the basic research or undertaking the clinical trials;
- public relations specialists — the company will be sensitive about its image and take advice on how best to present itself both to its clients and to the public;
- trade associations and professional societies — colleagues in the professions and in the industry;
- publishers and conference organizers;
- current and potential shareholders;
- banks;
- accountants and auditors;
- lawyers and patent agents;
- landlords, builders, contractors and service personnel;
- insurance companies — to insure people, buildings, cars, etc., and to obtain liability and indemnity cover;
- safety authorities — the law is becoming increasingly rigorous;
- travel agents;
- government departments and granting agencies — they set the regulations and are sources of R & D funding;
- local authorities — biotechnologists want to be good citizens;
- international agencies — they, too, may influence regulations and are also possible sources of funding.

Managing activities as complicated and interwoven as these is not easy. It requires much specialist experience coupled with an ability to integrate a group of diverse people into a team — a molecular geneticist may have little professional experience in common with an accountant but both are likely to be essential to the company that employs them, united by a common interest in its well-being and success. The role of management is as much leadership and inspiration as it is administration, and since no one person can do everything there has to be delegation. Textbooks and courses in management help in learning how to do it but it has often been said that management is best learned by example from a good manager.

Marketing Biotechnology

In its early days in the 1970s and 1980s, the "new" biotechnology of genes and enzymes was very science-driven, even in the commercial sector. Novel and exciting products, mostly in healthcare, were promised but many promises were premature and the markets largely unexplored. That climate has now changed dramatically, away from R & D towards bringing the products to market. Financing has become much harder as investors have grown more

Exploiting biotechnology

canny and realized that a speedy return on investment is less likely than it might once have seemed. At the present time few firms, especially those engaged in developing and marketing therapeutic products, have sufficient resources themselves to undertake the research and development, fund full-scale clinical or field trials and regulatory submissions, develop, optimize and validate large-scale manufacturing processes and facilities, and establish the necessary sales force. It has become necessary to spend less on research and more on addressing the markets.

Marketing is thus likely to be the main driving force in the biotechnology of the future. It means:

- recognizing what the market wants and is ready for;
- making sure the company is able to provide products and services to meet those needs;
- informing potential clients and customers about products and services before and as they become available;
- organizing the means for delivering the products and services to the clients;
- determining pricing policies to ensure adequate and ongoing profitability.

Close cooperation between what technology can deliver and what the market will buy is essential to ensure success. Neither "technology push" or "market pull" alone is satisfactory: the former tends to generate over-sophisticated products the market does not need (and will not buy) while the latter may result in a multitude of items for sale, none of them profitable.

The factors influencing competition in the biotechnology markets do not differ greatly from those in other technology-based industries. Patent cover and compatibility with government regulations are major factors in establishing a market position. Competitive advantage depends above all else on being able to produce the right product at the right time in the right market and for the right price. The "right product" means one able to do the job well, offering quality and reliability. The "right time" implies beating one's competitors in getting the product to the market when the buyers are ready; try to sell an unwanted product is not going to help the profit-and-loss account and, unless a new product is significantly better or cheaper than its predecessors, it will be difficult if not impossible to displace them once they are established. The "right market" is one in which the product readily finds willing buyers, a market which is not already dominated by a successful competitor or saturated with many competing varieties. The "right price", of course, is also important but it is not necessarily the lowest one. Determined partly by the costs of development and production, and partly by actual demand and competition in the marketplace, it is a price which will ensure an adequate return to the seller and satisfaction on the part of the buyer.

Financing Biotechnology

All new ventures require capital: for buildings, equipment, supplies and to meet salary, utility and other expenses before revenue begins to flow. The rapid development undergone by biotechnology since its birth in its present form in the mid- to late-1970s has consumed vast amounts of capital; it was fortunate that both the cash and the interest in investing it in this new activity were available when the needs of the budding biotechnologists were at their most acute.

Public sector investment

This type of finance, available mainly for research and some development in universities and research institutes, originates largely from government and international agencies in the form of long- and short-term grants and contracts, many for 1–3 years but others for longer. Some countries offer partial government support for joint university/industry projects if the industrial participants are willing to contribute matching funds in an agreed ratio. Governments in some countries have made direct capital investments into biotechnology companies, either taking equity in a private sector start-up or wholly capitalizing what is in effect a nationalized company.

In the private sector

Investment from private sources takes a variety of forms. For established companies a number of routes are available:

- they may decide to fund activities from their own reserves. Even if they are large, such resources are invariably limited and there are always competing demands upon them. The protagonists (i.e. champions) of particular developments, biotechnological or not, have to make their case in competition with others;
- in a *limited partnership*, funds are raised from investors for a specific developmental purpose to last a stated number of years with the resulting products and technology belonging to the partnership; while the generous tax deductions which may be allowed to them constitute an attractive inducement, the risk is actually taken by the investing partners. For the duration of the arrangement the company has a licence from the partnership to exploit the technology and an option to purchase rights to it at the end. In return for their investment the partners not only own the technology but may be offered favourable warrants to purchase the company's stock for a defined period at a fixed price. The investors are thus very concerned that the project should be successful, generate intellectual property and enhance the value (and hence the share price) of the company. This style of funding was especially popular in the US for the decade after 1975 but the Tax Reform Act of 1986 markedly reduced the benefits;
- with a *bank loan* the risk is taken by the borrower (unless there is a good chance that the company will fail, in which case the bank should not be lending). However, except for the interest and capital repayments, the gains are also retained wholly by the borrower. The bank looks primarily to receiving interest on the loan and the return of its capital. The success of the project is likely to be regarded as a secondary issue;

Exploiting biotechnology

- an option for a public company, one whose shares are traded on the stock market and available for anybody to buy, is a *rights issue* in which additional shares are offered to existing shareholders for the purpose of raising more capital. There are no residual payment obligations but, if some of the shareholders choose not to respond, the value of their existing holdings may be diluted while the risks are shared by all of them, new and old. A rights issues can also be offered by a private company, one whose shares are not publicly traded but are held by private individuals (often the founders) and investment institutions (see the next section).

Raising Funds for Start-ups

For new, start-up companies the situation is very different. Such companies are often started by an individual or a group of scientists who are alert to the possibilities of commercially exploiting their results but have neither the money nor the experience to do it by themselves. They have no commercial track record so bank loans are not usually available and, in any case, would probably be too risky. Stock market rules prohibit the sale of their shares directly to the public so all the funding has to come from private individuals (including perhaps some of it from the scientists themselves) or from those whose business it is to invest in such ventures: they are the *venture capitalists*, people and firms providing capital funds for new ventures and who have funded most of the biotech. start-ups in the last two decades.

Venture capitalists specialize in various types of investment: earlier or later developments, in the retailing, property, service, manufacturing or high-technology sectors, and so on. There are many funds in North America and Europe, some of which go for biotechnology start-ups. In that area the perceived risks and rewards are shown in Table 6.1. Because their investments

Table 6.1
Investment strategies in biotechnology*

RISK		
high ↑	unsuccessful biotechnology research companies	successful biotechnology research companies
↓ low	established companies with low biotechnology exposure	established companies with high biotechnology exposure

low ← REWARD → high

* *Reproduced by permission of Peter Laing*

are known to carry a high risk of failure, investors in a start-up must see the possibility of very high returns together with an exit route, the way in which they are going to sell their shares some years in the future when the price, they hope, will have risen many fold. The options will probably be to sell to other private investors, sell the company as a complete entity or in parts to other companies, or "go public", i.e. float the company on the stock market so that the shares can be publicly traded. The risks taken by venture capital, and hence the presumptive increases in share price needed to generate acceptable rates of return, are illustrated in Tables 6.2 and 6.3. In the 1970s the average performance of venture capital funds generally was equivalent to an annual compound rate of more than 20%, far higher than the stock market. However, the investor receives no interest on his investment and may have to wait several years before a suitable exit route is forthcoming. And the risks are high as Table 6.3 shows: a rule-of-thumb

Table 6.2
Venture capital investment*

Target annual rates of return and multiples of original investment			
Compound rates of return →	30%	40%	50%
Number of years before investment is realised	multiples of the original investment eventually received		
3	2.2×	2.7×	3.4×
5	3.7×	5.4×	7.6×
7	6.3×	10.5×	17.1×

* *Reproduced by permission of Peter Laing*

Table 6.3
Return on capital investments*

Average for 218 US investments	
Multiple of original investment recovered	Percent of total funds invested
More than 10 times	3
5–10 times	8
2–5 times	20
up to twice	29
partial loss	25
total loss	15
total	100

* *Reproduced by permission of Peter Laing*

Exploiting biotechnology

estimate is that 10% of the investments are very successful, 20-30% fairly successful, with 60-70% of them more-or-less write-offs. That may not be quite as bad as it seems because losses can be written off against profits for tax purposes.....nevertheless, profits are always best.

Raising the money

The proposal is the first tangible sign the venture capitalist has of the new company. It might come as a very informal document or even just a discussion, with the venture capitalist helping the proposers to put it into proper shape to interest prospective investors. What they want to see is a business plan which sets out in detail:

- a statement of the general background outlining the relevant science and technology, and indicating the perceived commercial opportunities and likely competition;
- the aims of the new company;
- its specific technology base, with strengths and weaknesses;
- details of intellectual property owned by the proposers and to be transferred to the company;
- the career backgrounds of the proposers, the scientists as well as any businessmen who may already have agreed to participate in managerial capacities;
- a financial plan to include a realistic estimate, on both optimistic and pessimistic bases, of the funding needed (and how it will be spent) before positive cash flow and ultimate profitability.

If the venture capitalist is satisfied that the proposed company will operate in interesting markets he undertakes due diligence (i.e. checks all the statements made in the proposal) to ensure that the proposers really have the requisite ideas, skills and competence. Several investors might cooperate at this stage; one of them becomes the lead investor and does the due diligence investigation on behalf of them all:

- recognized experts in the field will be commissioned to review the science/technology and perhaps to make market surveys;
- the investors will want to satisfy themselves about patents and a clear lead over possible competitors;
- they will, of course, meet the people in the company to make sure that everybody can work together and that there is the right mix of business and science plus entrepreneurial dedication: this is going to be a business venture, not an academic exercise;
- they may suggest that more specialist skills should be recruited;
- plans will include shareholdings and options for the founder members to ensure their continuing dedication;

The job of a biotechnology manager

- the prospective investors must feel that their investment is in good hands and being properly managed. They know that the innovative scientists who started the project probably do not have the skills and experience to drive the business forward so they have to find someone who does — he or she becomes the Managing Director (also called the 'Chief Executive Officer' or CEO).

Help and hindrance "from the Government"

One of the issues to be decided is where to establish the new company. Different countries, states, regions and districts may offer inducements and assistance to new businesses which they would like to have in their areas, businesses which will increase local employment and eventually broaden the tax base. They might include low-interest loans, start-up support grants and various sorts of collaborative funding for individual projects. The tax regime itself might be an important determinant and from time to time legislation may be enacted to provide partial relief from the various taxes on business, or a limited "holiday" from local taxes. On the other hand, legislation may hinder: particularly tight regulations about conducting genetic experiments and releasing genetically modified organisms, or heavy taxes on employment, may prompt a company to site itself in a less stringent location.

Raising more money

A new biotechnology company is likely to go through several rounds of funding, illustrated in Table 6.4. It starts with the founders, the people who had the original scientific ideas and some concept of the possible products or services which might result. Because they rarely have enough money themselves to start the company, and are short of rich relatives, they go to the specialists for help. The sequence of events that follows may be something like the one shown in Table 6.4.

A small "seedcorn investment" gets them started; it helps them develop their ideas and build the business to the point where it can attract further and more significant investment. The Managing Director is appointed — his early role is to give shape to the business and plan for the next round of funding. Money in the company is very tight at this stage and about 90% of such companies fail. However, if it does succeed, the original investors will probably still own about 25% of the company several years later, each of their shares being worth many times the price they paid for them.

Hopefully, a year or so after the seedcorn comes the first round funding. This is likely to raise several times more money but resources in the company will still be pretty tight. If the company survives to maturity, the investors who subscribe at this stage will own some 30% at about year six and may see their investment rise 30 times but the risk of failure is still very high at around 65%. The new money gives more security and the company can get down to some serious product development.

Exploiting biotechnology

Table 6.4
Investment stages*

Stage	Provides	Chance of Failure (%)	Investment ($m)	% Control (year 6)	Cost/Share at year 6 ($)
I: year 0 Founders	Original concept; patents	90+	0.1 No liquidity	25	
II: year 1 Lead venture investors	1st Round capital; helps to structure management team	65	0.5 Poor liquidity	30	0.17
III: year 2/3 Additional venture investors	More capital; possible board membership advice	40	2.0 Poor liquidity	25	0.80
IV: year 5/6 Public stock issue	Additional capital for expansion	20	10.0 Good liquidity	20	5.00

* Reproduced by permission of Peter Laing

By about year two or three the company is beginning to appear fairly healthy and ready to expand: it goes back to its venture capital partners for a second round. Capital is needed for expansion; products are now in sight. The investors at this round will own 25% of the company at year six, the value of their holdings going up perhaps five or six times by then. The company will now have been in existence for several years and the risk of failure will have correspondingly diminished (although still, on average, around 40%); the entry fee to join in (i.e. share price) has naturally gone up because some of the risk has disappeared.

With luck and good management, six years or so after being founded, the company will be looking like a real operation of some permanence, actually selling products and/or services. The risk of failure is falling fast (no more than, say, 20%) and a decision may be taken to apply for that precious stock market quotation which allows members of the public to buy shares — and provides the original investors with their exit route because they are now free to sell to whomever will buy. There is a bit more elbow room as far as money is concerned and the company management can cautiously begin to think of the things they would like to do, not just the minimum that has to be done. The company has arrived.

The Managing Director

The person who occupies the position of Chief Executive is obviously key to the success of the whole operation. It is his job to coordinate the variety of skills represented and controlled by the departmental directors. He will certainly already have had experience of commerce, business and management or he would not have been offered the job. He needs at least some understanding of the technical side of the business; it helps if his education included the biological sciences and chemistry, and it helps even more if he also appreciates something of engineering. Because he is "the decision maker of last resort" in the company, he must show considerable judgement in order to avoid being tempted to make his decisions on the basis of his own perhaps relatively limited technical understanding rather than listening closely to his experts in the field and weighing their evidence carefully. What is important is his skill in leadership, experience in management and ability to gain the confidence and trust of his colleagues and his Board of Directors.

The Managing Director therefore needs to formulate a clear strategy for where the company is going and how it is going to get there. He must carry his management team with him (and, through them, all the other employees). His word and decisions are difficult to overrule if he cannot be convinced by argument — only the Board can do that, something to be regarded as very serious and expressing a distinct lack of confidence. The relationship between the Managing Director and the Board may not survive much overruling.

Obviously a relatively gifted and unusual individual, he may be fairly hard to find and will command a correspondingly high remuneration package of salary, benefits and stock options, usually the highest in the company. The loss of the Managing Director is likely to have a more rapid and serious effect on the company in the short-term than the departure of any other employee or Board member. But a serious effect is not a necessarily a bad one — biotechnology companies, like others, do sometimes find that they have the wrong person in post and decide he has to go. The company often survives the shock.

7 Healthcare: prophylaxis, diagnosis and therapy

The complexities of the human condition, in both its healthy and diseased states, are enormous in extent and immense in their importance: the more than 6 billion people around the world are, at various times in their lives, acutely aware of the significance that healthcare has for each one of them. Such a level of interest and its obvious consequence — a willingness, variously expressed, to spend money — has made medical and other healthcare applications the most attractive commercial expression of biotechnology. It fits well with a recognition of the universal need for medical research and the view of many biological scientists and others that the basis of human function poses the most intriguing questions in all of science.

Clearly this chapter cannot be comprehensive about what is going on in the biotechnology of healthcare. It is intended rather to provide an overall understanding of the most important and rapidly developing lines of progress, a framework into which other ideas and information can be fitted. Some readers might find it the most difficult chapter in this volume, with too many long and unfamiliar words. There are no apologies about the vocabulary: these are the words normally used and, always with the proviso that they are properly explained, they must become at least minimally familiar to anybody wishing to understand the subject matter.

Dealing With Disease

Very generally, the treatment of human disease can be considered under three heads:

- *Prophylaxis* is directed towards avoiding the occurrence of disease and disability by using a range of preventive measures. One whole aspect is general healthcare: avoiding smoking and excessive quantities of animal fats in the diet are prophylactic measures every bit as important as administering preventive medication. To counter certain specific risks, drugs, like those used to protect against malaria, may be taken routinely before, during and after periods of potential exposure. For some infectious diseases, vaccination might partly or wholly protect an individual.

    In this account we are not concerned with general dietary practices and habits of life style nor, for the most part, with particular drugs for specific conditions — although biotechnology has much to do with the development of new ones; drugs new and old will be addressed in more general terms, mostly in this chapter and partly in the next. But vaccination and the development of new vaccines are some of the most important biotechnological contributions to healthcare and they will need more extensive discussion.

Exploiting biotechnology

- *Diagnosis* is the recognition of a disease or other medical condition. Rapidity and precision may be critically important and in some circumstances either or both might be difficult to achieve. Many conditions defy easy recognition; biotechnological methods, as well as other new approaches, are becoming increasingly important in helping to identify them quickly, accurately and reliably.
- *Therapy* encompasses all ameliorative and curative treatments. Biotechnology already has much to offer therapy, even in some surgical procedures as the discussion on new materials in Chapter 8 will show. But much of its importance lies in the development of new drugs: new antibiotics and new vaccines for infectious diseases; the effective manufacture of human hormones (Chapter 8) and other products for the treatment of deficiency conditions which may originate in an individual's own genetic defects; and the progressive development of a whole range of anti-cancer preparations, many of them proteins of human origin which have a natural role in suppressing the growth of cancerous cells.
- Another major area of biotechnological opportunity is *gene therapy*: dealing with genetic problems not by administering corrective drugs but by providing the patient's own tissues with the missing genetic information to allow him — just like a normal healthy individual — to produce for himself all the chemicals his body needs.

THE IMMUNE SYSTEM

The importance of vaccines in all three aspects of medicine, and hence in the role of biotechnology, is so great that we must at this point briefly take a look at what vaccines are, how and why they are made and what they can be used for.

The immune response is one of the most remarkable attributes of the animal system. It is a defence mechanism which has evolved primarily, it seems, to protect the host from the ravages of infection by foreign organisms. It follows from the basic discussion of microbiology in Chapter 2 that bacteria and fungi are opportunistic in their growth habits: if one of them finds itself in a conducive environment it will inevitably use the available chemicals in its surroundings to grow and consequently to produce whatever waste products are implicit in its metabolic activities. The internal environments of animals provide rich dietary opportunities for any microbes which might gain access fortuitously either through a wound or via natural orifices.

During the course of evolution many different types of microorganism have evolved to take maximum advantage of these opportunities. The effects they have on the host may be adventitious and incidental as far as the invading microbe is concerned, but for the host the infection can be more serious. The microbial cells themselves, the products they release, or both, often react unfavourably with the victim's biochemistry; collectively such materials, even if their chemistry is not yet determined, are referred to as toxins. (In a loose way, the collection associated with a particular infective microbe is often simply called that microbe's toxin.) The advantages of protection are obvious and during their evolution animals have developed mechanisms to protect themselves by countering the effects of microbial infections and eliminating the invading microbes.

Antigens

The immune system is based on the existence of large numbers of cells called *lymphocytes* and *monocytes*; there might be a trillion of them in every human being. Made in the spleen, thymus and lymph nodes, they are located primarily in the lymph vessels and, when they are mature, are also present in the blood. There are two main types: *B-lymphocytes* ("B" because they mature in the bone marrow) and *T-lymphocytes* ("T" for thymus, the site of their maturation). These cells respond to the presence of *antigens*, comprising proteins (and some polysaccharides) not found within the body of that individual. Thus, a protein, perhaps from an invading microbe, which is recognized as being "foreign", produces a response not evoked by the individual's own proteins. In the case of invading organisms the antigenic proteins will be those sited on the microbial surface or those released by the invader into its local environment.

Not all proteins are equally antigenic; that is, they do not all evoke similar levels of response and it is clear that some aspects of protein structure are more antigenic than others. Furthermore, the *antigenicities* of particular proteins are often modified by the attachment to them of certain specific carbohydrates. The sensitivity to differences between different proteins is very great but not absolute. The human immune system has no difficulty in recognizing a bacterial protein as foreign: it will be sufficiently different from all the human proteins for no confusion to result. But a protein from a close relative might fail totally to act as an antigen even though it does differ a little in structure; between identical twins all the proteins will, of course, be literally identical. An antigen can even arise internally; for example, a cancer cell which develops within an animal or a person might carry on its surface proteins different from any normally found in that organism and may as a result stimulate an immunologic response.

Antibodies and B-Lymphocytes

The immune system functions by producing special proteins of its own called *antibodies* which combine directly with antigens to counteract and inactivate them. Here again there is great specificity: each antigen evokes the production of a specific antibody.

Antibodies are produced by B-lymphocytes which carry on their surfaces receptor proteins able to recognize and bind to particular antigens: the two interacting proteins possess complementary structures which allow a precise fit and hence confer specificity. The form of the receptor protein on the B-cell is essentially that of the antibody itself. Each B-cell has no more than one type of antigen-receptor protein but every animal and every human being possesses enormous numbers of different types of B-cell, each with its own specific

Exploiting biotechnology

receptor/antibody. When the right antigen binds to it, the B-cell responds by dividing to produce large numbers of progeny *plasma cells*, each one of which actively secretes that same antibody. A population which originates from a single parent is a clone: a culture of bacteria arising from one single cell is also a clone and so is a population of plasma cells deriving from one B-cell. Thus, a single event, the binding of a particular antigen to the complementary receptor of a B-cell, produces a large number of derivative plasma cells of the same specific type, all secreting the antibody; the first enunciation of this concept was called the *clonal selection theory*.

Each antibody is a Y-shaped protein made up of two long "heavy" (h) chains and two short "light" (l) chains:

$$
\begin{array}{cc}
\mathbf{l\,h} & \mathbf{h\,l} \\
\mathbf{l\,h} & \mathbf{h\,l} \\
\mathbf{l\,h} & \mathbf{h\,l} \\
\mathbf{l\,h\,h\,l} \\
h\,h \\
h\,h \\
h\,h
\end{array}
$$

Part of the structure of each antibody (represented in bold in the diagram) is variable and specific for the antigen to which it will bind in a lock and key manner as shown in the next diagram:

$$
\begin{array}{cc}
\overline{\{A\}} & \overline{\{A\}} \\
\mathbf{l\,h} & \mathbf{h\,l} \\
\mathbf{l\,h} & \mathbf{h\,l} \\
\mathbf{l\,h} & \mathbf{h\,l} \\
\mathbf{l\,h\,h\,l} \\
h\,h \\
h\,h \\
h\,h
\end{array}
$$

$\overline{\{A\}}$ = antigen

While there are several classes of antibodies, only the part in bold type varies; the rest of the molecule is constant within each class.

Because each antibody binds to two antigen molecules, and some structures — such as whole microbial cells — possess many copies of the antigen, antibodies and antigens are likely to form meshwork complexes neutralizing the antigens and resulting often in insoluble *immune complexes* which may be engulfed by *scavenger cells* in the blood and tissues.

T-Cells

The role of T-cells is rather different. Some are called *helper cells* because they coordinate the immune response. On contact with an antigen they secrete a class of compounds called *lymphokines* which, among other actions, promote antibody production from B-cells; some of them might turn out to be important in therapy and will reappear later in this chapter. Other T-cells are *killers*: they attack cells infected with viruses, bacteria or other pathogens and may destroy both the cell and the infective agent. Some types of cancer cells are also vulnerable to attack by T-cells.

The Immune Response

How does the system work? Suppose an infection occurs for the first time in an individual's life. Antigens on the surface of, or released by, the invading microbe activate the appropriate B-cells to produce, in the course of a few days, their progeny of plasma cells which secrete the specific antibody necessary to neutralize the invaders' antigen: this is the *primary immune response*. Caught up in an immune complex, the invaders are consumed by the patient's scavenger cells and killed; the attack has been repulsed. Or has it? It takes time for the plasma cell population to develop and for sufficient antibody to be made by them. A high level of antibody production is not reached, nor is it well sustained, unless a second dose of antigen is received.

Meanwhile the infection might be proceeding apace as the invading organisms multiply on all the good chemical substrates in the blood and tissues of the host. The plasma cells may be too late; the host may already have succumbed. But if he survives he is likely to have a high degree of immunity against another attack by the same or a closely related microbe. This is because *memory cells* remain in his tissues, so called because they "remember" the earlier contact with that antigen and respond with the production of the required antibody much more rapidly and in much larger amounts than was the case on first contact: this is the *secondary immune response*. Natural infections, of course, involve a whole series of antigens associated with the infective microbes and the immune system generates a correspondingly complex set of antibodies in response; this mixed bag of antibodies circulating in the patient's blood is generally called *antitoxin*.

Acquired Immunity

There are two ways in which such desirable immunity to a particular antigen can be conferred on an individual. Because prior contact with an infectious agent can result in prolonged immunity to a subsequent invasion, that first contact can be arranged by design: the person to be immunized is actually given a minor bout of the disease in order deliberately to evoke an immune response, a procedure called *active immunization*. It may, of course, be dangerous to

Exploiting biotechnology

give people diseases, but this problem can often be avoided by using strains of the infectious organisms which have deliberately been so weakened that, while retaining enough of their antigenic properties to stimulate the immune response, they have lost their ability to cause the disease symptoms or to resist attacks by scavenger cells.

But there may nevertheless be risks: the offending microbe may not have been adequately weakened, or some individuals in the microbial population might still remain virulent. That risk may be avoided by not using living microbes but either dead ones (no risk if they have indeed all been killed) or antigenic proteins derived from the microbes, proteins which by themselves have no ability to cause disease.

Alternatively, antibodies (antitoxins) might be prepared in advance and stored against the need to use them to control the infection if it should occur. Such preparations are made in animals, usually horses, which are actively immunized in the way just described. The antibodies are injected directly into the patient who thereby acquires a *passive immunity*. Such antitoxins are used to treat cases of diphtheria, botulism and other diseases when they occur in unprotected patients because the course of these diseases can be so rapid and so dangerous that time cannot be lost in waiting the several days for the patient's own immune responses to neutralize the invading toxins. The risk here is not the first contact with the antitoxin but what may happen if the procedure is repeated after an interval, when the patient will "remember" those horse proteins and react vigorously and perhaps dangerously against them. So repeating the treatment with vaccines prepared in horses must be used with great care.

Active immunity is widely used for tetanus, diphtheria, whooping cough, polio, measles, typhoid, yellow fever and other diseases. For some infections it is less effective, either because the infectious agents frequently change the antigenicity of their surface proteins (influenza, sleeping sickness and others), because (like tuberculosis and leprosy) they invade tissue cells without killing them and lie within protected from contact with B-cells and T-cells, or even because they attack the immune system itself (the AIDS virus attacks T-cells). In malaria the pathogens spend much of their time inside cells; although antibodies are made, the intermittent periods when the parasites are present in the blood appear not to be long enough for the antibodies to be effective against them.

Incidentally, the conferring of active immunity is often called *vaccination* since it was first noticed 200 hundred years ago that milkmaids showed a lower than normal incidence of smallpox. This was because they often became infected with the cowpox virus, the Latin name for which is *Vaccinia*. Cowpox is a related but mild disease caused by a virus antigenically similar enough to the smallpox virus for anti-cowpox antibodies to be able to react with the smallpox virus. Active immunity against smallpox can thus be acquired by injecting cowpox and this has little or no risk for humans.

Self and Non-Self: Is That Antigenic Protein Mine?

It is obviously important that the antibodies we make do not combine with our own proteins; human proteins are certainly antigenic in other species as theirs are in us and some are even antigenic in other people. How does our immune system "know" which is which?

The answer remains obscure but the pattern appears to be established in the fœtus before birth. Somehow, at some time, the fœtal immune system takes stock of its own proteins and decides that all those then present are "self" but that any appearing later are "foreign". If a foreign protein is introduced into a fœtus before birth that individual may forever recognize the protein as self.

Sometimes, however, the system goes wrong and tissues may become extensively damaged by antibodies produced against self proteins. Rheumatoid arthritis is one of the conditions in which that happens.

BIOTECHNOLOGY AND ACQUIRED IMMUNITY: MONOCLONAL ANTIBODIES

The discovery ("invention" might be a better word) of monoclonal antibodies has had and will continue to have a profound effect on the way the advantages of the immune response can be exploited. The ideas were first developed in a UK public sector research institute. Presumably because their commercial opportunities went unrecognized by scientists and administrators unused to thinking in commercial terms, patent protection was not sought and commercial exploitation, largely outside Britain, has been unhampered by the payment of royalty or licence fees to those who did the groundwork.

Conventional Production of Antitoxins

The extraordinary value of monoclonal antibodies for diagnosis will appear later. First we must consider the value of antitoxins for treating infections by the process of passive immunization. Antibodies, as we saw before, are produced in horses, which some have called a "farmyard technology". To obtain, say, tetanus antitoxin, graded doses of tetanus toxin, or material derived from it, are injected into a horse which accordingly makes the corresponding antibodies. Several days later the animal is bled; an antiserum containing a crude mixture of antibodies and native horse serum proteins constitutes the antitoxin which is then administered to the patient. Note that the recipient perceives the horse antibodies as foreign proteins. Next time such an antiserum is used not only will the human patient react by producing his own antibodies against the horse serum proteins, he will also do so against the horse-derived antibodies, thereby antagonizing the very therapeutic agents intended to help him! This is a less than satisfactory situation: it would be much better to use an antitoxin completely devoid of horse proteins and containing only those antibodies which are necessary at the time.

Monoclonals for Making Single Antibodies

Before the monoclonal concept, single antibody antitoxins could not be produced either in the laboratory or industrially. To do so the correct B-lymphocyte clones would have to be

Exploiting biotechnology

isolated and cultivated; the difficulty lies in the ability of such clones to undergo only a limited number of cell divisions before they die. The growth of cells on a scale necessary for generating usable amounts of antibody requires them to be effectively immortal, able to go on growing and dividing, and secreting their particular antibody, as long as essential nutrients are provided for them.

While ordinary cells cannot do this, cancer cells can. The scientific breakthrough in antibody production demonstrated that antibody-producing cells can be fused in the laboratory with cancerous lymphocytes (*myeloma cells*) of a similar basic type which are readily available but which have lost their own ability to generate their antibody. The resulting fused cells contain the nuclei — and hence the genetic information — of both progenitors: some of them are immortal and produce the antibody. Such cells, called *hybridomas*, are often stable and can be stored until needed; they may be cultured on a large scale in a manufacturing process for antibody production.

This development is a form of genetic engineering which does not directly involve recombinant DNA technology. The antibody-producing cells needed to make the hybridomas are usually obtained from mice which are treated with the appropriate antigen in order to stimulate the growth of the relevant clone and the cells are isolated from their spleens. But the isolated cell population will contain more cell types than just the one required. Separation is necessary, a difficult but possible operation. Essentially unlimited amounts of an antibody can in principle be obtained from a selected hybridoma which may be stored and used as required. It is called a "monoclonal antibody" because it is produced as a single, pure substance from a single clone of cells.

Human Monoclonals

Valuable though mouse-based monoclonal antibodies undoubtedly are, human ones would presumably be even better. If used in the treatment of human patients, mouse products elicit a number of undesirable side effects, including vomiting, rashes, shortage of breath and irregular heartbeats. Mouse antibodies also display the disadvantages mentioned earlier with regard to horse serum antitoxin: they are foreign proteins for humans who will, given time, produce their own antibodies against the mouse antibodies and thereby nullify their effect.

While recent work has extended the mouse technology to the production of human monoclonal antibodies, a number of technical medical difficulties remain unresolved even with human products because of the variation between individuals. Although the relationship between people is obviously much closer than between people and mice, different people do show protein variations and individuals do not readily accept tissue transplants from others unless the relationship between them allows for compatibility. Thus, monoclonal antibodies made by cells from a particular individual might not be without side effects in others. There is not yet enough experience to know.

The importance of monoclonals cannot be exaggerated. We will encounter them again several times later in this chapter.

Healthcare: prophylaxis, diagnosis and therapy

BIOTECHNOLOGY IN PROPHYLAXIS

New, Safer Antigens

The risks of using whole microorganisms as antigens to confer active immunity have already been mentioned. Individuals may acquire such immunity naturally by trivial contact with the disease organisms, resulting at the time in infections so mild that the patient suffers no symptoms. Active immunization seeks to achieve the same effect by design: conferring immunity without risk, and with a minimum of symptoms, by injecting a weakened preparation of the microorganism in question. For some diseases, including smallpox and tetanus, there appear to be few difficulties. Others, like typhoid, do sometimes cause slight discomfort and distress. In yet other cases, whooping cough for example, serious difficulties with long-term consequences occasionally arise. For some diseases the risks are unacceptable because, as with the AIDS viruses, there may not yet be a way of attenuating the natural viruses to make them safe enough to use as vaccines. It would be more acceptable if these forms of immunity could be acquired with no risk.

Another set of problems derives from the variability or the low antigenicity of the microbe's antigens as perceived by the host's sensing systems. It helps little to acquire active immunity to one or more microbial surface antigens if the next time contact is made with that microbe its surface proteins have different structures and fail to elicit the secondary immune response. The viruses responsible for influenza and AIDS readily mutate their surface antigens, while the protozoon responsible for sleeping sickness is able to shuffle its complement of surface proteins around so readily as to allow it to escape much of the host's immune response.

Artificial Vaccines

One way of avoiding such problems might be to make use of certain weakly-antigenic proteins which the viruses and other parasites also possess: because they do not provoke strong responses, and are largely ignored by the patients' immune systems, there is little advantage in the invading organism being able to change them easily and so they are likely to remain as constant features. Using genetic engineering, their genes could be cloned into bacteria from which the proteins themselves might later be isolated and used directly as provoking antigens in amounts which the host system could not ignore. If such proteins can be made in quantity, the antibodies to them — produced via acquired active immunity — may be sufficient to bind them into complexes and allow the organism of which they are a part to be eliminated.

An example of this approach was reported recently. Vaccines, being protein-based, must usually be injected directly into the blood or into a muscle because proteins ingested by mouth are broken down in the digestive system and do not enter the bloodstream and tissues. A new concept for oral vaccines is being developed, making use of bacteria which sometimes gain entry to the gut and cause food poisoning. The bacteria are modified genetically so that, while still able to colonize the intestine, they can no longer cause the disease. Further genetic

Exploiting biotechnology

manipulation places the genes for one or more of the tetanus bacterium's surface antigens into the modified intestinal resident in such a way that the latter actually makes the tetanus proteins and places them on its own surface. At this stage the modified intestinal bacteria are taken by mouth and evoke a primary immune response both to themselves *and also to the tetanus proteins.* If a real tetanus infection subsequently occurs the rapid and massive secondary response is triggered and the tetanus infection dealt with more effectively than if the body had to start again from the beginning.

Major Outstanding Problems — Malaria and AIDS

One of the world's most important health problems, malaria affects hundreds of millions of people resulting in several million deaths each year. The disease is caused by a protozoon parasite carried by mosquitoes; while effectively controlled in the early days of DDT and other modern insecticides, mosquitoes have lately become increasingly resistant as have the malarial parasites themselves to the insecticides designed for their control.

The difficulty of preparing conventional vaccines for protection against malaria derives in part from the complex life cycle of the parasites as they migrate between the liver and the blood stream, and in part from the varying antigenic signals which the immune system has to recognize. Thus, one vaccine would be needed to attack the so-called *sporozoites* injected into the blood by the mosquito bite. To be effective, the victim's immune system would have to respond within about thirty minutes of the bite, after which time the sporozoites enter and are protected by the liver cells. A separate vaccine would be needed for the *merozoites* which leave the liver, attack blood cells and provoke the symptoms of the disease. Natural immune responses simply cannot take effect fast enough. Several protective vaccines have been prepared but immunizing animals and humans with artificial antigens made by genetic engineering have largely not been successful. The importance of the disease ensures that efforts will continue but it is impossible to predict when and how they will succeed.

The viruses causing AIDS (*acquired immune deficiency syndrome*) present another particularly difficult set of problems. Called *human immunodeficiency viruses (HIV)*, they attack the T-cells of the immune system itself because a protein (gp-120) on the outer surface of the virus has a particular affinity for one (CD4) in the membrane of T-cells; the immune system is thereby inactivated. HIV belong to the category of retroviruses and DNA copies of the RNA bearing their own information can become incorporated into the host's own DNA and lie there dormant for long periods. During the period of dormancy no symptoms are displayed but in a proportion of cases some stimulus eventually provokes the virus to cause the disease.

Antibodies are at first produced but fail to eliminate HIV. One reason might be that the lymphocytes to which the viruses bind fuse with other lymphocytes and spread the viruses without their ever being released into the bloodstream where the antibodies are located. Another is that HIV appears to alter the surface properties of infected T-cells so that they can no longer destroy cells harbouring the virus. A third seems to be that the viruses mutate so rapidly that the nature of the surface antigens is constantly changing and the immune response mechanism cannot keep up.

Healthcare: prophylaxis, diagnosis and therapy

Hope for vaccines lies either in making antibodies to gp-120 directly or in incorporating gp-120 into the vaccinia virus. The latter would then be used as an antigen in the expectation that, while such a modified vaccinia virus cannot cause AIDS, the host will produce antibodies to gp-120 and react strongly to any later infection with genuine HIV. A possibility for treatment might lie in using genetic engineering to make quantities of the CD4 protein with the intention of using it to swamp the CD4 binding sites on the virus particles so that they can no longer attach to the CD4 located on T-cells, which would thereby remain active.

Other ideas being tried are based on the notion of making artificial, non-infectious HIV with enough of the native surface architecture to evoke an effective immune response. For instance, it might be possible to construct the hollow protein coat but leave out the RNA so that the particle, while displaying antigenicity, is not a complete virus and so cannot multiply and integrate.

Antibodies as Antigens

Still another way of preparing vaccines composed of pure antigens, and devoid of other protein contaminants, is to use antibodies to the original antigens as antigens in their own right; these can then be employed to evoke a second generation of antibodies which ought to resemble the original antigens. But sometimes things do not go quite as planned, as the following set of diagrams illustrate — in this model, all the structures composed of solid squares are behaving as antigens while the antibody each raises is printed in hollow squares.

Suppose this is the original antigen, with the antigenic site itself represented by the protruding triangle:

The antibody produced in response will have a matching site, the recessed triangle. But antibodies are not simply reflections of antigens; they have their own distinctive structural characteristics and the one raised to this antigen might also have other potentially antigenic sites, like the protruding rectangle:

```
     □□□□□           □□□□□
    □□□□□□          □□□□□□
□□□□□□□□□□□□ □□□□□□□□
□□□□□□□□□□□□□□□□□□□□
        □□□□□□□□□□□□□□□□
        □□□□□□□□□□□□□□□□
        □□□□□□□□□□□□□□□□
```

113

Exploiting biotechnology

If this antibody were now to be used as an antigen:

the next generation of antibodies might, at least in some important properties, be expected to resemble the original antigen:

However, instead of raising an antibody to its recessed triangle, the first generation antibody might instead evoke one to its protruding rectangle: thus, the second generation antibody, the intended artificial antigen, may not resemble the original antigen or have the desired properties:

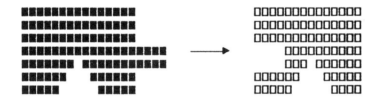

Vaccines Against Cancer

The technique of using antibodies as antigens shows considerable promise though, as we have just seen, problems remain to be solved. While perhaps more properly included under therapy than prophylaxis, it is appropriate at this point to mention the possible value of antibody-antigens for anti-cancer therapy. One of the difficulties to be overcome with cancer is, of course, the fact that cancer cells arise spontaneously from within an individual; because they have so many of their proteins in common, it is not easy to find ways of attacking the cells of the cancerous tissue without also damaging healthy ones. Traditionally, drugs have sought to make use of the fact that cancer cells grow and divide more rapidly than the normal cells from which they are derived. By attacking *growth processes* (for instance, DNA replication) an attempt can be made to direct a chemical attack against the faster-growing cancer cells while leaving other cells relatively unscathed. The problem is that ongoing growth processes are necessary elsewhere, as in the continuous production of new blood cells to replace those destroyed as part of a natural cycle. Serious consequences may follow any interruption of cell growth and division, with undesirable or dangerous effects. (Hair is another tissue which grows continuously, which is why people on courses of chemotherapy treatment not infrequently become bald, but their hair grows back when the treatment stops — hopefully the cancer will not.)

Nevertheless, cancer cells really are different and, while these differences are subtle, the very remarkable discriminatory powers of the immune response might be able to spot them and generate specific antibodies to "cancer proteins". Positive results have been reported with a number of tumours, including some sarcomas, bladder tumours, lymphomas, colon carcinomas and melanomas, but other complex effects are also observed and further development is clearly necessary.

Other Uses for Antibodies

When discussing product purification in Chapter 3, mention was made of using the way in which specific binding agents, attached to solid matrices, can be used to extract and concentrate complementary chemicals from mixtures. The words used then were "key" for the binding agent attached to the matrix and "lock" for the free chemicals so bound.

Antigens and antibodies can form ideal "lock" and "key" pairs because they recognize one another so precisely. Antibodies can thus be used both in the laboratory and in production processing to isolate and purify proteins such as human products manufactured by microbial systems (Chapter 8). Conversely, of course, antigens can be used to purify antibodies. The production of usable quantities of monoclonal antibodies has helped to make these purification techniques feasible for larger-scale industrial operations.

Allergies

Some secondary immune responses are so vigorous as to merit the description *hypersensitivity*. Allergies fall into this category and constitute severe and occasionally fatal reactions to the presence of a wide range of possible antigens in foods, hair and fur, house and other dusts, and pollens. The antigens react with cells of the immune system causing them to make certain *immunoglobulin* proteins; these trigger the release of pharmacologically-active agents such as *histamine* which in turn cause tissue damage and other undesirable and unpleasant effects.

Treatment of acute allergic conditions is based on suppression of the symptoms with *anti-histamine* drugs, while cures are attempted by a process of desensitization: over a long period of time, the patient is challenged with gradually increasing quantities of the antigen, called an *allergen* in this context, in the hope that tolerance will eventually develop. Cures are not always successful.

Exploiting biotechnology

A new approach is to create a novel protein fragment to block the mechanism releasing histamine. A ten-amino acid peptide has been artificially synthesized which recognizes the histamine release trigger and binds to it in a way which does not cause histamine release but which does prevent the binding of the immunoglobulin.

BIOTECHNOLOGY IN DIAGNOSIS

One way in which biotechnology is beginning to play a significant part in diagnosis is in the development of new sorts of sensitive instruments called *biosensors*, designed to detect and measure the concentrations of certain chemicals characteristic of specific diseases. For example, they can be used to measure glucose in blood or urine as an indicator for diabetes. Biosensors are dealt with in more detail in Chapter 12; for the present our discussion takes us into other diagnostic procedures.

Diagnosis with Antibodies

Antigens which prompt the production of antibodies are not all proteins. Some are quite small molecules, and the antigenic nature of a particular protein can be altered by combining it with such a small molecule. By suitable manipulations of this sort, antibodies can be used to detect the presence of a wide variety of substances which may be indicative of disease or have other medical interest; some of them are proteins, others not.

The visible signs of reaction between antigens and antibodies depend particularly on the nature of the antigen. Each antibody molecule, as we have seen, binds to two molecules of its antigen. If those two antigens are themselves joined in some way, perhaps by residing on the surface of the same microbial cell, a whole meshwork of joined-up antigens and antibodies develops which is manifest as a precipitate of visible particles. Pure proteins, however, which may have but one antigenic site on each molecule, are less likely to precipitate when mixed with the corresponding antibody.

Many techniques have been developed to detect antigen-antibody interactions. This is not the place to describe them in detail but it is worth listing a few of the more common ones:

- *radioimmunoassay (RIA)* and *fluorescence immunoassay (FIA)*, which use an antibody labelled with an easy-to-detect radioactive element or fluorescing chemical as the detector for the reaction. ("Radioactive" is explained in the next section);

- *enzyme-linked immunoassay (EIA)*, in which the immune reaction is coupled to an enzyme catalysis;

- *enzyme-linked immunosorbent assay (ELISA)* is a refinement of EIA with either the antigen or the antibody first bound or adsorbed onto a surface.

Healthcare: prophylaxis, diagnosis and therapy

Monoclonal Antibodies for Diagnosis

Just as in prophylaxis, the advent of monoclonal antibodies has enabled the use of antibodies in diagnosis to expand enormously. Because of their very great specificity and ready availability, monoclonal antibodies have facilitated the development of precise methods for distinguishing between closely-related microbes, between drugs as similar as morphine and heroin, and between a variety of steroid hormones differing only marginally, but physiologically importantly, in their molecular structures. The levels of hormones and other proteins circulating in the bloodstream can be monitored as indicators for characterizing a patient's condition, as can the concentrations of drugs administered during treatment. None of this would have been possible with the mixture of antibodies present in a typical horse antitoxin serum. Monoclonal antibodies are one of the major success stories of contemporary biotechnology and a view of their commercial significance will be taken at the end of this chapter.

DNA Fingerprinting

This is diagnosis of a very different sort. It is primarily concerned not with recognizing disease conditions or physiological states but with exploring and establishing relationships between individuals. In order to understand how it is done, a little more science is necessary and two additional concepts are explored.

The nature of radioactivity

The atoms of each of the elements known to chemistry coexist in slightly different forms. Hydrogen, for example, has three sorts of atoms. The lightest and simplest in structure weighs one unit. A second version incorporates an additional sub-atomic particle, bringing its weight up to two units; in all respects other than weight it behaves exactly like the first version. The third possesses yet another particle and weighs three units. Chemically it is still every bit hydrogen but this new configuration is unstable and tends to disintegrate into an atom of another element, helium, which weights three units. The disintegration is accompanied by the emission at very high speed of a tiny, virtually weightless, particle called an *electron*, which carries a single negative electric charge. In some respects this electron behaves like a ray of very energetic light; for example, it will fog a photographic film. It is because of these "rays" that the term "radioactive" is used for such disintegrating atoms. Every one of the 105 elements has some radioactive varieties (called *radioisotopes*) among its atoms.

Repeat sequences in DNA

The second concept relates to the organization of DNA in animal and human genomes. In addition to the bulk of the DNA which codes for the many proteins necessary for the person's

Exploiting biotechnology

or animal's biochemistry, there are also present large quantities of DNA which apparently code for no proteins at all. Why they exist is a matter of considerable debate. Much of this non-coding DNA is in the form of short sequences, each of a few tens of bases ("letters" in the models used in Chapter 2), repeated over and over again, maybe tens, hundreds or thousands of times.

In genes coding for specific proteins it is very important that the sequences remain unchanged from generation to generation; if they were to vary, the amino acid sequences of the proteins for which they coded would also change and most probably inactivate or otherwise impair protein function. Thus, individuals of any one species show very little or no variation in the base sequence of the genes coding for particular proteins. But the repeat sequences, with no apparent coding roles, are not limited in this way; it matters little if bases are exchanged because no proteins are made from them. Random base alterations in the repeat sequences inevitably occur from time to time. Not being involved in the formation of proteins, they are neutral in their effect; conferring neither advantage nor disadvantage, they persist in the population, accumulate in the chromosomes and are transmitted from parent to offspring. Not only do repeat sequences between individuals differ with respect to base changes in particular positions, they differ also in the number of repeats in certain of the sequences.

In the following oversimplified example, the likelihood is calculated of two individuals (other then identical twins) having the same pattern of repeat sequences. Human cells possess 23 pairs of chromosomes, the sites of the repeat sequences as well as of the genes coding for proteins; one chromosome of each pair originates from a persons' mother, the other from his father. If one maternal chromosome pair is designated A-A', and the corresponding paternal pair a-a', an individual and his siblings might have any one of four combinations of that pair: **A-a, A-a', A'-a and A'-a'**. The chances of two offspring of the same parents sharing identity for *any one of their 23 pairs of chromosomes* is thus one in four. Multiply up that probability for all 23 pairs and the probability of total identity reduces to 1 in 70 trillion. Compound that likelihood further by the fact that mutation and variation is taking place actually within an individual's lifetime, and the probability of two people showing the same pattern diminishes even more. A person might even display differences in his own pattern over a period of time! Do note, however, that in practice the number of different patterns is markedly lower than theory would postulate.

DNA probes

DNA molecules are enormously long. It is extremely arduous to locate any particular region of the DNA molecule by analysing it chemically. But there is another way, a clever way. Pieces of DNA complementary to a specific region of a very long DNA molecule can readily be constructed artificially. To illustrate the point, let us return for a moment to the models of Chapter 2. The simple message in its double-stranded form was:

Healthcare: prophylaxis, diagnosis and therapy

```
···T-H-E-C-A-T-S-A-T-A-N-D-A-T-E-T-H-E-R-A-T-A-N-D-R-A-N-O-F-F···
   : : : : : : : : : : : : : : : : : : : : : : : : : : : : : :
···T-H-E-C-A-T-S-A-T-A-N-D-A-T-E-T-H-E-R-A-T-A-N-D-R-A-N-O-F-F···
```

Let the two strands be separated and the lower one discarded, leaving:

···T-H-E-C-A-T-S-A-T-A-N-D-A-T-E-T-H-E-R-A-T-A-N-D-R-A-N-O-F-F···

Using bases ("letters") containing radioactive atoms, a piece of DNA is made corresponding to just part of that message, say:

A-N-D-A-T-E-T-H-E-R-A-T-A-N-D

Such a piece of radioactive DNA is called a *probe* because, when mixed with any DNA containing the sequence ···A-N-D-A-T-E-T-H-E-R-A-T-A-N-D···, the probe will bind to it in register as a result of the pairing tendency. Thus it binds to:

```
···T-H-E-C-A-T-S-A-T-A-N-D-A-T-E-T-H-E-R-A-T-A-N-D-R-A-N-O-F-F···
                  : : : : : : : : : : : : : : :
                  A-N-D-A-T-E-T-H-E-R-A-T-A-N-D
```

and just as well to:

```
···O-N-E-D-O-G-R-A-N-A-N-D-A-T-E-T-H-E-R-A-T-A-N-D-S-A-W-R-E-D···
                  : : : : : : : : : : : : : : :
                  A-N-D-A-T-E-T-H-E-R-A-T-A-N-D
```

DNA to which the probe has bound thereby becomes radioactive and its presence can be detected by its ability to fog photographic film. We are now ready to explore DNA fingerprinting.

Comparing DNA from different individuals

Samples of DNA are obtained from tissue or body fluids. The DNA is broken down into small pieces with a restriction enzyme of the type that cuts DNA at defined positions as described in Chapter 2:

Exploiting biotechnology

```
                    ↓
···?-?-?-?-?-?-?-E-D-U-D-E-?-?-?-?-?-?-?···
   : : : : : : : : : : : : : : : : : : :
···?-?-?-?-?-?-?-E-D-U-D-E-?-?-?-?-?-?-?···
                    ↑
```

There are many sorts of restriction enzymes and a convenient one is chosen to snip out the repeat sequences which display variable length distributions in different individuals. Each sample of cut DNA, dissolved in water, is loaded as a narrow band into a slab of a special jelly, with samples to be compared placed next to one another. An electric voltage is applied across the length of the slab and, as the pieces of DNA carry electric charges, the voltage difference between the two ends of the slab causes them to move. The distance each piece travels is governed by a combination of its size and weight, and the amount of electric charge it carries. Thus, the different lengths of repeat sequences from an individual source will travel different distances.

After sufficient time for the DNA fragments to be spread along the whole length of the slab as a series of as yet invisible bands, the current is turned off, the native double-stranded DNA fragments, still in the slab, are separated with chemicals into single strands and the single-stranded fragments transferred to a filter without disturbing their pattern or the distance each has migrated. After drying, the filter is baked to prevent the DNA fragments from moving when the whole filter is next treated with a solution containing the radioactive DNA probe. Binding takes place wherever the probe finds a piece of complementary DNA on the filter; any residual unbound probe is washed away. The filter is again dried and pressed against a sheet of photographic film which is later developed to reveal the location and quantity of the radioactivity associated with each DNA fragment. The result looks like a supermarket price bar chart comprising some 50 bands. Comparisons are made between the adjacent samples run on the same slab in order to establish degrees of similarity. Often the comparison between samples is performed in triplicate to provide three independent sets of complementary data and so reduce the chances of error.

Most of the procedure is accomplished quickly but, because the time taken to fog the photographic film depends on how much radioactivity is present on the filter, that part of the analysis might take several days or even weeks.

Uses for DNA fingerprinting

The technique is used for medical, forensic and research purposes. Medically, it can help to establish the genetic relationships between individuals in order to trace patterns of disease inheritance and predict the susceptibility of individuals. Forensically, it is now widely used in criminal cases to determine whether or not recovered samples of blood, semen or tissue originate from a particular suspect. It is also employed by immigration authorities to determine family relationships. In research, there are many applications both in medicine and more widely in population biology to explore the way populations and individuals are related to and have interacted with one another.

Healthcare: prophylaxis, diagnosis and therapy

Amplifying DNA sequences

When no more than very tiny samples of DNA are available for comparison, the sensitivity of fingerprinting and other genetic tests can be improved by repeatedly replicating selected segments from a complex DNA mixture. The DNA is snipped with a restriction enzyme, separated into single strands by heating and mixed with a large excess of a probe, designed to bind to one of the ends of the cut DNA pieces; the probes will act as primers for the synthesis of replicate single strands of DNA. A supply of the individual nucleotide units ("letters") from which DNA is made is added, together with the enzyme (*DNA polymerase*) which links them together. The individual nucleotides spontaneously line themselves up on the existing single strands of DNA (as we saw in Chapter 2) and, starting at the probes bound to the sections of restriction enzyme-snipped DNA, the enzyme adds new units in the correct order, eventually making new pieces of DNA complementary to the snipped to the sections of the original. The mixture is heated again to separate the old strands from the new ones made on them and, on cooling, another probe attaches to each piece of DNA, new and old, more individual nucleotides line up and the enzyme again joins them up starting at the probes. By repeated heating and cooling, the original DNA can be amplified as much as a millionfold in a few hours. This technique, called the *polymerase chain reaction*, has become very important in genetic manipulations in biotechnology; its inventor was awarded a Nobel Prize in 1993.

Diagnosis of Microbial Infections

The speed with which a microbial infection is diagnosed can be of critical importance to the patient. In urgent cases the physician does the best he can from an examination of symptoms and starts the treatment he deems most appropriate. It would often be a great help to be more certain of the diagnosis within a short period of the patient presenting.

Recognition of microorganisms on an individual basis has hitherto been very difficult if not impossible — drawing a small sample from an ill patient for microscopic examination is unlikely rapidly to confirm a diagnosis beyond doubt, or whether a particular infection is resistant to certain antibiotics or likely to be responsive to a specific therapeutic vaccine. To find that out the pathogen has to be cultured under nutrient conditions, normally for a minimum of 24–48 hours, until a large enough population has grown for a battery of biochemical and other tests to be applied. The patient could be dead by then.

In recent years a number of rapid testing methods have been devised exactly with the objective of confirming within a few hours at most the identity and such critical properties as antibiotic resistances and sensitivities, and antigenic properties. Some methods are intended to make the microbes easier to see under the microscope among cells and debris in samples of patients' blood or tissue while others depend on chemical and physical measurements:

- the sample from the patient is concentrated by collection on a very fine filter membrane and the bacteria specifically stained (coloured) with a fluorescent dye before they are counted. When illuminated by an invisible beam of ultraviolet light, fluorescent particles are easily seen as bright points against a dark background;

Exploiting biotechnology

- making use of certain electrical properties of microbial cells, a suspension of the microbes is pumped through a very fine hole in the wall between two adjacent glass vessels. Each vessel contains an electrode and a voltage applied across the hole results in a certain current flowing. As each microbial cell passes through the hole it momentarily alters the electrical properties of the fluid — the change in current flow can be registered as a pulse;
- all living organisms contain a certain chemical (*ATP* — its full name is "adenosine triphosphate") essential to their energy mechanisms. ATP is used by fireflies as their source of energy for generating light; extracted from the fireflies, the relevant enzyme will produce light flashes in the test tube in the presence of ATP. The technique can be used to recognize living microbes under the microscope: dead microbes will not flash;
- when microbes are inoculated into a growth medium, new sensitive instruments are able to detect population increases within a few hours, much earlier than by traditional methods. The instruments measure electrical or optical properties of the population, or the minute quantities of heat generated by the cells as they go about their chemical business. Using radioactive nutrients, the carbon dioxide produced in oxidation or fermentation can be measured in tiny amounts. Radiolabelling can be used for unequivocal identifications by revealing the proportions of different proteins in extremely small samples of microbial matter;
- DNA probes use specific nucleic interactions for the recognition of particular microbial strains;
- immunoassays permit very precise identification of microbes and microbial products by their reactions with specific monoclonal antibodies.

Many of these methods have been commercialized and are available in well-equipped medical facilities.

BIOTECHNOLOGY IN THERAPY

There are many physical and mental problems to be treated and therapy is accordingly a large subject. It encompasses all measures relevant to the alleviation and cure of disease — the use of drugs, surgical procedures and a host of other measures. Biotechnology has little to contribute to some of these and its present involvement in surgery, for example, is very limited though not totally absent: Chapter 8 includes a discussion of new bone replacement materials. It is largely through the use of drugs, new developments in vaccines, and the identification, amelioration and rectification of inherited genetic disorders that biotechnology has most to offer. The possibilities are obviously vast and the sections which follow can do no more than sketch the general scene.

Healthcare: prophylaxis, diagnosis and therapy

Natural, Artificial and "Designer" Drugs

While many therapeutic drugs come from natural sources, the growth of the chemical industry and the realization that chemicals made for other purposes sometimes have medical applications has led to the development of a myriad of entirely artificial preparations of value in medical practice. In the contemporary context, as scientific understanding of the molecular mechanisms underlying disease processes increases rapidly, the possibility of designing and making drugs for very specific purposes becomes ever more real.

Natural products as drugs

The early drugs used in therapy originated empirically in folk remedies based on observing, without understanding, the beneficial effects of a variety of natural products. As chemical skills progressed it became possible to identify the active agents in crude natural preparations, sometimes improving their performance by isolating and purifying the product, learning how best to use it and perhaps also to modify some part of its chemistry to make it more effective.

The antibiotics are well-documented examples of drugs which have developed like this. *Penicillin* (discussed in more detail in Chapter 8) was discovered some 60 years ago as a microbiological phenomenon. More than a decade elapsed before it was purified, longer still for its chemical structure to be elucidated and not until 30 years or more after its discovery did systematic chemical modification of the natural molecule bring a marked improvement in performance.

Many of the old folk remedies have been re-examined in the light of modern technology and new, potentially valuable products have emerged. The ability to find such new products depends on the scope and discrimination of the screening procedures. Using methods designed to identify a wider range of potentially valuable properties than has hitherto been possible, some commercial companies are instituting systematic searches for new therapeutic substances and re-examining old ones for new and hitherto unrecognized properties.

Biotechnology is also having a major effect on the *availability* of certain natural products, particular those from human sources. Patients with hormone deficiency diseases, for example, usually need to receive the missing or inadequate hormone; for some conditions an animal product may be fairly satisfactory but the human version is likely to be best — even if difficult or impossible to obtain in sufficient quantities. Therapy for heart attacks includes methods for dissolving blood clots and a natural human protein, *tissue plasminogen activator*, is one way of doing so — although some bacterial enzymes like *streptokinase (fibrinolysin)* are reported to work just as well. The new techniques of genetic engineering allow such products to be made relatively cheaply in large amounts using microbial systems, another matter dealt with in more detail in the next chapter.

Exploiting biotechnology

Artificial products as drugs

R & D in the chemical industry is continually generating new compounds which are put through a battery of tests to reveal useful properties for a variety of purposes, including medical ones. Hundreds of thousands, perhaps millions, of new chemicals have been evaluated, sometimes demonstrating quite unexpected benefits for therapy. Derivative versions of compounds which have already been found to be useful are also explored in the hope of obtaining better formulations. These explorations are likely to continue indefinitely though, as more and more information is accumulated, the probability in any particular case of discovering anything new and important must be expected to decline.

New drugs by design

The objective of drug therapy has always been to employ drugs which have specific beneficial effects on the diseases being treated without causing any untoward consequences for the patient. That is not always a tall order: in the early days of penicillin usage, before widespread microbial resistance became a reality, it really did seem to be a perfect drug. Miraculously, infections by susceptible microbes were easily controlled and cured with few patients showing unfavourable side effects.

Penicillin was nevertheless discovered by chance and, while later antibiotics were found by systematically screening natural microbial populations, they were all obtained empirically. They can be seen as some of nature's own successful experiments in drug design. Increased understanding of the detailed and complex chemical structures of proteins and other biochemical macromolecules now begins to offer the possibility of imitating nature's example and designing compounds to fit precisely with defined receptor sites. Mention has already been made of specific targeting in connection with alleviating allergic reactions. The concept will certainly be extended to other diseases characterized by the occurrence of molecular interactions which, in the normal state, either do not take place at all or are quantitatively different.

To take just one possibility, imagine a disease resulting from the excessive production of a hormone binding to certain receptor proteins located on cells at various places in the body. Several therapeutic alternatives might be considered. One might be to reduce the rate of hormone production by physically removing part of the producing gland, although surgery is always best avoided. Another would be to interfere chemically with hormone production or block its interaction with the receptors; understanding the biochemical mechanisms controlling hormone production or the detailed chemical configuration of receptor sites might allow the use of specific drugs to affect just that hormone system and no other aspect of body chemistry.

Recent advances, particularly in elucidating the structures and modes of operation of increasing numbers of physiologically important proteins, will extend virtually without limit the opportunities for designing drugs to order.

Healthcare: prophylaxis, diagnosis and therapy

Methods of drug delivery

Traditional routes for drug administration are by mouth or by injection. Oral ingestion cannot be used for proteins, including the many immune system products to be discussed later in this chapter, because they are broken down in the digestive tract. Injection, on the other hand, might be unsatisfactory for patients not under direct medical care. Furthermore, both traditional routes deliver the drug at a concentration which is high at the time of entry, gradually falling as the drug is eliminated or degraded. Convenient means of delivery as well as steady circulating and tissue levels of a drug are clearly desirable.

It can be a problem to move a drug across a membrane barrier. Encapsulation within *liposomes*, tiny capsules bounded by fatty membranes, facilitates access to cells covered by proteins carrying their own fatty components because the two sorts of fat tend to bond to one another. Injected liposomes might accumulate at inflammatory sites, or in the liver or spleen, where they can persist for weeks, gradually releasing their encapsulated drug. Alternatively, a water-soluble polymer may be used either to enfold the drug or to provide a surface on which the drug is deposited. The drug is liberated within the tissue as the capsule dissolves, or it is released progressively and in a controlled fashion from its surface attachment. A further refinement is to use a tablet constructed as an *osmotic pump*. An outer insoluble polymer coating allows water to enter the interior and dissolve the drug at a controlled rate from the internal surface while a tiny hole drilled in the outer casing permits leakage of the dissolved drug from the interior.

Yet another approach is to attach the drug to a natural carrier molecule able to bind to a specific receptor and so deliver the drug exactly to the right place. An example of this type of system, in which a toxin is bound to an antibody, will be described in a later section of this chapter.

Modifying the Immune Response

As the workings of the immune system become better understood, so new possibilities emerge for using its properties to advantage. *Cytokines* promote growth and cell division and it might be possible to use them to enhance the immune response through the greater proliferation of B- and T-cells. Certain cytokines have received a good deal of publicity in recent years for their potential as anti-cancer and anti-viral drugs. *Adjuvants* are substances which enhance the immune response when administered together with an antigen. Exploring new ways of using all of these is expected to lead to major improvements in immunotherapy.

Interleukins

Among the lymphokines, substances made by helper T-cells, is a category of proteins called *interleukins*: several have been identified and together they stimulate the growth of T-cells and

Exploiting biotechnology

the production of antibodies by B-cells. Interleukins are in effect hormones of the immune system, secreted by T-cells and interacting with specific receptors located in other parts of the immune system. They can be compared with *endocrine hormones*, one of the two main internal communication mechanisms (the nervous system is the other). Released by producing cells often located in special glands like the thyroid or adrenals, the function of endocrine hormones is to bind to other cells elsewhere in the body and provoke a specific response from them. *Insulin, thyroxine* and *adrenalin* are well-known examples; there are many others.

As part of the natural mechanism for promoting B-cell growth and antibody production, interleukins are being developed for use directly as therapeutic agents, for instance as *immunostimulants* in diseases such as certain types of leprosy in which the patient's unaided immune mechanism is relatively ineffective. The performance of interleukins is likely to be improved by deliberate modification via protein and genetic engineering; interleukin-2 has been extensively worked upon from this point of view. Its structure is now partly characterized and derivatives have already been prepared which interact with the interleukin-2 binding site without evoking the normal response. This offers hope of blocking the natural reaction in conditions resulting from aberrant lymphokine-receptor effects, including *autoimmune diseases* (in which the body attacks its own tissues — it fails to recognize "self"), the generation of certain tumours and some forms of diabetes. Engineered versions of interleukins have been made to bind to interleukin receptors on recently activated T-cells without affecting those on T-cells which remain quiescent. Interleukins might thus be used as *immunosuppressive agents*, killing T-cells responding to a graft or transplant and thus minimizing rejection responses.

Another approach to interference is via the interleukin cell-surface receptors. Although the receptors are present in small numbers and difficult to isolate and characterize, parts of their protein structures have now been analysed and several of their genes cloned. It seems certain that in time some, at least, of the interleukin receptors will become available in quantity in a soluble form. They can be used to bind and inactivate excessive amounts of circulating lymphokines before they can attach to their natural cell-surface receptors. Malignant cells, too, may have interleukin-2 receptors; blocking circulating interleukin with added receptor molecules might help to slow down or stop the growth of such malignancies.

Interferons

It was discovered more than 30 years ago that these small proteins were produced by cells infected with viruses. They appear to interact with neighbouring cells, stimulating them to make anti-viral proteins and protect themselves against viral infection. Three types of interferon are known: all induce the production of anti-viral proteins while one of them has additional effects on the immune system.

Viral infections are much more difficult to treat than those of bacterial origin. Bacteria are complete living organisms, with their own characteristic metabolism and distinctive biochemistry. The differences between them and the cells of a human patient potentially offer many opportunities for attacking the bacteria without harming the human. Two brief examples will suffice. Both humans and bacteria in the course of their metabolism make use of

Healthcare: prophylaxis, diagnosis and therapy

a molecule called *folic acid* which is a participant in certain essential enzyme reactions. Bacteria make this chemical for themselves from simple starting materials; humans cannot do so and need it fully formed in their food. Thus, substances which prevent bacteria making folic acid have no effect on humans and such chemicals were the basis of the anti-bacterial *sulphonamide* drugs which were one of the mainstays of bacterial disease therapy before antibiotics; they are still in use today. Penicillin is another case: it prevents some types of bacteria from building their cell walls but, as human cells do not have walls in the same sense, they are not affected.

Viral diseases are quite different: viruses have no independent metabolism but integrate themselves closely with the host cells, making attack difficult without also harming the host. *Interferons* act against viruses forming double-stranded RNA (as distinct from double-stranded DNA) during their replication process. It stimulates the production of an enzyme which is particularly active in the presence of double-stranded RNA and attacks those single-stranded RNA molecules which act as intermediates in the formation of viral proteins. (In the Chapter 2 model, single-stranded RNAs are printed in lower case letters: -t-h-e-c-a-t-s-a-t-a-n-d-a-t-e-, etc.). Many clinical trials have been undertaken to evaluate the suitability of interferons for treating viral diseases.

The value of adjuvants

The effectiveness of immunization programmes notwithstanding, only about 20 human vaccines are commercially available and, while the versatility of biotechnology promises to extend that number significantly, it is not a foregone conclusion that each new vaccine will evoke the desired intensity of protection. That may depend on the use of adjuvants, substances which, when administered at the same time as the antigen, stimulate and extend antibody synthesis.

There are many different types, originating mostly from bacteria and plants, although apparently the only widely approved adjuvants are the hydroxide and phosphate of aluminium — entirely inorganic and therefore nothing to do with any form of living organism. Not much is known about the way adjuvants work and they have mostly been found empirically. Some, while effective, are toxic and the search for better ones continues. People make them, too; in a sense, interleukins are the human body's own adjuvants and we have already noted the efforts being made to harness their action in therapy. A good adjuvant can be the critical factor in a vaccination procedure and they are important commercially for the production of effective vaccine preparations.

Therapy for Cancer

Cancer is very different from an infection with a foreign organism. The problem is that tumour cells are so similar to their normal neighbours. Agents which damage one will probably attack the other and it is difficult to discriminate between them; difficult, but fortunately not impossible.

Exploiting biotechnology

All attacks on tumours seek to maximize their differences from the normal cells which surround them. Even surgery does that: this piece of cancerous tissue is to be excised, but not that normal tissue. From a chemical and physiological point of view cancer cells exhibit a number of characteristic properties:

- they grow relatively quickly;
- they invade other tissues;
- unlike normal cells, which can divide no more than a limited number of times and hence are constrained in their ultimate growth, cancer cells are essentially immortal. In a conducive environment they can, in theory, go on multiplying for ever.

All these properties, dangerous and potentially lethal for the unfortunate individual afflicted, nevertheless represent special characteristics which in principle offer opportunities for selective attack.

Attacking cancers chemically

Because most cancer cells grow and divide faster than their normal counterparts, one area of vulnerability is their growth processes. Certain drugs as well as radiation are able to prevent DNA replication and so to inhibit the development of cancer cells. Unfortunately, tumours are not the only fast-growing and hence susceptible tissues: blood cells are continuously being formed in the bone marrow, the intestinal lining is constantly renewed, hair and nails grow, and this necessary growth is also inhibited by radiation and *chemotherapy*.

Chemicals which interfere with cancer cells often also have unpleasant and perhaps dangerous side effects. New chemicals are routinely screened for anti-cancer properties; it has been reported that, of 800 000 compounds tested in the past 50 years, only 40 have proved clinically safe and effective. Just as with microbial resistance to antibiotics, cancers become resistant to chemotherapeutic agents. The parallel with microbes is close: not all the cells of a tumour are equally sensitive and from the beginning some are likely by chance to be resistant. Treatment with the chemical at first kills a high proportion of the sensitive cells and remission appears to have occurred. But eventually the residual population of resistant cells takes their place and the tumour has then become either less sensitive or quite resistant to treatment with the original drug. Using a combination of drugs might help to overcome this problem but is also more likely to affect healthy tissues. The search for new compounds will certainly go on.

Attacking cancers immunologically

The idea that immunotherapy might be applicable to cancer has been stimulated by repeated observations that severe bacterial infections contracted by patients also suffering from cancer may cause tumour regression. The effect was presumably based on the fact that cancer cells are in some respects immunologically different from normal cells, although the differences are clearly not great enough for most patients to rid themselves of the tumour by their own unaided immune responses. However, when strongly provoked by a bacterial infection, the immune system also reacted to the tumour antigens. As long ago as the 1890s papers were published on anti-cancer therapies based on the injection of bacteria. But such treatments tended to be by-passed in favour of others, particularly the use of radiation.

The idea of immunotherapy has made a partial comeback in recent years. It has been known for a long time that some microbes act as powerful non-specific stimulants of the immune system and that, in both animals and humans, the injection of certain bacteria produced tumour regression, especially if the tumour was not well developed and had not spread. The early vaccines consisted of whole bacteria which often caused serious toxic side-effects. This prompted workers to try to fractionate the microbes in order to separate the immune stimulators from the toxic components. Something of this problem has been mentioned earlier when discussing the ways in which to improve vaccines and reduce their toxicity.

One example of a *bacterial immunomodulator* is BCG, the bacterium responsible for bovine tuberculosis and originally called *Bacille Calmette-Guerin* after its discoverers. The living bacterium, no longer in its virulent form, enhances immunity generally and is fairly effective in preventing tuberculosis in humans; it has been in use for that purpose for decades, particularly among high-risk groups, but may cause local inflammation and other problems. Recent efforts have been directed to isolating a number of cellular components from the bacteria in the hope of separating immunoactive from toxic factors. Dose quantity and scheduling has a major effect on the outcome and, while trial results are variable, tumours such as bladder cancer often respond well with partial or total remission in a majority of cases.

Cytokines again

It seems likely that interferon and interleukin-2 may have important anti-cancer properties in addition to their other immunological benefits. By encouraging them to express their own characteristic antigens, interferons might make several types of tumour cell more susceptible to attack by killer T-cells and the drug may also act directly by stopping the growth of cancer cells. Clinical data show that interferons, when administered singly, are not remarkably effective against tumours even in high doses. However, they do work in some cases of which the most outstanding is the 70-90% partial or total remission in hairy cell leukaemia achieved by α-interferon. Success has also been reported with other forms of leukaemia but in general the family of interferons has not yet shown the promise expected of them ten years ago.

Interleukin-2 is one of the most intensively studied of the cytokines. It stimulates the action of killer T-cells — if human blood is incubated for several days with interleukin-2, T-cells able to attack a variety of tumours are generated. Impressive tumour regressions have been obtained in extensive clinical trials, particularly with melanoma and renal cell carcinoma.

Tumour necrosis factor (TNF) is yet another cytokine produced in response to a variety of factors, especially certain bacterial toxins. It causes disintegration of tumour cells, perhaps by interfering with the blood supply to the tumour mass, but clinical trials have not yet

Exploiting biotechnology

shown it to be an effective therapeutic agent. In an interesting new development, a retrovirus, engineered to carry the relevant TNF genes, is used to infect T-cells which have been removed from patients with melanoma and then grown in a laboratory culture. The idea is to re-inject into the patient the virus-infected T-cells — now carrying the TNF genes — in the hope that they will migrate to the site of the tumour and kill the cancer cells.

With even the human versions of these cytokine drugs there is a tendency to toxicity, leading to fever, hypotension and depression — not something to be expected from one's own molecules! These difficulties may be caused by the relatively large amounts which are injected, a problem shared with the administration of many "natural" drugs like insulin and other hormones. In a normal person these compounds are present in low concentrations in balance with a large number of other components. In order to avoid the awkwardness of continuous administration, they are usually injected only at intervals and in quantities which are large at the time of introduction. Such high concentrations can cause problems, even though the body gradually gets rid of what it perceives as excessive quantities. The ideal answer would be a convenient means of slowly but continuously introducing a modest amount of the drug, a matter addressed in Chapter 12.

Specific vaccines for specific cancers (and maybe for specific people)

Some tumour cells display cell-surface antigens quite distinct from any others in the body. When used as antigens, such tumour cells will raise an immune response in animals; perhaps soon it will become possible to immunize in advance against cancer.

Other ways of attacking tumours

A number of avenues are being explored. One of them might involve monoclonal antibodies. This approach could be particularly useful when, as in the case of B-cell lymphomas, each tumour makes its own specific antibody. A monoclonal antibody could be raised against the lymphoma antibody used as an antigen (making antibodies to antibodies was described earlier). The monoclonal antibody could then be coupled to a toxin, the combination *immunotoxin* being administered to the patient. The tumour cells would bind the monoclonal antibody and be killed by the accompanying toxin.

Another way might be to use agents directly to control the growth and spreading of tumours. Cancers are known to induce their own blood supply and recent studies have demonstrated that many factors control the development of the new blood vessels. It might in time prove possible to manipulate them to starve a tumour. Further points of attack might be directed to controlling the way in which tumours spread by blocking motility or antagonizing the enzymes which the tumour cells use to break down barriers and invade tissue. All of these ideas are likely to be explored and, while it is unlikely that all will work with every kind of cancer, it seems not improbable that effective treatments will be found for many malignant conditions.

Healthcare: prophylaxis, diagnosis and therapy

Xenotransplantation

The prospective benefits of genetic manipulation seem almost endless. Quite recently a new one has emerged which could offer major advances for organ transplants.

When an organ is transplanted from one individual to another, differences in their protein, and hence antigenic, structures prompt the recipient's immune system to recognize the transplanted organ as non-self; it responds by producing antibodies to neutralize the foreign invasion, which eventually results in a rejection of the transplant. This severe immune response has to be countered by the administration of immunosuppressive drugs but they have the secondary effect of seriously weakening the patient's defences against infections. Only transplants between identical twins (who share an identical genetic and hence immunological constitution) avoid this problem.

In practical terms the whole situation is exacerbated because of the shortage of available human organs — the number of human cadavers with healthy organs is in any case limited and relatives are often unwilling to authorize their use for transplants. As a result, tens of thousands of potential recipients are waiting for suitable donor organs and attention is turning to the prospect of using animals as donors. Their size, and their acceptability because they are already bred for food, make pigs the most likely source.

The rejection problems are, of course, even more severe in pig-to-human transplants than they are in human-to-human operations. Many of the immunological incompatibilities arise because porcine and human proteins have different types of sugars attached to them, sugars which give the proteins much of their antigenic individuality. One possible way of dealing with this problem is to inject sugars of the pig variety into human recipients in the hope that the antibodies will bind to the free sugars and leave the transplants relatively unaffected. The trouble is that, in the quantities needed, those sugars may be poisonous. Alternatively, pigs might be modified genetically so that their organs are no longer so antigenic to humans, possibly by inactivating the pig genes coding for one or more of the enzymes that make the sugars.

There may be another way. The mechanism of antibody rejection is to activate host complement, a group of proteins which promotes an attack on the foreign tissue and leads to its destruction. As a precautionary measure against action by their own immune system, cells also possess a protein which blocks the action of complement. This protein, however, is species-specific: pig cells can protect themselves against pig complement but not against the human variety. Could this be changed — could the genes for protection against human complement be introduced and expressed in pigs in such a way that would not harm the pig but would protect its organs from immunological attack when transplanted into humans? The answer seems to be an encouraging "yes". By itself this will not solve all the problems of interspecies transplants. There are other difficulties to be overcome before pig organs are routinely available for human needs but some experts are confident that it is only a matter of time and effort before clinical xenografting becomes widely available.

Exploiting biotechnology

Gene Therapy

Perhaps the most daunting of all challenges facing healthcare biotechnology is the correction of defective genetic inheritance. A healthy human body depends on the right information being both present in a person's genes and properly expressed. With so much complexity, all manner of errors and omissions inevitably occur; some parts of the body (they might be individual enzymes or other proteins and therefore not necessarily result in visible defects) fail to be made or have the wrong structures and either themselves malfunction or are not correctly integrated into the whole.

There is no prospect within the next few years of genetically correcting those gross physical disabilities which result from a defective genome. Individuals born with parts of their bodies not fully developed or actually malformed are usually helped either by surgical procedures or prostheses. But in those cases in which the genetic defect is confined to a single malfunctioning enzyme there is at least hope that gene therapy might solve the problem rather than palliatively administering a drug to make up the deficiency or counter a defect, which is all that is possible at present.

An example of a defective enzyme

Consider the inherited disease *phenylketonuria*, where the defect lies in the liver enzyme system which normally converts the amino acid *phenylalanine* (one of the twenty present in proteins) into another called *tyrosine* (also a component of proteins). Phenylalanine accumulates in body fluids, producing derivatives, including *phenylpyruvic acid* which is excreted in the urine, a condition called *phenylketonuria*. (Phenylpyruvic acid belongs to a class of chemicals called *ketones*; "phenylketonuria" means a phenyl-type of ketone present in the urine.) While there are minor physical symptoms of the condition, the most serious consequence, for reasons not understood, is mental retardation. It is therefore important to recognize the condition at birth and treat it within the first year by confining the infant to a diet which contains just enough phenylalanine for normal growth.

Even were it available, there would be no point in injecting the missing enzyme into the bloodstream because it would almost certainly not be taken up into the cells which needed it. Having identified the disease, ideally it could be cured by extracting the defective cells, supplying them with the correct gene so they could make the enzyme for themselves and then putting them back into the patient. The genetics might be possible; what cannot yet be done is functionally to insert the modified cells into the liver.

Doing the genetics

Genetic manipulation of human cells requires the removal from the person of the cells to be cultured under artificial conditions in the laboratory; currently this can be done in a routine

Healthcare: prophylaxis, diagnosis and therapy

manner only with cells from bone marrow and skin, although it is reasonable to look forward to similar possibilities for other tissues. Genes may be transferred into some cultured cells by using relatively innocuous retroviruses which have been derived from mice and altered genetically to make them suitable vectors for conveying the target genes into cultured human, rather than mouse, cells. The technique is difficult and many of the practical problems are far from resolution, but promising results have been achieved for some genes and some types of cells.

Fitting the genetically engineered cells back into the right place

Partly because techniques for transplantation are already successful, the tissue most seriously being considered for cell replacement at the present time is bone marrow. Marrow cells might be removed from the patient and a laboratory culture established in preparation for the genetic manipulation. When the genetics had been successfully accomplished, most or all of the bone marrow cells remaining in the patient would be destroyed by radiation or other means and the genetically modified cells re-introduced. Originally from the same patient, the altered cells would presumably trigger no immunological response; Figure 7.1 shows how it is done in mice. Another possible transplant of this general type is a treatment for diabetes: a million or so insulin-producing cells obtained from the pancreas of a non-diabetic donor would be implanted into the liver of the recipient, hopefully to continue producing insulin under normal physiological control in the right quantities when needed.

A simpler alternative technology?

If every case of gene replacement therapy were to entail virtually a research programme of its own, the costs would be unsustainable and the concept unlikely to become of major medical

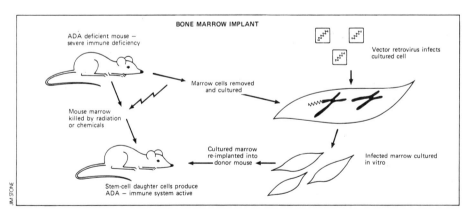

Figure 7.1 Diagram of bone marrow transplantation in mice. (*Copyright Bio/Technology*)

Exploiting biotechnology

benefit. But a remarkable recent research finding suggests that there might be a much simpler (and hence cheaper) alternative, provisionally called *gene therpeutics*.

It was found that injecting a simple aqueous solution of genes into mouse muscle resulted in at least some of the genes being integrated into the muscle cells and their specific proteins expressed for a protracted period. If gene therapeutics turns out to be as simple with human tissue, the feasibility of correcting inherited disorders takes on a totally different dimension. It might be possible to treat muscular dystrophy, cystic fibrosis, Parkinson's disease and even high circulating levels of cholesterol (considered to enhance the risk of heart attacks) by administering the relevant genes via an appropriate route to the target tissue — trials are already under way to do this with cystic fibrosis. A new and safer form of vaccination might also be possible if, as has been reported, genes from foreign organisms injected into muscle stimulate the production of antibodies against the proteins for which they code. Some workers anticipate this as a way of avoiding injection altogether and administering the genes orally.

The Human Genome Project

A few years ago the international molecular biology and biotechnology community floated the idea of determining the complete sequence of bases ("letters" in our model) in all the human chromosomes. Governments and others were persuaded to provide much of the funding; in 1989 the Human Genome Organization (HUGO) was set up to facilitate global collaboration and coordination, and in 1990 the Human Genome Project was born as an international cooperative activity to last 15 years at a total cost of $3 billion. That is a lot of money and it has to be said that not even all biologists think so much should be spent on this particular project, to the detriment, as they see it, of possibly more exciting options.

The human genome contains an estimated 30,000–90,000 genes comprising about 3 billion base pairs. Doing the job by conventional methods might take 100,000 man-years of effort but new techniques will make the whole analytical process much more efficient, with each base pair identification costing only $1 and perhaps falling in time to 50¢. The potential benefits are enormous: a profound understanding of the way the human organism works with profound implications for being able to deal with genetic defects and perhaps other disease conditions as well. It is not unreasonable to expect that a great deal of light will be shed on cancer and age deterioration as well as on the more obvious genetic defects we have discussed above. People foresee novel diagnostic tests for genetic diseases, new strategies for making designer drugs and technical developments in laboratory equipment. Another implication of the programme is the development of computer and database technology in order to communicate, analyse and correlate the prodigious amounts of information which are to be generated.

Since funding will come from charities and private corporations as well as from governments and international agencies, inevitably the question arises of the ownership of information and the benefits to be derived from it. Private companies are not charities and normally invest only when they see the prospect of returns; they will expect to receive intellectual property rights (patents and licences) or other advantages. And it is not only the private sector that is worried about intellectual property rights. To gain public funding support, protagonists of the project have emphasized all the economic and medical benefits that can be expected; that has served to pinpoint the question of who gets the commercial benefits. Government agencies, arguing that taxpayers' money had already been spent on the fundamental science as well being committed for the project itself, reserved their position and some said they would refrain from publishing results until ownership questions were resolved; applications for patent protection began to be made. People have asked whether work paid for by their nation's taxpayers should be released to countries which "fail to live up to their international responsibilities". A good deal of heat as well as genetic information is likely to be generated by the Human Genome Project!

MARKETING HEALTHCARE BIOTECHNOLOGY

There can be little doubt of the commercial potential for biotechnology in the whole field of medicine and healthcare. Individuals (and governments — the latter, perhaps, with less enthusiasm) spend more and more on their own or their citizens' physical well-being as their net disposable income becomes greater; societies can apparently absorb medical and healthcare services without limit.

Enormous market potential will continue to be provided by new drugs of major therapeutic value, particularly those which offer successful ways of dealing with cancer and cardio-vascular disorders. Difficulties arise both with development costs and with competition. Finding a potentially valuable drug is only the start of the matter. Its efficacy and acceptability must be thoroughly demonstrated or it will not receive a licence for use. Major new drugs for human use may take ten years to perfect, possibly entailing an investment in production and marketing facilities of $100 million or more, a very large financial risk.

Competition is an ever-present threat, especially if a successful new drug for a particular application is likely to capture the market and exclude both its competitors and possible successors. (But this is not always so: a recent study has suggested that tissue plasminogen activator — at some ten times the price — might actually be less satisfactory than streptokinase, the alternative older product of bacterial origin.) Competition is keen because biotechnologists in any R & D activity share virtually the same intellectual climate, all with access to the same database of scientific research. They have probably all had similar educations and read the same journals. Even if confidentiality is maintained, a new idea on the part of one person or group may not for long retain its novelty because others are likely to be thinking similar thoughts. While originality might buy a few months' headstart, that may not be enough to guarantee success in the light of a possible ten-year development period.

Minor developments are commercially less sensitive. Monoclonal antibodies intended for specific local diagnostic and other uses not involving administration of drugs to patients, will be very much cheaper to develop and may be entirely viable when produced and marketed on a comparatively small scale. They may therefore offer better opportunities for small businesses. The existing commercial pattern of major drugs coming from large corporations with adequate

Obtaining Regulatory Approvals

New drugs must be judged safe and efficacious before they can be used. Different countries differ in their procedures for granting approval but the intentions are generally similar. Approval by one country may greatly assist a grant of permission in others but not always: interleukin-2 is an anti-cancer drug which was being developed by an American company in the late 1980s. It had already received marketing approval for the treatment of advanced kidney cancer in nine European countries and the company sought similar approval for the US. However, an advisory committee of the US Food and Drug Administration, in reviewing the application, decided to take no action — they neither approved nor disapproved, but suggested working with the company to resolve various outstanding issues regarding the application. That failure to gain approval was quite unexpected, particularly in view of the many European approvals. Obviously it was a serious setback and may have contributed to the company's merging with another a couple of years later. In spite of using their best judgement, the company got it wrong and perhaps paid the price by losing its independence.

New therapeutic drugs are first tested in animals. If they are successful, human clinical trials proceed through three stages:

- in Phase I, a small number of healthy human volunteers enable assessments to be made of safety, pharmacological effects and metabolism;
- Phase II uses a larger group, now of patients suffering from the disease. These tests, often done in university or government medical centres, evaluate effectiveness and side-effects;
- Phase III extends testing to thousands of patients in a variety of locations and must show a low incidence of adverse effects.

If the drug satisfies all the criteria a licence is issued for specific applications. It is this testing procedure which is so lengthy and expensive: periods of up to 10 years and sums as high as $100–200 million are spoken of. So much time and such a lot of money is an enormous hurdle, particularly for a drug intended to treat a disease for which there may not be many sufferers and for which the market is too small to justify all the costs. "Orphan drug status" in the US may be granted if the prospective market is smaller than 200 000 patients; it confers market exclusivity for seven years. Sometimes competing drugs may each be given orphan status, while a disease which initially affects comparatively few people may grow in significance, giving opportunities for large revenues: AIDS is a case in point.

Approval for diagnostics is simpler but evidence is nevertheless required of their safety and efficacy. Approval for new diagnostics may take two or three years but, if it can be shown that the diagnostic is "substantially equivalent" to something already on the market, the time may be reduced to 90 days.

Healthcare: prophylaxis, diagnosis and therapy

Paying for Biotechnological Developments

All the advances discussed in this chapter are intended, one way or another, to be sold to the public. Whether private sector initiatives or developments resulting from public policy, each one will cost money and somebody has to pay, either in advance through taxes or insurance premiums, or when the bills for treatment come in, or perhaps both. Who will pay and how will payment be secured?

Some patterns are already developing. Diagnostic methods are intended to ensure faster and more accurate recognition of disease. As new methods become available they will inevitably come into use, partly because their costs can be transferred to the patient, directly or via private or public insurance schemes, and partly because medical practitioners will want to avoid any charge that they have failed to do the best for their patients. Either way, the cost falls on the patient or the taxpayer (the same person, anyway: all taxpayers are at some time patients, and all patients taxpayers). To the extent that they have a choice, and so long as they feel they can meet the cost, patients will go to those doctors whom they think have the most up-to-date methods. People who are less able to pay will nevertheless hear about the new methods and sooner or later demand equal access.

No doubt similar considerations will hold for prophylactic and therapeutic procedures. It will not be considered ethically acceptable to withhold protective vaccination from patients on the basis of their inability to pay personally. Responsible governments are in any case likely to make sure that such vaccination is available at public expense to the population at large. But what about expensive new drugs to treat potentially serious disease, drugs which might daily cost hundreds of dollars or pounds or marks or whatever? Will public health services be able and willing to meet the cost? Will the public be prepared to pay higher taxes? What happens when people who can afford the drug do receive it and those who cannot have to do without? Will there not be outrage and accusations that some people are left to suffer, or even die, because they are too poor to pay for treatment?

Yet medical and pharmaceutical development seems never to have come to a halt because of a concern about how it will be paid for. As societies grow richer and people apparently become relatively sated with material goods and the amount of food they can eat, there is an increasing willingness to pay for more services, healthcare among them. Perhaps there is no need to worry — just so long as the advance of all the new drugs and medical services does not seriously outstrip society's ability to distribute their cost on a broad and fair basis.

8 Chemicals, enzymes, fuel and new materials

BIOTECHNOLOGY IN THE CHEMICAL INDUSTRY

The chemical industry converts feedstocks into valuable products. Each production process is based on a sequence of catalysed chemical reactions engineered into a viable production technology. Thus, a feedstock is converted, often through several stages, into a number of products, perhaps only one of them ultimately destined for the market place; those intended for sale must usually be separated from unwanted by-products and purified. Biotechnology fits neatly into this scene; it offers additional ways of effecting chemical change and differs from conventional chemical methods mainly in the nature of the catalysts and the ways in which they are used.

The primary advantages of biological catalysis will already have become clear in Chapter 2: they are specificity and efficiency. Enzymes, very restricted as to the range of reactions each will catalyse, speed up specific chemical interactions to thousands or even millions of times the rates they would achieve if left to themselves. Their specificity means that a particular enzyme, while limited in the variety of feedstocks (or *substrates*) with which it will interact, generates an equally limited range of products. In appropriately designed procedures, the use of enzymes rather than non-biological catalysts can greatly simplify product purification.

Biocatalysis in Production Chemistry

With some exceptions, products of the chemical industry can be divided very broadly into those with biological significance in their origin or application, and those without. Aside from artificial pesticides, the former (biological) category mainly comprises compounds present in or made by living organisms; it includes products for direct human or animal use (food, drugs, toiletries, etc.), substances involved in the processing of biological materials and certain chemicals used for industrial purposes. The second group of products are those not of biological origin, and not designed and intended for biological use. Among them are petrochemical products (fuels, plastics and others), metals and inorganic chemicals, and a host of artificial organic chemicals which are based on carbon (and hence, by definition, "organic") but which are products of the chemical laboratory and not found in nature.

The world of biology is in a continual state of biochemical flux, the sequence of birth, growth, death and decay giving rise to an endless recycling of the chemicals of which living organisms are composed. At various times in the distant past, carcases of animals, plants and microbes have been buried in silt or mud before they had time to decay and, over long periods of time, have been turned into peat, coal or oil. But when no burial takes place, all

Exploiting biotechnology

biological chemicals are recycled. This plethora of chemical activity depends on biological catalysis: enzyme-catalysed reactions are involved in the synthesis of all compounds of natural biological origin and those compounds serve in turn as substrates for other enzymes in other organisms which degrade them, so completing the cycle.

Every living cell contains thousands of different types of enzymes, each with its own specificity. Because much of biochemistry is common to all cells, a high proportion of the enzymes from one organism have similar catalytic functions to those from other sources, although their precise physical structures are rarely if ever identical. However, some organisms have enzymes which are not shared by most of the others, or possess particular enzymes in large quantities, or in some especially favourable physical form — for example, an enzyme from a bacterium normally living in a hot environment will probably be more resistant to heat than one from a relative inhabiting a cooler location. The opportunities are legion for acquiring from nature enzymes with defined functions and properties; the technology for using them in many different ways for industrial purposes is developing rapidly.

Although a few are released into their external environment, most natural enzymes are normally present either inside cells or bound to their surfaces; the cells generating the enzymes might be simple unicellular organisms like bacteria or form part of the complex structure of a plant or animal.

Enzymes employed for industrial purposes mostly originate from microbial sources because microbes are cheap and easy to culture in the quantities necessary, and the extraction and purification of enzymes from their cells is relatively simple. However, enzymes do not necessarily have to be purified in order to serve as industrial catalysts; providing certain operating criteria are met, catalysis can often be achieved by using whole microbial cells. It will be instructive to look briefly at the different ways enzymes can be used.

Catalysis with living, growing cells

Enzymes tend to be rather fragile molecules, easily degraded; retaining them in their original environment within the cells where they belong is the most effective way of extending their operating lives. Using whole, living microbes as catalysts has many advantages:

- they are cheap and easy to obtain;
- in many cases they are more robust than enzymes which have been extracted and purified from them;
- often they will use simple, cheap feedstocks and synthesize high value products;
- some production processes are complex, requiring the collaboration of several enzymes as well as smaller, non-enzymic compounds (*coenzymes*). The spatial relationship of these components is an important factor in determining efficiency and their natural configuration, inside the cell, is likely to be the most active one. It might prove impossible to reconstruct that configuration if the enzymes and other components are removed from the cells. Furthermore, using whole living cells prevents the coenzymes from being lost to the surrounding medium.

But there are also disadvantages:

- An enzyme on the surface of the cell will easily be reached by the substrate and the product readily released into the environment, making subsequent recovery and purification fairly easy. But if the enzyme is located inside the cell, protected chemically from the external environment by the cell's covering membrane, the substrate might have difficulty reaching it and the conversion to product will accordingly be slow. If, furthermore, the product does not readily escape, purification might involve smashing the cells to release it. This can be difficult (and hence expensive) and is likely also to liberate all manner of cellular components which complicate subsequent purification of the desired product.

- As we noted in Chapter 2, a living microbial cell is an integrated system adapted to making the best use of its resources for growth and the production of offspring: its goal in life is not to make chemicals as cheaply as possible for us. In addition to using much of their feedstock for growth and the formation of biomass, living systems may also unavoidably convert part of the substrate into excretory products, the generation of which is essential to maintaining their life processes, particularly under anaerobic conditions. In some cases, it is true, the accumulation of those excretory products (like alcohol or acetone) might be the purpose of the production process, but in many cases it is not. Not infrequently the chemical(s) we want are produced in low yields, which is all the microbe might need to make for its own purposes. It then takes the combined skills of the microbiologist to modify the cells' genetics and biochemistry, and the biochemical engineer to design such production protocols as to leave the cells no option but to do what the operator wants and produce more of the desired product. This type of strategy is often successful but it does sometimes fail. A microbial culture contains enormous numbers of cells; some of them may mutate spontaneously and escape the restraints imposed to force high production yields. If the mutants grow well, they may become the dominant population in the reaction vessel and, if they generate much less of the product, may subvert and ruin the production run. Some microbial systems are less stable in this regard than others.

Catalysis with living, non-growing cells

For some procedures it is possible to use cells which are intact but unable to grow because a particular chemical essential for growth is omitted from the medium. This has the advantage of retaining the internal cellular and biochemical architecture for maximum catalytic efficiency while ensuring that the feedstock is used for product synthesis and not simply to generate more microbial mass. The production of some valuable chemicals, however, is so tightly linked to the growth process itself that the deployment of non-growing cells is not possible. Processes dependent on non-growing cells suffer little risk of ruined production runs because of mutation but they may be vulnerable to contamination by extraneous microbes.

Exploiting biotechnology

Catalysis with dead cells

Death among the microbes is not quite what it is in familiar human terms. People beyond reproductive age are not normally regarded as dead if they continue to exhibit the conventional signs of human life. But a microbe that cannot reproduce is certainly not viable and lack of viability might be little different from death. However, even if a microbial cell is "dead" in the sense that it cannot grow and reproduce, some of its enzymes may still be active and useful for manufacturing purposes. Death must nevertheless inevitably imply some measure of organizational degradation and disruption so that dead cells are probably most useful for single enzyme steps and less effective for integrated, multi-enzyme reaction sequences. Like other non-growing systems, they are not subject to mutation.

Catalysis with isolated enzymes

Some production procedures lend themselves to the use of isolated enzymes. Their advantages lie in the simplicity of the reactions, leading to few or no by-products, unimpeded access of substrate and release of product, and perhaps easier product purification and recovery. The problems come with the relative fragility of many isolated enzymes and their frequent requirement for coenzymes as well as for rather precisely defined environments in which to function; depending on their source, they might be difficult and expensive to isolate.

Configuring Biocatalysis for Production

Processes run in batches, like brewing beer and making cheese, use microbes free-floating in the reaction vessel. The vessel having been loaded with feedstock, microbes, enzymes, etc., the reaction is allowed to run under defined and controlled conditions until the most cost-effective conversion to product has been achieved. The reaction mixture is then removed from the vessel for product purification, when appropriate, and packaging. The vessel is cleaned, sterilized if necessary, and recharged for the next batch.

When the production run is complete, the microbes may be separated from the fermentation broth (as in beer) or left in place actually to form part of the product (cheese and yoghurt). Microbes are often easily separated by filtration or centrifugation; sometimes they flocculate (i.e. form clumps), which makes their removal easier yet. A proportion of the microbes from each batch may be used to inoculate the next production run simply by putting some of the last product into the next batch (yoghurt). Recovering enzymes is more difficult because enzyme molecules are minute compared even with the tiny microbial cells: they cannot easily be separated from the medium by filtration or centrifugation and they do not settle out of solution. Often it will not be worth trying to recover an enzyme and a fresh quantity will be used for the following batch.

An alternative to batch production with living microbes is to run in a continuous mode. In this configuration, feedstock is admitted continuously into the reaction vessel which already contains an actively growing microbial population. The excess liquid overflows into a receiver from which it is channelled through the separation and product purification procedures. Back in the production vessel, the microbes which have been removed in the overflow are replaced by growth of the remaining population.

Continuous operation has the advantage of maintaining more constant production conditions than the batch mode and of not having frequently to strip down, clean and recharge the vessel between runs. For some purposes, however, batch production is necessary even with the disadvantage of equipment downtime: certain microbial products are made just when the cells stop growing as they run out of food and it would therefore not be helpful to keep them growing continuously, always with an adequate supply of nutrient.

Immobilized systems

Another way of using microbes and enzymes in a continuous mode is to immobilize (i.e. attach) them onto a solid support from which they cannot escape. Separating the microbes or the enzyme from the product is then very easy indeed: the liquid reaction mixture is simply drained away from the solid support bearing the catalyst.

Immobilization facilitates continuous production. The immobilized catalyst resides permanently in the production vessel; a stream of feedstock solution is introduced continuously into the vessel and a stream of product emerges. No catalyst has to be separated and the production vessel does not have to be cleaned, recharged and perhaps sterilized at the end of each run. This results in a more efficient use of equipment and, for systems where such a configuration is possible, a lowering of operating costs. Some immobilized catalysts continue to function for months on end and the fact that immobilized cells are usually not growing means that feedstock is not used to make biomass, and risks of production failure due to mutation are eliminated. But, as always, the benefits of an immobilized system have to be viewed in the light of the additional costs of immobilization as well as relative performance compared with free-floating batch and continuous systems.

Whole cells are immobilized by entrapping them in a chemical jelly which readily allows the penetration of chemicals, attaching them to a solid support or chemically linking the cells so that they clump together in *flocs*. Enzymes can also be immobilized by entrapment in a gel matrix, although other methods employ adsorption onto certain large polymeric compounds or attachment by chemical linkage to the surfaces of various polymers, plastics, glasses or ceramics. The immobilized forms of both cells and enzymes are usually fashioned into beads which are either contained in a cartridge through which the feedstock solution flows, or allowed to float freely in the reaction liquid from which they are readily recovered by simple settlement or filtration.

Exploiting biotechnology

Pros and Cons of Biotechnology for Product Manufacture

Before reviewing the present and possible future roles for biotechnology in the chemical industry, it might be helpful to summarize some of the operating considerations.

Biotechnological procedures are useful because:

- they operate at ambient pressure, thus obviating the need for expensive pressure vessels;
- they operate also at or near ambient temperature with little or no fuel requirements for heating, but sometimes with a need for cooling because of heat generated in the reaction vessel;
- in some cases they have a high rate or a high efficiency, or both, of converting feedstock to product;
- they are able to generate very complex products, many beyond the capacity of conventional chemical skills;
- except in the case of food, many biotechnological products have high added value;
- they can use a variety of agricultural and petrochemical feedstocks.

Limitations include:

- lack of robustness in some cases;
- risk of contamination with extraneous unwanted organisms or mutants arising within the operating population — hence the need for costly sterilization and other precautions to exclude contaminants;
- downstream processing may be elaborate;
- feedstocks for some specific reactions are expensive;
- biological systems are water-based, with a consequent need for extensive dewatering of most products. There might be problems in both the use of water-insoluble feedstocks (such as oils) and the generation of water-insoluble products because of the difficulty and expense of achieving good mixing.

One last general thought — at the present time manufacturing biotechnology can be used economically only to make difficult-to-isolate natural products or compounds closely resembling them, not totally artificial materials. (A different situation prevails in microbial mining, oil production and environmental remediation technologies. In none of those cases is the microbial product itself of primary value — the economic targets are the extraction of metals, recovery of oil or cleaning of a polluted environment. Chapters 10 and 11 will explain.) While in principle available from existing sources, these natural products are made with biotechnology either because supplies of the conventional sources are too limited or product isolation from them too difficult (i.e. too expensive). Almost all human proteins are examples of compounds for which isolation from natural sources is not economically and practically feasible; employing biotechnological procedures to make them in microbial systems was touched upon in Chapter 7 and will be discussed again later in this chapter.

Chemicals, enzymes, fuel and new materials

CHEMICALS

The range of chemicals already routinely made with biological methods is impressive. Some are "traditional" products which have been manufactured for millennia, for much of that time, of course, in total ignorance of any biological involvement: alcoholic beverages, vinegar, soy sauce, cheese and yoghurt are all examples. Biological methods, such as the retting of flax and hemp (a partial microbiological decomposition of the plant material to release the best fibres), have also been used traditionally for some processing purposes.

Within the last century, our understanding of microorganisms, their biochemistry and genetics has made enormous strides and, as a result, microbiological procedures in the manufacturing and processing industries have developed apace. Probably the largest of the processing operations in terms of sheer volume is waste management, especially the disposal of garbage and other wastes in landfills and the treatment of sewage, both activities addressed in Chapter 11. The first uses of specific microbes and enzymes for product manufacture under precisely controlled conditions date from the second decade of the 20th century. While the earliest products were used for explosives manufacture and subsequently in the painting of motor vehicles, microbial fermentation expanded into a wide range of products during the inter-war years. It was the dramatic mid-century growth of antibiotic production, however, that stimulated much of the development which has led to the contemporary sophisticated production technologies now included under the heading of "biotechnology". Notwithstanding the dependence of biotechnology on enzyme catalysis for manufacturing purposes, some of its significant products are themselves enzymes, sold as such and used for catalysing a variety of industrial processes.

The biotechnological production of chemicals has shown a steady upward trend which is likely to continue, but progress is not constant in all sectors. The advent of the modern petrochemical industry undermined the cost-effectiveness of some biological manufacturing although, as will become apparent in the discussion of converting biomass feedstocks into fuels, swings go in both directions.

Market Sizes

The biological production of chemicals is big business. Rough indications of some worldwide annual sales during the first part of the 1980s are:

	$ million
• antibiotics for medical and agricultural use (more than 70% of them penicillin and related compounds)	1400
• riboflavin, a vitamin	70
• citric acid, used as a flavouring agent, etc.	360
• high fructose syrup, a sweetening agent	1400
• enzymes for use in detergents	140
• all enzymes together	400*

*estimates vary; some authors go as high as $900 million

Exploiting biotechnology

Range of Chemical Products

Antibiotics

Since the pioneering work with penicillin in the early days of the Second World War, the growth of the antibiotics industry has been startling. The idea of using specific drugs to treat particular diseases has for long been part of folklore and, throughout history, trial and error has given rise to a host of remedies based on natural materials. It was often thought that the cure for a disease would be found near its cause. For example, rheumatism is a condition associated with low-lying and damp locations. Willow trees grow on river banks and from them was extracted a substance which was developed chemically to form the basis of modern aspirin, still one of the more effective and safer drugs for the long-term treatment of rheumatism and arthritis.

As microbiology and biochemistry progressed, so did novel concepts of treating disease, particularly infections. Early in the 20th century, Paul Ehrlich conceived the notion of what nowadays might perhaps be called *designer chemicals*, specifically tailored to interfere with the biochemistry of a disease-causing microbe while leaving the host's activities unimpeded. Ehrlich referred to such chemical compounds as "magic bullets". The most famous of the substances he looked at was No. 606 in his sequence. Active against syphilis, it did unfortunately have severe side-effects; it was used under the name of *Salvarsan* (because it offered salvation from syphilis and contained arsenic!).

In 1928 Alexander Fleming noticed that a fungus with which he was working inhibited the growth of certain bacteria. The fungus was a species of *Penicillium* and, although at that time Fleming did not know the reason for the antagonism, he realized that a chemical made by the fungus was preventing growth of the bacteria. He called that unknown substance "penicillin" and his discovery led 15 years later to the development of the first antibiotic production process.

Antibiotics are natural versions of Ehrlich's magic bullets. They are compounds made by certain microorganisms which interfere with particular aspects of the biochemistry of others, either killing the target species or stopping their growth; in nature they presumably confer a competitive advantage on the synthesizing organism by curtailing the activities of some of its neighbours. In human contexts they give the body's own defences a better chance to eliminate an infection. The effects of antibiotics are neither random nor universal — some microbes are susceptible to individual antibiotics while others are not. Nor are all antibiotics by any means useful in medicine. Even when they act against disease microbes in laboratory tests, they might fail to do so in an infected patient — or side effects might preclude their use in practice. The most valuable antibiotics in medicine are those which specifically affect the microbes causing the disease but do nothing untoward to the patient.

In the years following 1945 a systematic search was made for natural antibiotics and more than 2,000 microbial products have been found to possess antibiotic activity. Like all natural biochemicals, antibiotics are susceptible to biodegradation; the reason why some microbes are not sensitive to particular antibiotics is often because they possess enzymes which break them down. It turns out that in nature DNA can be passed from one sort of bacterium

Chemicals, enzymes, fuel and new materials

to another; this phenomenon leads to the possibility of a bacterial species sensitive to an antibiotic acquiring genetic information from a resistant form, so enabling it to make the "resistance" enzyme and become resistant in its own right. The more widely a particular antibiotic is used for therapy, the more likely that the sensitive bacteria will be eliminated from the human population to leave the field clear for resistant forms to take over. Thus, disease bacteria resistant to antibiotics have tended to become more common as time passed, leading both to a need for new antibiotics to which they are not (or at least not yet) resistant as well as caution in the indiscriminate use of existing antibiotics.

In addition to the well-known penicillin, most antibiotics, including *tetracycline, erythromycin, neomycin* and *streptomycin*, are active against bacteria but some, such as *nystatin* and *griseofulvin*, inhibit certain disease-causing fungi. As time goes on it becomes harder and harder to find new natural antibiotics. Screening of natural microbial populations on a random basis will mostly throw up species already known and characterized, with the probability of finding new ones diminishing progressively. One way of extending the potency of existing antibiotics is to purify the natural substances obtained by fermentation and then modify them chemically (i.e. non-biologically) to produce derivatives which retain their anti-microbial properties but are no longer readily biodegradable and hence do not easily give rise to resistance. Such an approach has resulted in a family of modified penicillins which has markedly consolidated and extended the effectiveness of this group of antibiotics; we will meet them again later in this chapter.

Other pharmacologically active substances

A number of microbial fermentation products are known to have pharmacological actions, among them antidepressant, hypotensive, anti-inflammatory, anticoagulant and coronary vasodilatory effects. For the moment, however, they are probably not being produced on a commercial scale.

Amino acids

While these compounds are, of course, the building blocks for making proteins (Chapter 2), markets do exist for the some of the individual amino acids themselves. They can be extracted from natural sources of protein by chemical procedures but in practice it is usually cheaper to make them by microbial fermentation, and production processes are already in use for some of them. They are discussed again in Chapter 9.

Organic acids

These form another category of industrial fermentation products. Among those produced by fungi, the best known is probably *citric acid* once obtained from lemon and lime juices but now

Exploiting biotechnology

made by fermentation using a mould called *Aspergillus niger*; annual production worldwide is of the order of 300,000 tons. It is used both in soft drinks and confectionery to give a lemon flavour, as well as in metal processing and cleaning. *Gluconic acid* and its derivatives are also used for metal complexing, particularly in some detergents. It, too, is recovered from an *Aspergillus niger* fermentation but its preparation from the sugar glucose is biochemically simple and future production might be based instead on an immobilized enzyme system to catalyse the oxidation of the glucose to this product. *Itaconic acid*, yet another fungal product, has a role in improving the properties of vinyl and other polymers in fibre production and paints.

Other types of bacterial fermentation produce *tartaric acid*, used both as a food acidulant and as a retardant in the setting of plaster; *lactic acid*, the souring agent in cheese and yoghurt production, which also has medical and industrial applications; and *acetic acid*, the basis of vinegar and an important industrial chemical, a large fraction of which is, however, normally produced chemically from petroleum and not by biological means.

Water-soluble polymers

There are many requirements in the food industry and elsewhere for agents which "thicken" water, that is make it more viscous. This requirement has traditionally been met by polymer "gums" extracted from plants and seaweeds, but many microorganisms are also able to synthesize viscous polymers and they can be used to develop manufacturing processes based on microbial fermentation. These polymers, with names like *dextran, xanthan, scleroglucan, alginic acid* and *pullulan*, are based on long chains of sugar units of the type described very generally in Chapter 2. Of these, dextran and xanthan are the most important.

Microbial polymers produced by fermentation have several advantages over comparable products extracted from vegetable sources:

- they provide a range of potential products from which to choose;
- they offer the opportunity of a more constant product than one obtained from harvesting a natural crop;
- they avoid seasonal fluctuations in supply;
- in some cases the properties of the product can be varied by changing the conditions in the fermenter;
- recovery and purification procedures are relatively simple.

Several microorganisms produce water-soluble polymers which, when dissolved in water, display valuable rheological properties. One of them is *viscosity*, making the solution syrupy and resistant to flow. Another is called *pseudoplasticity* in which viscosity is dependent on shear rate so that the harder the viscous solution is shaken or forced through a narrow orifice the less viscous it becomes; viscosity is restored when the shaking stops or the shear falls on the far side of the constriction. (This is the property which allows tomato ketchup to be poured from the bottle after it has been vigorously shaken and to set firm again on the side of the

Chemicals, enzymes, fuel and new materials

plate next to the fish and chips without running all over the tablecloth.) A third property is *thixotropy*, the tendency of the fluid under constant rate of shear to become thinner with time.

The oil industry is an important consumer of biological polymers, especially xanthan. When an oil well is drilled, the drill bit penetrates through rock, grinding it into small pieces which must be removed from the hole and carried up to the surface. This is achieved by pumping a fluid down to the drill bit through the hollow drill string and returning it to the surface where it is either discharged to waste or the rock fragments it carries are stripped out before recirculating the fluid back down to the drill. The fluid serves to cool the bit and contains chemicals which act as lubricants as well as polymer to make the fluid sufficiently viscous so that the rock chippings do not simply fall straight to the bottom of the hole but are held in suspension long enough for them to be carried to the surface.

Holding particles in suspension is also important in rock fracturing. In some oil production procedures it is necessary to crack the underground rock in order to improve drainage into the wells. This is done by forcing water into the ground at such high pressure that the rock actually cracks. Unfortunately, when the pressure is released the rock simply settles back to where it had been previously and seals the crack. Measures must therefore be taken to keep the crack open and this is done by including sand, carborundum or other hard *proppant* particles in the injected fluid to keep the fractures open and prevent the rock settling back to its original position. The injected fluid must be sufficiently viscous to ensure the proppant particles are suspended for long enough to allow them to enter deep into the fractures; this is accomplished by making the injected fluid viscous with polymer.

Another petroleum production use of polymers is in tertiary oil recovery. As we shall discover in Chapter 10, one way of improving the recovery of crude oil from natural underground reservoirs is to introduce water through an injection well in order to sweep the oil out of the porous rock and collect it some distance away via a production well. As crude oil is often rather viscous, the water has to be thickened with polymer in order to sweep oil efficiently.

The food industry has many requirements for polymers (see Chapter 9) while there are also many industrial uses, including chemical and biochemical processing and the manufacture of absorbent wound dressings for medical use.

Other chemical products

The chemicals already mentioned by no means exhaust the list of potential biotechnological products. Further examples include pesticides against certain plants, insects and soil-inhabiting worms, more flavour enhancers, vitamins, and surfactants or detergents — on which there is more discussion in Chapter 10 in relation to oil recovery and in Chapter 11 with reference to cleaning up oil spills.

Biotechnological systems employing whole microbial cells, or enzymes extracted from them, are also used to modify certain biochemical compounds in ways which are very difficult to achieve either by conventional chemistry or by using organisms other than microbes; such procedures are called *biotransformations*. The reactions start with complex molecules

Exploiting biotechnology

originating from plant or animal sources, or derived from microbial fermentations. Specific changes in the complex molecules are then effected by the transformation processes. It may eventually be possible to extend the range of biotransformations by employing cells obtained from whole plants and grown, like microbes, in culture vessels. Such cells possess valuable enzymic capacities but they grow very slowly and at the present time are difficult to use. Improvements in technique will eventually make them more amenable as production vehicles.

Biotransformation reactions are of particular significance for the synthesis of a range of *steroids* which, in plants and animals, act as hormones regulating various metabolic activities. With respect to human medicine, biotransformation of steroids generates such therapeutic materials as male and female sex hormones, corticoids for the treatment of arthritis and the synthetic hormones used in contraceptives (the active ingredients in "the pill"). Annual production of steroids probably exceeds 2,000 tons, worth some $1.5 billion. About 85% of the volume is contributed by corticoids but the 5% of market volume accounted for by the contraceptive products equates to 35% of the value.

Some chemicals are particularly interesting. *Indigo* is a blue pigment used to dye fabrics, including blue jeans. It originates from a plant called *Indigofera* and in 1983 it was estimated that the world market, served by four main producers, was more than $100 million at a unit cost of $10–12 a pound. Genetic engineering enables indigo to be produced by a common bacterium, one used extensively as an experimental organism for studying bacterial biochemistry and genetics. Could market share be secured by bacterial indigo if the selling price were as low as $3–5 a pound? How would the existing suppliers react? The answer is not clear; the response of the present producers and their clients is not predictable and the fashion for blue jeans may change.

Making Foreign Proteins

One of the early advantages perceived for genetic engineering was the ability to produce cheaply and in essentially unlimited quantities proteins of great therapeutic potential but ones often very difficult to obtain from their human sources. Human proteins are unique to humans, although in some cases animal versions might work well for therapy. However, for many purposes animal products are unlikely to win approval either because they are not active in human beings or because they evoke undesirable and unacceptable side reactions.

The manufacture of foreign proteins using yeast and certain bacteria include some which are derived from other microbial sources and plants as well as those of human origin. Among the human ones are:

- *growth hormone* for the treatment of dwarfism and other growth defects;
- several varieties of interferon for the control and treatment of virus infections and their possible involvement in the onset of certain cancers;
- the potential anti-cancer agents *interleukin-2* and *tumour necrosis factor*.

Insulin was one of the first proteins to be considered in this context. Made in the pancreas, it is an important hormone which controls sugar metabolism; a deficiency leads to one form of *diabetes*. Discovered in the early 1920s, a type of insulin almost but not quite identical with the human variety can be obtained from pig pancreas available from slaughterhouses; it has been used to control diabetes for nearly 70 years. Although largely successful, it does have adverse reactions in some patients. While, of course, it has always been impossible to use human beings as a source of insulin protein in quantities other than as small samples of experimental material, it was easy to obtain enough human genetic material to insert the genes for human insulin into bacteria. This offered the possibility of microbial synthesis. Many technical hurdles had to be overcome, but overcome they were and "human" insulin, identical in every respect with the natural product save that it is made in bacteria, is currently generally available for treating patients.* Indeed, advances in genetic procedures during the past decade now make the requirement for a sample of original human DNA unnecessary for the production of a human protein in bacteria. Knowing the sequence of amino acids in the protein is sufficient to allow the manufacture in the laboratory of a substitute gene which can then be fitted into a production bacterium.

ENZYMES

As well as the critical role they play in all forms of biotechnological product manufacture, enzymes are themselves industrial products of no mean significance. Their benefits as industrial catalysts lie in the mild conditions in which they work, their virtual absence of side products and our ability to select enzymes with just the right catalytic properties. Those benefits must, however, be weighed against the fact that enzymes are intrinsically unstable and their actions are easily inhibited by a host of extraneous chemicals. Problems can also arise because substrates or products may have low solubilities in water, the medium in which enzymes normally operate. Nor are enzymes necessarily cheap: their price per kilogram ranges from as little as $3 for comparative crude industrial preparations to as much as $300,000 for highly purified enzymes needed for research purposes; few are available in bulk and any needed for specialist applications would probably have to be isolated and purified for the purpose. Without detailed analysis it is difficult to judge quantities and hence cost; those depend, among other factors, on the catalytic activity of particular preparations and the time needed in the overall procedure for accomplishing the catalytic step.

---

* The real-life story of insulin is not that simple. Competition duly appeared from a number of sources. One rival company used a different microbial system to produce genetically engineered "human" insulin while another made use of the fact that pig insulin has only one of its 51 constituent amino acids different from the human version; the latter company found an enzymic method for changing the different porcine amino acid into the human one. But treating patients with insulin is a complex matter because their responses to the different human and animal preparations vary and doctors often prefer to stick with drugs whose effects they know well. So at least five types of insulin are now on the market: pig, beef and three sorts of "human" — one made with genetically modified bacteria, one with yeast while the third is the one with the modified amino acid.

Exploiting biotechnology

Sources of Enzymes

Enzymes are present in profusion in all living organisms but for each type of enzyme there is usually a preferred source which is either the most readily available or technically the most convenient, or both. Slaughterhouse material is obviously a convenient starting point for enzymes present in tissues from cows, pigs and sheep but the range of suitable ones is rather limited, the enzymes are often not very stable and purification is sometimes a problem, as indeed it is from plant leaves and fruits, other potentially prolific sources. Many industrially-produced enzymes originate from microbes and are used in various manufacturing processes to break down large natural molecules; some of them are considered in more detail later. Microbes are good sources: they are cheap to obtain, there are very many varieties from which to choose, and it is often possible to manipulate them genetically and physiologically to improve yields of the desired enzymes.

Enzymes are marketed in solution, as solid powders and immobilized onto solid matrices as described earlier in this chapter. Various chemical additives may be present to stabilize acidity and to prevent microbial attack on the enzymes, all of which are biodegradable and hence potentially food for microbes. When enzymes were first used in laundry detergents, people using the products sometimes suffered skin rashes which were at first thought to result from contact with the enzymes. Although that was probably not the cause, enzymes were for a while omitted from washing powders and they later reappeared in a granular form which avoided any problems of enzyme dust.

Commercial Uses of Enzymes

This section will explore some of the uses of enzymes for catalysing and facilitating a number of process operations.

The starch industry

The main uses are for enzymes called *amylases* to break down starch (a polysaccharide polymer) into glucose, the sugar units of which it is composed. (Note, incidentally, that although many biochemical names are long and complex there are some useful tips about their use. The ending *-ase* is a good indicator that a name refers to an enzyme. Similarly, many sugar names end in *-ose* and amino acid ones in *-ine*.) Starch can also be broken down to glucose by the use of hot acid, but that requires fuel to heat the mixture to the temperature required and has a deleterious effect on a proportion of the product molecules. Fairly heat-stable amylases permit the operating temperature to be lowered from 140–155° for acid to 95–100° for enzymic treatment, although the contact time might be longer.

Amylases and related enzymes are used for sizing and desizing of textiles (covering the fibres with a protective coating of an adhesive starch-based polymer to stop them breaking during weaving, with later removal of the size) and of paper: sizing the fibres or coating the surface of paper contributes to stiffness, strength and some other desirable properties. These enzymes also find a number of uses in the food industry, for example in helping to remove the cloudiness in fruit drinks. The elimination of "haze" in fruit preparations, as well as in the manufacture of jams and jellies, is also aided by a group of plant enzymes called *pectinases* which attack *pectin*, another common plant polysaccharide.

Proteinases

Also called *proteases*, these are enzymes which help to break down proteins. They are very valuable in detergent formulations because many stains (from blood and other bodily fluids, food, etc.) have a high protein content and it is the protein which often makes the stain stick tenaciously to the fabric. There is an interesting relationship between biotechnology and cultural practices. In the UK it is said that people traditionally insist on washing their clothes in very hot water so, if they are using enzyme-containing detergents, the enzymes must be tolerant of high temperatures and that presents a number of technical problems. In the US, however, cold-water washing has been acceptable for some time, permitting the use of more heat-sensitive enzymes which are probably cheaper.

Protein-degrading enzymes are used in leather production to remove hair, a more satisfactory way of doing so than traditional "dehairing" with lime and other chemicals, yielding a better quality product and fewer problems of noxious fluid disposal. Proteinases are employed in brewing to improve the release of nutrients from the barley and prevent cloudiness in the beers, while they improve the grain and texture of bread and other baked products. Additional varied uses include removing remnants of waste protein from carcases in abattoirs (the scavenged protein being used in soups, canned meat products and animal feed supplements), recovering silver from photographic emulsion by dissolving the gelatin base (gelatin is a protein) and in various cleaning preparations, coarse ones for drains and more specialized ones for instruments and delicate equipment.

Lipases

Another category of enzymes, these degrade fats (or *lipids*). They have a number of uses in the food industry for flavour development and texture modification and, often together with proteinases, are contained in cleaning preparations designed for treating drains blocked with accumulations of grease, fat and hair.

Speciality Enzymes

Most of the enzymes discussed so far have been used in rather a general sense — employing a lipase/proteinase preparation to clean a blocked drain cannot be expected to depend on a

Exploiting biotechnology

biochemical analysis of the particular fats and proteins trapped in the pipe. Other enzymes are used much more specifically for defined stages in the synthesis of particular products.

Penicillin acylase

This is an enzyme used in the production of penicillin derivatives. Penicillin produced by fungal fermentation is altered catalytcially by this enzyme and the product of the reaction serves as the starting point for chemical modifications to make derivative antibiotics. This process was briefly mentioned above.

Glucose isomerase

High fructose sweetener for the soft drinks and other areas of the food industry has become a major industrial product. Its production involves this enzyme; Chapter 9 will tell the story in more detail.

Prospects for the Future

Modern biotechnology is still to have a major effect on industrial enzymes; change has tended to be confined to small-scale activities leaving the major industrial processes largely unaffected. That pattern will probably continue but there are some grounds for thinking that the pace of change might increase as the underlying science advances to offer enzyme catalysis in a wider range of applications and industrial users gradually recognize the potential benefits. A few of the likely technical advances are worthy of some consideration.

Thermostable enzymes

Many processes benefit from running at fairly high temperatures approaching the boiling point of water, but most enzymes are not stable in such hot environments. However, there are microorganisms which normally live in hot places and which might serve as sources of thermostable enzymes but their range is limited and not all needs can be met with enzymes from those organisms. Increasing understanding of protein structure will lead to an ability *de novo* to design new enzymes with defined high temperature and other desirable properties. Coupling that skill with modern genetic technology for making artificial genes to code for such new enzymes, and getting the genes working to make the enzymes in convenient host microbes, is one way in which new catalytic tools will become available for industry.

Chemicals, enzymes, fuel and new materials

Enzymes out of water

We have stressed many times that biology is water-based; one of the limitations of using enzymes in production is that many sought-after products are not water soluble and much chemical processing (for instance, of petrochemicals) is not water-based. Recent scientific advances show that enzymes can function in liquids ("organic solvents") which do not mix with water. They may do so by being coated with a water film so thin that contact with chemicals in the surrounding non-aqueous medium is not impaired by the existence of a boundary between the water and the organic solvent. Efficiency demands dispersal of the water-coated enzyme molecules throughout the non-aqueous environment; there are several ways of doing that and the prospects look good for enzymic catalysis of reactions in organic solvents. Enzymes can also work in gases. One immobilized microbial catalyst, flushed with a gas mixture of propylene and oxygen, will make propylene oxide, an important industrial product.

New specificities

While prodigious numbers of alternative amino acid sequences can exist among proteins (indications of this were presented in Chapter 2), recent work has shown that the number of different molecular shapes is nowhere near as great as the number of different sequences. For a protein with a chain of 100 amino acids, the number of sequences is theoretically 10 followed by 130 zeros but experiments suggest the existence of no more than 100–200 million different shapes. Large though this number seems to be on paper, in molecular terms it is actually very small. The number of molecules in a level teaspoonful of that same protein with 100 amino acids in its chain would be 10 followed by 20 zeros, more than a trillion times more than the 100 million different shapes; even a tiny speck of the material will contain more molecules than the number of all the possible shapes.

Such a finding implies that it will not be impossibly difficult to extend enormously the range of protein molecular recognition interactions on which enzyme catalysis is based: 100 million different ways of binding small molecules to enzymes offers huge practical opportunities for new sorts of catalysis. So making novel enzymes to fit new and different molecular shapes, and hence to catalyse new reactions, is now a much clearer possibility than it might have seemed just a short time ago.

BIOMASS INTO FUEL

Biomass, all the organic matter forming the bodies of plants, animals and microbes, those still living as well as those in the process of decay, is clearly a major source of chemical feedstock. What part of it can be used in industry and what are the products which might result?

Exploiting biotechnology

Estimates place the total amount of biomass at about a trillion tons, some 10% of which is renewed annually through the process of plant photosynthesis. Plant material forms the bulk of this potentially usable material; calculated in human terms, that adds up to about 20 tons of new plant growth per person per year. The two major components of plant material are *cellulose*, forming about 40% of the total, while *lignin* comprises 20–30% of more woody plants and smaller percentages of others; straw, for example, may contain 10-15% lignin depending on the plant of origin. Wood, a prolific source of complex, energy-rich organic compounds beckons as a feedstock for possible conversion into fuels and other chemical products. Of course, many forms of plant material are already used as fuels with no chemical treatment at all — they are burned to provide heat and many people take the view that probably that is the most cost-effective way in which they can be used as fuel sources. But there are other views.

Lignin

A major component of wood, lignin is a very strong material with a heterogeneous chemical structure. In principle biodegradable, it is in practice remarkably resistant, especially when kept dry; it is the combination of its strength and resistance to decay which makes it such a valuable structural material. Comparatively few microorganisms are able to attack woody plants and those that do tend to act rather slowly — as anyone who has noticed how long it takes for fallen trees to decay will have realized.

That resistance of lignin to both biological and chemical degradation also makes it rather an unfavourable feedstock for chemical processing. Pretreatment of the wood, either physically by milling or shredding, exposure to steam at very high temperatures or steam explosion (a process in which a chamber containing the wood is filled with high pressure steam and the pressure suddenly released, causing the wood to disrupt), or chemically with hot acid, is usually necessary before further processing can take place. Enzymic degradation using microorganisms is possible although slow and, while several attempts have been made to speed it up in order to make lignin more useful as a feedstock for industrial processes, many difficulties remain. The industrial uses of lignin are thus mainly prospective: it can be used as an adhesive in plywood and particle board manufacture, as a surfactant, rubber reinforcer and asphalt extender, but its greatest use at the present time remains as a fuel to be burned directly. It does not appear promising as a feedstock, partly because of its complexity and variability, partly because of pretreatment costs and also because there is no particular shortage of alternatives.

Cellulose

The other major chemical component of plants is somewhat more promising. Cellulose is composed chemically of chains of glucose units and the polymer is fairly susceptible to being broken down to those individual glucose molecules either enzymically or by the use of acid.

Acid treatment is generally regarded as the more satisfactory. There is an inverse relationship between the strength of the acid and the temperature necessary to achieve breakdown: the stronger the acid, the lower the temperature. Breakdown with strong acid yields a cleaner product for subsequent use as a feedstock for microbial fermentations. However, for cost effectiveness it is necessary to recycle the acid and this requires expensive, corrosion-resistant equipment. Enzymic breakdown entails the cooperation of a number of enzymes which occur in a range of microorganisms and a good deal of effort in recent years has gone into optimizing enzymic degradation of cellulose with a view to making it a more cost-effective industrial process.

The amount of cellulose available from plant sources, and its decomposition to glucose (an excellent substrate for many different microbial fermentations), makes it worth thinking about cellulose as an inexpensive feedstock for the manufacture of a range of products on a medium-to-large scale. For the moment, however, the technology for utilizing cellulose as a source of glucose is still not satisfactory and, like lignin, the best way to use it as a fuel is to burn it.

Starch and Sugar into Alcohol (*Ethyl Alcohol or Ethanol*)

These, however, are two chemicals which are much more immediately suitable as microbial fermentation feedstocks, and it is to them that we must turn for a view of what large scale microbial production might imply.

These two substances are not present in large amounts in all plants but their content is very high in some agricultural products: starch, particularly in maize and cassava, and sugar in the stems of sugar cane and the roots of sugar beets. Cane sugar is directly fermentable by a range of yeasts and other microbes while starch is readily digested by acid treatment, or by enzymes present in many plants, to yield glucose as a fermentation substrate.

The fermentation of sugar to make beverage alcohol has been known since antiquity and, until it was largely displaced by production from petrochemical sources, industrial alcohol, too, was made by fermentation. There was accordingly a wealth of experience in this technology and not only for beverages and solvents: even before World War II some petrols for powering vehicles contained 20% alcohol.

With the steep rises in crude oil prices in the 1970s, a number of countries began to take seriously the possibility of using agricultural sources of starch (in the US) or cane sugar (in Brazil) to make alcohol, either to replace or extend the use of petrol. Though some problems do exist, alcohol does this rather well with no more than minor technical modifications to vehicles. Extensive programmes were undertaken in both countries for parallel but certainly not identical reasons. The US had both a very large agricultural production of starch, with the capacity to expand, and extensive facilities for milling and processing food grains. There was political concern at the high price of crude, which was expected to go on rising, as well as the ever-growing dependency on foreign oil, some of it from politically unfriendly and potentially unstable areas. Brazil had a large sugar cane industry which could easily be expanded. The

Exploiting biotechnology

country had little domestic oil production and a shortage of foreign currency with which to buy crude on the world market. A number of social benefits from an alcohol-for-fuel programme were perceived.

In different ways, both countries developed extensive alcohol fermentation and production facilities. In the US, *gasohol* was promoted, gasoline containing 10–12% alcohol; in 1983 sales reached 3-4% of all petrol sold. Ten years later, gasohol seems entirely to have disappeared: the price of crude oil fell in the mid-1980s, technical problems were encountered in many installations and comments have been made about poor profit margins, poor management and a rush to take advantage of tax concessions without thinking through the implications of the necessary investment in equipment and processing.

In Brazil the programme was more extensive. As early as the 1930s, laws required the addition of alcohol to imported petrol in order to protect the domestic sugar industry. Plans formulated in 1975 under the *Programma Nacional do Alcool* called for 20% substitution of petrol by alcohol in 1980 and 75% by 2000. Benefits were to accrue not only in terms of foreign currency saved but also in a reduction of income disparities between regions and between social groups, and in the expansion of employment opportunities. Cars were designed and manufactured to run solely on alcohol and a programme of converting existing vehicles was put in hand; pump prices of alcohol were kept below those of petrol and various prohibitions on the use of petrol were introduced.

Technically the Brazilian programme probably succeeded: alcohol was produced on a large scale and most new cars do run on it. Economically the value of the initiative is more doubtful and social benefits are a matter for debate. Crude oil had to be imported anyway for making diesel fuel for trucks and buses, and to make liquefied petroleum gas for cooking, generating petrol as a by-product. But with so many cars using alcohol, the country has a surplus of petrol which must therefore be exported. On the other hand, subsidies to sugar producers were cut for political reasons. The producers turned to other agricultural products and exports of sugar were encouraged by rising world sugar prices; sugar into alcohol declined in importance. Alcohol production is now said to fall 10% below requirements, the shortfall being made up with methanol. Unfortunately Brazil cannot produce enough methanol and large amounts will need to be imported.

NEW MATERIALS THROUGH BIOTECHNOLOGY

Many traditional materials for cloth making and building construction are biological in origin:

- cotton and linen, based on fibres of cellulose (a polysaccharide) from plants;
- wool and silk, fibrous proteins from animals;
- animal skins both for clothing (furs and leather) and tenting, composed of protein fibres in a polymeric matrix;
- wood, a complex of mainly fibrous plant materials made of lignin, cellulose and some other substances;
- various forms of thatching materials, all largely cellulose and derived from plants.

Chemicals, enzymes, fuel and new materials

Working from a knowledge of the structures of these natural substances and an understanding of why they perform so well, biotechnology is beginning to offer the possibility of new materials incorporating some of the best properties of natural products and extending them into novel specialist applications. This chapter will look at four:

- bacterial cellulose;
- new types of fibrous proteins;
- *polyhydroxybutyrate*, a bacterial "plastic";
- bone-replacement materials for surgical use.

Bacterial Cellulose

Not only plants make cellulose — a number of bacteria also do so and produce a very pure and fine product which, while inevitably more expensive than the best plant cellulose in the finest cottons, has value for a number of special uses. Recent reports speak of the successful commercialization at relatively low cost of "Cellulon", a bacterial cellulose produced in a fermentation with genetically-modified microbes. The bacterial product differs from conventional plant cellulose in having extremely thin fibres with 100–200 times more surface area than the usual material, giving it very high absorptive powers both for water and many other chemicals. Since it is non-toxic, uses are perceived in medicine for artificial skin grafts and wound dressings, and as dietary fibre, bulking agent and texturing agent in food. In the paper and processing industries it might be used for separation membranes as well as a thickener in paints, dyes, glues and cosmetics.

New Proteins, Fibrous and Sticky

Among modern synthetic materials, the family of nylon polymers are extremely successful products with properties which make them suitable both for high tensile strength fibres and where light-weight, wear-resistant characteristics are needed as in bearings, gear wheels, etc. In terms of their chemical chains, nylons are reminiscent of proteins and, not surprisingly, various people have looked at the range of natural proteins with a view to developing new industrial products.

Silk produced by silk worms, the larvae of the silk moth, is an obvious candidate for microbial synthesis. Silk has long been recognized as a desirable fibre but the comparative difficulty of production and correspondingly high cost have limited its use. In one case, the silk protein gene has successfully been transferred into bacteria while, in another, an artificial protein has been made, a synthetic gene constructed to code for it in microbial systems and the new gene inserted into a suitable recipient bacterium. So far, however, no cost-effective working technology appears to have been developed.

Exploiting biotechnology

An attractive group of natural proteins which would find many industrial uses is that of silks produced by spiders, especially the dragline silk which forms web frameworks and also provides a safety rope for the spider when it is climbing. Some of these protein silks have very high tensile strengths coupled with very high extensibilities, better than those of natural silk and approximating, and in some cases bettering, the properties of high-strength artificial fibres like *Kevlar*. There are reports that the gene for dragline silk has been isolated and transferred to bacteria, potentially for a production process. At the very least, applications of the high breaking energy are seen for flak jackets and other bullet-proof clothing and there seems little doubt that the valuable properties of spider silks would find many other uses if it were available in quantity at reasonable cost.

Yet other valuable proteins are the glues; animal-based products have a long history in glue manufacture. Many more remain to be exploited, or exploited more extensively, the mussel glues among them. They have the interesting property of being applicable to a wide range of substrates under wet conditions, since mussels live in water. Already available in small quantities directly from its natural source, the gene for mussel glue protein has, like those for various silks, been transferred to bacteria for potential production. It might become possible to modify the natural protein for wider purposes and, via artificial gene synthesis — some of which has already been accomplished — variant glue proteins might be made by bacteria. Application is foreseen in small quantities for surgical procedures and, if large amounts could be made at an acceptable cost, many engineering uses might become possible.

Polyhydroxybutyrate (PHB)

Most people think of plastics as entirely modern artificial materials, not associated at all with biology. That used to be true, but no longer. Certain bacteria accumulate PHB as a storage material, their equivalent of starch; it can actually form the basis of an extremely valuable biotechnological plastic.

PHB is accumulated in the cells of those bacteria in the form of small granules. For industrial production it is made in conventional fermenter vessels, the bacteria being persuaded to accumulate the material as a reserve by first allowing them to grow normally and then later depriving them of certain nutrients essential for further reproduction, while allowing them to continue to make PHB until it constitutes around 80% of the total cellular dry weight. The cells are broken open to release the PHB which is recovered either by solution into organic solvents, in which the cell debris is not soluble, or by using enzymes to digest the cell remains and leave the desired product. Pigments and plasticizers are added as necessary and the powder dried, extruded and granulated into a form convenient for the manufacture of products by injection and blow moulding, the formation of films and fibres or incorporation into other plastics formulations.

When PHB was first developed industrially, at the time of the crude oil price increases in the early 1970s, it was seen as a possible replacement for plastics based on petrochemicals, but the subsequent fluctuations of the oil price — often downwards in real terms — undermined that view. Nevertheless, it was found to have valuable speciality properties, among them its very high degree of toleration by human and animal tissues, leading to great promise for certain types of surgical implants and other high added-value medical applications. As a biological product, with a normal susceptibility to biodegradation, it clearly had potential as a biodegradable plastic for such products as disposable containers, with consequent benefits for minimizing long-term environmental contamination.

Further exploration of its biotechnological implications showed that it was possible to make not just pure PHB but also plastic materials which combined PHB with related chemical compounds. By introducing various process modifications, the properties of the original substance could be altered in defined ways and a new range of applications became possible in fibrous and film configurations with various degrees of tensile strength, clarity and permeability to gases. It is not out of the question that PHB and its derivatives will in time be only the first of a new and diverse range of biologically-generated plastics.

Bone Replacement

The relationship between biotechnology and new substances is not confined solely to their manufacture from natural biological products: it is also concerned with the use of non-biological materials for specific purposes in living systems. Materials for bone replacement are just such an instance.

Severe problems arise in the implantation of orthopaedic prostheses, perhaps the best known being hip joint replacement in patients with disabling arthritis. Bone is a living tissue, constantly in a state of change and reacting to the stresses and loadings put upon it. A consequence of bone implants might be to alter the loading and stress patterns in such a way as to result over time in a loss of bone substance with a consequent loosening of the implanted prosthesis. Although all patients are liable eventually to suffer loosening, the process is greatly retarded in more elderly recipients and the operation performed in people over 65 years old may well last 10-15 years. In younger people, however, with more active tissue bone responses, problems may arise in less than five years.

There are a number of difficulties encountered with existing bone implant materials (alumina, cobalt-chromium alloy, stainless steel, titanium alloy, polyethylene and polymer-based bone cement) because of the ways in which their properties differ from those of bone itself. Incompatibilities promote bone resorption and weakening of the bond between the implant and the residual bone. Following a detailed analysis of the response of bone structure to implant stresses, a new material was designed which seeks to avoid these problems by using existing substances to match as closely as possible all the relevant properties of bone. Natural bone is formed of *hydroxyapatite*, a mineral form of calcium phosphate, embedded in a matrix of protein fibres. To mimic this structure and maximize compatibility, the new material is structured as a composite of hydroxyapatite powder (a readily available material used in the manufacture of bone china) in a matrix of polyethylene. Once processing problems were overcome,

Exploiting biotechnology

a usable material resulted which is expected markedly to reduce bone resorption and the loosening of implants. Four years of use have shown that the composite performs well.

Manufacturing New Materials

The food industry apart, biotechnological manufacturing is confined largely to low volume, high added-value products. The cost-effective use of microbes to manufacture the sorts of new materials discussed in this section is likely to require scaling up to a minimum of multi-tonnage production coupled with good process control, as in existing microbial production, to ensure high and constant product quality. It will take time and experience to develop large-scale manufacturing technologies; for the near term, these new materials are likely to find rather specialized uses, requiring no more than a relatively modest scale of operations. Nevertheless, improvements in microbial quality and performance, achieved by traditional means as well as by the newer ones of genetic manipulation, must be expected to combine with developments in production engineering progressively to widen the commercial and industrial opportunities for using novel structural materials.

CONCLUSIONS

Biotechnological methodologies have enormous scope for the production of high added-value products on a small- to medium-scale. Genetic engineering and other technical advances may confidently be expected greatly to extend the range of such products and significantly to improve production methods and reduce costs.

Once more with the exception of the food processing industry, the picture may be less buoyant with respect to the large-scale production of commodity chemicals. Aside from sewage treatment and the disposal of wastes to landfill (Chapter 11), the two largest microbial fermentation projects in bulk terms, single-cell protein from hydrocarbon and alcohol for fuel from sugar and starch, have both been successful technically but less so economically. A number of authorities nevertheless look forward to the growth of biotechnology in various parts of the chemical industry. In Japan, one of the leading countries in this field, a whole host of chemical procedures is being explored to see what advantages biological catalysis might have to offer.

When thinking of the future, it is worth remembering that the "new" biotechnology of genetic engineering and enzyme modification is not much more than 15 years old. With so much scope for new development in such a wide range of activities, most of them so far with barely a scratched surface, it is hard to imagine that the next 15 years will not witness changes in production technology every bit as dramatic as those which have taken place in laboratory research.

9 What we eat: agriculture and food

Right at the beginning of this book we touched upon the problems of defining biotechnology, noting how very difficult it is to draw absolute boundaries between all the different commercial expressions of the biological sciences. Biotechnology cannot just mean genetic engineering, or genetic engineering plus microbial technology, or indeed, any limited set of combinations — one can always think of more examples which are clearly biotechnological but fall beyond arbitrary definitions. Perhaps in this chapter we should therefore logically discuss every aspect of agriculture and food provision, each of them clearly commercial uses of biology. But the story has to end somewhere and so this account will have to be selective, concentrating on recent and current developments with no more than an occasional reference to traditional experience.

Originally hunters and gatherers, obtaining food where they could, people nowadays in industrial countries retain fishing as the only large-scale foraging activity; true, some hunting of birds and game still goes on but their total contribution to the human diet is minute. Ever since pre-history, mankind has become increasingly reliant on agriculture for primary food sources as well as for some of his raw materials. From time immemorial, farmers have learned by trial and error the practicalities of improving plant crops and animal herds, leading to greater reliability and better returns by increased yields, enhanced resistance to disease and extended ability to flourish in harsh environments. In parallel with agriculture, various types of processing have been used to make secondary foods like bread, wine and sour milk products. Impressive empirical skills in all these areas were acquired over the centuries but only in the last hundred years or so have they been understood in scientific terms. Those developments continue unabated but now, with the aid of a profound appreciation of the detailed underlying biological mechanisms, we do it with our eyes open, greatly improving the efficiency and effectiveness of what we seek to achieve. This chapter is about that interaction of science and technology in agriculture and food.

Of course, agriculture is not solely about the provision of food. A wide range of non-food products and raw materials is derived from plant and animal sources: fibres for cloth (cotton, silk and wool), paper (wood pulp) and rope (hemp, sisal, coconut and others), etc.; structural materials (mainly wood); adhesives (bones); perfumes (natural plant and animal oils and other substances); fuel (wood and dried animal droppings); decoration and ornamentation (flowers and shrubs), and so on through a long list.

THE AGRIBUSINESS

"Agriculture" is a pretty broad term which can include many forms of cultivation. All parts of this major human activity may in time be influenced by biotechnology and some of the

Exploiting biotechnology

main areas of current activity will be touched upon in this chapter; as far as food provision is concerned, only hunting and fishing are almost entirely excluded from discussion.

In agriculture, as in so much of biotechnology at the present time, most new products and services represent improvements in methods for solving existing problems which have been matters of concern to farmers for many years. No doubt wholly new developments will eventually emerge but in the relatively near-term — say within the next couple of decades — the most probable impacts of biotechnology will be in:

- the use of genetic and tissue culture procedures to produce modified plants with:
    - enhanced resistance to pests and disease;
    - improved tolerance to drought, flood and other environmental changes;
    - less sensitivity to the presence of heavy metals in the soil;
    - better yields;
    - products with improved flavour and keeping qualities;
    - perhaps reduced dependence on chemical fertilizers;

- genetic manipulation of crop plants enabling field cultivation, rather than microbial systems in fermenters, to be used for making therapeutic and other proteins (such as human and animal hormones) which are difficult or impossible to isolate from their primary sources;

- yet more genetics to make new sorts of shrubs and flowers, or flowers in new colours or combinations of colours, for the decoration and ornamentation market;

- diagnostics for plant diseases;

- new pesticides, particularly "biopesticides" which avoid the use of noxious chemicals;

- more widespread use of algae to generate a variety of food and other products;

- more positive control of microbiological interactions with agriculture, for example in the development of mycorrhizal associations, silage and legume inoculants and probiotics;

- improved vaccines for animal prophylaxis;

- diagnostics for animal diseases;

- use of animal hormones to improve product yields;

- the deployment of transgenic animals for reasons similar to those for seeking improved plants: better productivity per unit of input, better resistance to diseases and environmental conditions, as well as for the production of "bioactive" molecules.

Most of these topics merit exploration in this chapter.

Plant Genetic Manipulations

As we observed in Chapter 2, plants can be difficult to deal with compared with microbes when it comes to genetic manipulation. The reason lies partly in their cell walls.

All living cells, plant and animal as well as microbial, are bounded by thin, flexible membranes which confer little mechanical strength but are immensely important in controlling the passage of chemicals into and out of cells. In essence, the cell membrane delineates the boundary between the chemical and physical environment inside a cell and the world outside. Its role varies in detail in different types of organisms. Cells within the larger animals derive considerable mechanical and chemical protection from their "outside worlds" by being part of an organism much larger than each individual cell. Most animals have a strong outer coat, a "skin" of some sort, which protects its soft and vulnerable internal tissues from environmental damage. That skin also enables the animal within its own body to maintain a constant chemical and physical environment, one conducive for its constituent cells. Within that friendly environment the cell membranes maintain the characteristic internal compositions of the cells they surround (brain cells, for instance, are different from muscle cells), while all are bathed in a fluid (lymph in the case of brain and muscle) which provides just the right kind of chemical environment they all need.

Life is more difficult for bacteria living, perhaps, as single cells in a river, a much more hostile environment for bacteria than lymph is for brain or muscle. In the bacterial case the membrane constitutes a boundary between two very different fluids (the river versus the inside of the bacterium), much more diverse in their properties than lymph and the living matter in brain and muscle cells. Bacterial membranes are therefore more selective in controlling the passage of chemicals so as to keep the right ones inside their cells and the wrong ones (in the river) out.

Most plant cells are protected by being located in the leaf, root or stem of a large individual plant and, while not bathed in a fluid quite in the same way as animal cells, do nevertheless exist in a compatible microenvironment comprising other cells like themselves. In addition to its boundary membrane, each plant cell is normally enclosed by a physically tough wall composed of cellulose, a material used in almost pure form to make cotton fibres and some types of paper. Methods like microinjection and particle acceleration (microprojectile techniques), effective ways of getting DNA into naked animal cells devoid of such a tough cell wall, can work with plant cells only if the wall is first removed. Certain enzymes can dissolve the walls, allowing naked plant cells to be manipulated just like animal cells: they will undergo fusion, DNA can be introduced by microinjection (Figure 9.1), electroporation and particle acceleration, and so on. Given a chance, the cells will eventually regrow their walls but, until they do so, they are very fragile and need special care and attention if they are to survive.

Learning how to overcome all these difficulties delayed somewhat the application of recombinant DNA technologies to plants, but eventually most of the problems were solved and plant genetic manipulation is becoming an important procedure for generating new strains for many practical purposes. It is important, however, to bear in mind that genetic manipulation of organisms as complex as crop plants carries a high degree of uncertainty which can be resolved only by experience in the field under growing conditions.

Tissue culture and protoplast fusion

The elegant methods of gene splicing are directed to introducing relatively small and specific pieces of DNA, each carrying one or a few genes. An alternative means of genetic manipulation

Exploiting biotechnology

is to join together two cells with different properties to make a hybrid combining the characteristics of both. In animal systems this is the approach used in the production of monoclonal antibodies (Chapter 7).

The experimental protocols are rather empirical and rely on two laboratory techniques: the ability to fuse cells from different plants and the fact that it is often possible to grow a whole plant from a single cell. Thus, whole cells from two sources can be fused but, because their genomes each contribute a wide range of different properties to the hybrid, it is usually not possible to predict with any certainty in any particular hybrid cell which properties will

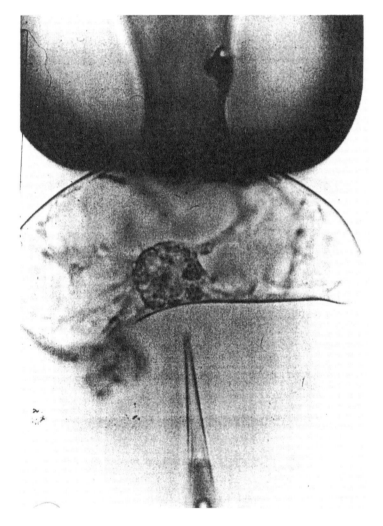

Figure 9.1 Microinjection allows the direct incorporation of DNA into the nucleus of a variety of plant protoplast cells. (*Photo by permission of Calgene*)

What we eat: agriculture and food

be expressed. Take a simple hypothetical example: suppose that from a plant which has blue flowers and is resistant to a particular virus, and another variety with red flowers but susceptible to infection, it is desired to produce a resistant plant with red flowers. Cells from each plant type are mixed and fused. The plants developing from some hybrid cells might have red flowers and viral resistance (useful), some will be blue and sensitive (undesirable and discarded), while yet others may originate from the fusion of two cells of the same sort and be just the same as the plants from which they came. From this mixed bag of hybrids it is necessary to isolate and cultivate those types which are wanted while rejecting the rest.

The properties of the individual hybrids can be recognized only when they are allowed to grow into complete plants. This is done by *tissue culture* or *micropropagation*, in which the hybrid cells are spread onto a jelly containing appropriate nutrients, often in the presence of light to allow photosynthesis; if the conditions are right, a proportion of the hybrid cells will develop into recognizable plants with their familiar roots and shoots. In some cases further individuals can be cultivated from portions of the growing plants to build up a large population. Eventually the new plants are transferred to soil and allowed to grow to full size. During this process the plants tend to be delicate until they enter their normal growth patterns and have to be protected carefully from infestation by pathogens.

Micropropagation is not restricted to generating plants from fusion hybrids. In an application called *clonal propagation*, single cells are obtained from the rapidly growing (meristem) zones of existing plants and cultivated by micropropagation to produce many individual clones, all identical because they have originated from one source. Clonal propagation can be used to avoid the variability which is always present in populations resulting from natural forms of sexual reproduction. As long as the costs are acceptable, it is a valuable way of obtaining plants of high quality and uniformity, free from pathogens and whose availability is not governed by the seasons; there is also the further advantage that space requirements are limited. These considerations do hold in many cases and clonal propagation has become important commercially, particularly for the provision of pot plants.

In spite of the cells in clonal propagation being derived from a single genetic source, there is some variation (perhaps about 0.1%) among the product plants; this presumably implies a corresponding variability among notionally identical cells in the source plant, a variability normally masked by the presence of the much greater number of all the other (identical) cells. In clonal propagation these small differences become manifest because a whole plant is derived from each single cell. The method has been used, for instance, to obtain tomatoes with high solids contents and better harvesting characteristics. Such an approach, termed *somaclonal variation*, is being used to improve strains of bananas, carrots, celery, maize, sugar cane and wheat. (The word *somaclonal* is derived from the concept that, in multicellular plants and animals using sexual reproduction, at least some of the reproductive cells are immortal in the sense that they give rise directly to the next generation but most of the body [the *soma*] eventually dies.)

One can go further and combine clonal propagation with mutant selection. As we saw in Chapter 2, it is possible to use chemical and physical procedures for making random changes to DNA of cells, giving rise to mutants in various ways different from their parents. With suitable selection techniques, the mutants can be recognized even in the presence of large numbers of the parental forms. Propagation/selection can be applied to plant breeding, for

Exploiting biotechnology

example to obtain herbicide- or disease-resistant varieties (Figures 9.2 and 9.3). (Herbicide resistance is a valuable property in crop plants, allowing sensitive weeds to be controlled by herbicide application without the crop being affected; it becomes economically justified when the weed cover exceeds about 10% of the crop area.) Cells in tissue culture are exposed to the herbicide or the disease agent and the survivors grown into complete plants, a proportion of which will show the resistance. This has been successful for obtaining herbicide-resistant maize.

A summary of these genetic procedures is shown in Figure 9.4.

Commercial Prospects for Plant Genetics

For thousands of years, farmers have been employing methods of plant breeding to develop hardier, more productive strains of crop plants which may also be easier to cultivate and harvest. Modern agriculture uses plants improved by centuries or even millennia of selective breeding and which are vastly more productive than the wild species from which they originated. But until now plant breeding has been largely an empirical matter seeking to promote the desirable traits of the breeders' plants (or animals) and minimize or eliminate the undesirable ones. Although the genetic manipulation of plants is complex and still in its early stage of development, real prospects are now in sight of defining sought-after new developments in genetic terms and achieving them with a precision undreamed of by earlier generations. One can now ask farmers what improvements in plants they would like to have and then set about

Figure 9.2 Agracetus scientists inspect the progress of young tobacco plants during the first outdoor test of a genetically engineered plant in 1986 near Madison, Wisconsin. The plants were altered so as to be resistant the crown gall disease, a bacterial infection. (*Photo Courtesy of Agracetus*)

What we eat: agriculture and food

actually doing something to meet their requirements. (Do note, however, that breeding new plants takes time, not just for the laboratory manipulations but even more for testing under growing conditions. Several seasons may be needed to evaluate yields and disease resistance in the field, all of which adds to the cost of new varieties.)

Improvements in crop qualities

Genetic engineering and other modern breeding methods can potentially be used to upgrade the quality and/or yield of the crop, extend its geographical range and improve its tolerance of drought and flooding. The future might also require adaptation to increased atmospheric carbon dioxide from the continued burning of fossil fuels, and high temperatures which some believe will occur in the process of global warming, as well as to an enhanced level of ultraviolet radiation if the ozone layer is progressively destroyed.

Figure 9.3 Glyphosate-resistant tobacco plants. The three on the left have not been transformed while those on the right have been genetically engineered to be resistant to the broad spectrum, non-selective herbicide. The front two plants have been heavily sprayed with glyphosate; those in the middle row were sprayed moderately, while those in the back were not sprayed at all. (*Photo Courtesy of Calgene, Inc.*)

Exploiting biotechnology

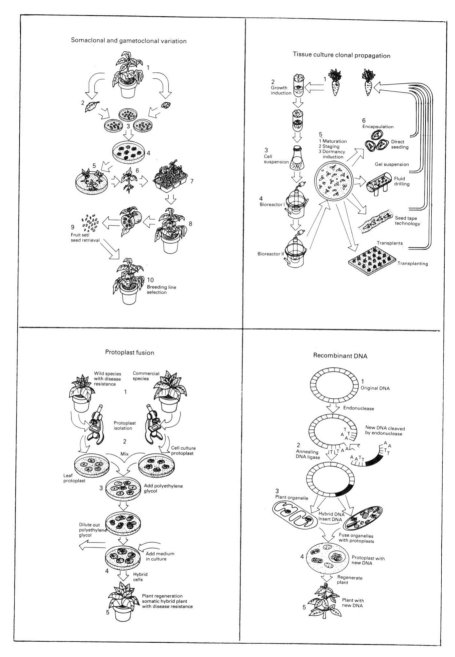

Figure 9.4 Molecular biological techniques as used in plant breeding. (*Reproduced by permission of R Jeffcoat*)

What we eat: agriculture and food

From an economic point of view, oil quality, total protein content and the proportions of particular amino acids in plant protein are all important. For instance, both soybean and maize are major global food sources but each is short of two of the amino acids essential for human and animal health; if they could be engineered to produce more of the deficient components their food values would clearly increase. Conventional breeding has so far failed to do this because plants which make more of the amino acids in question have lower yields; unless the amino acid-rich varieties offered price margins per ton high enough to offset the lower yield, farmers will not grow them. New attempts are even now being made, using genetically modified plants to increase yields, but it is too early to be sure of the outcome.

Since plants are such important sources of industrial raw materials, improvements in the products themselves, or in their processing, can have significant commercial implications. Such improvements are already leading to de-icing chemicals and biodegradable plastics from maize. Soybean is the source of new types of printing inks giving better colour and sharpness, less tendency to be rubbed off yet — because of their biodegradability — are easier to clean up and dispose of with minimal environmental damage.

Resistance to herbicides

Looking for resistance in tissue cultures with or without untargeted, random mutagenesis is not the only way of obtaining resistant plants. In some cases genes which encode resistance to particular herbicides can be introduced into the plant genome: one such gene, isolated from a bacterium able to grow in the presence of, and hence resistant to, the herbicide *bialaphos*, has been successfully transferred to alfalfa, oilseed rape, sugar beet, tobacco and tomato. An alternative approach is to identify the target molecule in the plant which is sensitive to the herbicide and make it resistant. This is being done with a chemical called *glyphosate* which interferes with a plant enzyme participating in the synthesis of a number of amino acids essential for protein synthesis. The strategy is to mutate the plant enzyme to a form resistant to glyphosate.

The downside of increasing resistance may include poorer yields per plant and an encouragement to farmers to use more herbicide. That in turn tends over time to lead to resistance spreading by cross-pollination to weeds related to the crop plant.

Defence against insect predators

While certain insects are, of course, essential for the pollination of some species, plants in general are prey to attack from insects, nematodes (roundworms, threadworms), fungi, viruses and bacteria. Many chemical insecticides, fungicides, nematocides and so on are known but there is increasing concern about the effects they might have on non-target organisms, including humans and animals who subsequently eat the plants. Can the consumer be sure that the chemical is harmless to him, or that all traces have been eliminated? In some quarters there is a greater willingness to accept the use of natural rather than artificial pesticides because they are often felt to be safer. However, this is a serious misconception because many natural substances (including a number of fungal toxins sometimes found in grain and other plant products stored under poor conditions) are very poisonous indeed for man and animals.

As part of their strategy for invading plants, insects and fungi produce proteinase enzymes which decompose plant proteins and assist penetration of the tissues. Different strains and species of plants show varying degrees of susceptibility to attack: many possess their own natural forms of resistance, often in the form of proteinase inhibitors, chemicals which antagonize the invading proteinases and halt their action. Modern agriculture, however, depends heavily on *monoculture* in which tracts of land, sometimes very large indeed, are dedicated to the cultivation of single crop strains. When a crop consists of essentially identical plants, disease spreads very rapidly once it takes hold because all the plants are equally susceptible. Enhancing resistance to disease might best consist of finding ways to boost the plants' own defences or, if they have none, to import a defence mechanism by transferring the appropriate genes from a resistant variety.

One way of doing so is illustrated by a toxin poisonous to butterflies and moths. It is made by a bacterium called *Bacillus thuringiensis* which infects these insects and causes paralysis or death via a general septicaemia. The simplest way of using the toxin is to produce large quantities in fermenters and spray it onto the leaves of plants, just like any chemical pesticide; ingestion by foraging caterpillars causes death from toxin poisoning. Related strains of the bacterium produce toxins affecting mosquitoes and blackflies while a number of other insect toxins are made by different species of microbes. In some cases, the toxin is most effectively administered within the live bacteria or their spores. From the farmers' point of view the best method would be one which entirely eliminated the pest while leaving some of the bacteria permanently present in the environment ready to infect again should the pests reappear. But that would not have too much appeal for the toxin suppliers — they would prefer that the pests were brought under control (but not wiped out) and that the bacterium then died out. Recurrence of the pest would require payment for a further application of the bacteria or toxin. Indeed, if repeated applications were not implicit, there might be reluctance to develop the toxin in the first place because it would not be seen as a worthwhile commercial investment. This is just one of many examples in biotechnology which demonstrate the crucial interplay of technical and commercial considerations.

An extension of this toxin approach to insect control is to transfer the toxin gene directly into the genome of the plant which would then be able to make the poison for itself without any further need for the bacteria. That has already been done for tomato and tobacco plants, conferring resistance to a number of insect pests. In time, more plants could be made resistant to commercially important insect pests like the Colorado beetle, boll weevil, and Asian rice gall midge, while alteration of the toxin gene itself might enable variant toxins to be produced with an extended range of insect targets.

Another way of tackling the insect problem is to interfere with their mating activities so that the species becomes locally extinct. This was done very successfully with the gipsy moth: large numbers of male insects sterilized in the laboratory were released into the wild where they competed with normal males for the females and, being sterile, greatly reduced the size of the next generation.

What we eat: agriculture and food

It is important to appreciate that insect control does not consist simply of killing susceptible insects. The use of a single pesticide control measure puts very great selectional pressure on the target insect population, inevitably giving rise to pesticide-resistant individuals which are then even more troublesome than their sensitive precursors. Managements systems are necessary to combine several methods, perhaps an insecticide plus insect-resistant plants.

Defence against viral, fungal and bacterial diseases

For many years cross-protection has been used mitigate the effects of viral infections in plants. This makes use of the fact that the presence of a virus may reduce or prevent symptoms appearing in a subsequent infection by the same or another virus. The problem lies in the inoculation of the protecting virus by spraying or other means, something not always acceptable to the public. An alternative is genetically to introduce the DNA coding for certain of the viral proteins into the plants, which then make the viral product as part of their own protein synthesis. By mechanisms not yet fully understood, the presence of viral protein inside the plant cells either entirely prevents infection by infectious virus particles or reduces their impact.

Plants defend themselves against bacterial and fungal invasion passively by structural barriers which cannot be crossed and actively by specific responses to the presence of a pathogen. Some of the barriers already contain antimicrobial chemicals to inhibit invasive action. Active response usually involves changes to the inside of the cell wall which prevent penetration. These changes may include the synthesis of proteins called *phytoalexins* which are made in response to wounding and help to antagonize bacterial and fungal action. Different plants show a variety of resistances to various potential pathogens; understanding in molecular terms the way these defence systems function may one day offer opportunities for transferring resistance genes to sensitive plants.

New flowers and shrubs

Plants for the decorative and ornamental markets will benefit from advances in disease protection just like crops in the field. Some cut flowers and pot plants occupy a very interesting market niche. People like novelty and are often prepared to pay for it. So a number of tropical countries have a lively trade exporting their floral products to the industrial world while enterprising biotechnologists think of new colours, new shapes and new odours — even of luminous plants! — to be produced by genetic engineering.

Diagnosing Plant Diseases

For the farmer, recognizing diseased plants is an important factor in crop management — it will tell him whether and when to use a pesticide spray and how much will be needed. He can

Exploiting biotechnology

decide to spray anyway on the chance that the crop is already or will become infected, use one of the computerized forecasting techniques to try to decide the likelihood of disease or even actually test plants in the field to find out; only the last choice gives real local information.

Antibodies are said now to be available for hundreds of plant diseases; many kits are on the market for carrying out reliable immunoassays with a minimum of fuss and bother. Crops and flowers can be kept disease-free (and hence worth more) if affected plants are recognized and destroyed at an early stage and the proper remedies applied; golf courses and gardens will always be nice and green and free from brown patches, just the way people like to see them.

Chemicals From Plants

Plants are the sources of many important chemicals. They include:
- pharmaceuticals — cardiatonic agents (like digoxin), antimalarial alkaloids (quinine), analgesic alkaloids (codeine, morphine), anti-inflammatories (colchicine), muscle relaxants (tubocurarine), antifertility drugs (diosgenin), antileukaemics (vincrystine and vinblastine), and lots of others;
- food additives — quinine, vanillin and many more flavouring agents;
- cosmetic and perfume components — jasmin, rose oil, and other plant oils each comprising many chemical components;
- chemicals for agriculture — nicotine, anabastine, pyrethrins and rotenone are insecticides. Some of the pyrethrins have low toxicity to mammals, short persistence in the environment and good powers of immobilizing and killing flying insects. They are thus suitable insecticides for maintaining public health, spraying areas where people are present;
- pigments — an example is indigo without which blue jeans would not have been blue!

Making plant chemicals in tissue culture

These chemicals usually have complex structures which render their chemical synthesis difficult and hence expensive. Yet they may also be comparatively difficult and expensive to isolate and purify from their original plant sources — concentrations may be low and purification protocols lengthy. Were it not for problems of maintaining high rates of productivity, making them in plant tissue culture could offer several advantages which might include greater control than for field crops, more consistent yield and quality, and independence from pests, climate and seasons. However, tissue culture is technically complicated: it depends on extensive research to provide a thorough understanding of the way the system works and how to get the best response, followed in the production phase by continuous monitoring and control. Nevertheless, recent advances have pointed the way to the successful exploitation of tissue culture systems. Already several alkaloids and other pharmaceuticals have been made on an experimental basis and their high market value coupled with reliability of supply may justify the costs of producing them by this route rather by extraction from plants.

What we eat: agriculture and food

One clearly successful example of tissue culture manufacture is the production in Japan of *shikonin*, a red plant pigment used both as a dye and as an anti-inflammatory for haemorrhoids. About 0.15 tonnes was formerly extracted annually from roots and sold for £3.3 million ($5 million) per tonne; in recent years half the total product has been made in a high-yielding root culture at about 80% of the earlier cost of extraction.

Making foreign proteins in plants

Chapter 2 explained the basic technology for transferring genes from one organism to another while Chapter 8 noted the production of human hormones and anti-cancer and anti-viral agents in genetically engineered yeasts and bacteria. Whether microbial cells (or even plant cells in tissue culture) will turn out to be the most cost-effective systems for making foreign proteins remains to be seen: genetic problems associated with the differences between prokaryotic and eukaryotic organization, difficulties with culturing procedures, complex equipment and buildings to accommodate it, the provision of feedstocks and processing, and the maintenance of sterility all add to costs.

Instead of engineering the genes for foreign proteins into bacteria or yeast, it might make more sense to put them into plants. Modified plants might make taste and smell chemicals, pharmaceuticals, pigments and a host of others. Studies are now in hand to express pharmaceuticals in tobacco and insulin in alfalfa. Imagine fields of crop plants as sources of insulin or of interleukins, gathered in with combine harvesters. How might their pros and cons relate to factory production using microbes or plant tissue culture?

Nitrogen Fixation and the Use of Fertilizers

The problem

Like all living organisms, plants need a supply of the element nitrogen. Available in some places as nitrate minerals, the obvious major source of supply is the atmosphere, nearly 80% of which is gaseous nitrogen. The problem is that, in common with most organisms, plants cannot directly capture nitrogen from the air but have to take it up from the soil in a chemically combined form via their roots. Any particular plot of soil contains only a limited amount of combined nitrogen which becomes progressively depleted by successive generations of plants. Some of it returns when the plants die in place but, if they are cropped, the soil will obviously become more and more deficient and at some point further plant growth ought to cease. In most places, however, plant cover continues indefinitely — where does the nitrogen come from?

It turns out that certain bacteria have the ability to convert nitrogen from the air into soil chemicals in the process of *nitrogen fixation*; not surprisingly, the microbes concerned are called *nitrogen-fixing bacteria*. They grow in nodules on the roots of leguminous plants like clover, alfalfa (lucerne grass), lentils, soya, beans and peas so those plants have no problems with their supply of nitrogen; the bacteria in the root nodules fix enough nitrogen both for

Exploiting biotechnology

their own needs and for those of the plants they have colonized. It is a good arrangement: the bacteria benefit from a cosy protected environment and the plants get the nitrogen they need. When the plants eventually die and decay, their nitrogen is returned to the soil for others to use.

This relationship between legumes and nitrogen-fixing bacteria is the basis of crop rotation invented in the early eighteenth century by Charles ("Turnip") Townsend. He suggested that the fertility of the soil could be maintained if every four years or so a field is not cropped but sown with a legume (such as clover) which is later ploughed in. Townsend's idea may be derived from the sabbatical year of Mosaic law in which the land was to be untilled every seventh year (Leviticus xxv, 1–5). We now know that the rejuvenation of fertility in the fallow year is the result of bacterial action.

The need for artificial fertilizers

Crop rotation works well but it does mean a loss of revenue by taking the land out of production every four years or so and the amount of nitrogen fixed in the fallow year is unlikely to be sufficient to support intensive agriculture during the intervening period. The application of nitrogen fertilizers is essential to maintain yields and keep prices down; it is a salutary thought that about one-third of the world's population depends on chemical fertilizer for adequate food production.

Many crops are therefore routinely supplied with nitrogen fertilizer made on an industrial basis by the *Haber process*: three parts of hydrogen gas and one part of nitrogen gas under very high pressure and temperature give two parts of ammonia which forms the basis of the fertilizer. The quantities manufactured are by no means trivial: a recent source suggested an annual world total of 40 million tons from industrial production compared with 175 million tons of nitrogen fixed biologically and 40 million tons obtained from nitrate mining. A Dutch study showed that, in intensive agriculture, about a quarter of the energy input into the consumable crop material comes from fossil fuel to make fertilizer, and that nitrogen fertilizer production accounts for 2-3% of the world's energy expenditure. Estimates in the 1980s gave $8 billion as the annual cost, with more than $1 billion for nitrogen for maize alone — those costs will be much higher now. These are important social and economic issues; the advocates of "organic farming" notwithstanding, many would argue that, without intensive agriculture and the extensive use of chemical fertilizers, the world could not support its present human population. Yet the consumption of fossil energy (with its attendant pollution) and the ecological consequences of treating vast tracts of land with nitrogen fertilizer are matters of acute public concern.

Could biological nitrogen fixation cope?

The advent of genetic engineering has prompted investigations into the possibility of extending the range of biological nitrogen fixation to plants which do not naturally harbour the bacteria.

What we eat: agriculture and food

In principle there are two ways in which it might be done (remember that the ability to fix nitrogen resides genetically with the bacteria, not the plants):

1. Modify nitrogen-fixing bacteria so that they colonize the roots of plants other than legumes. Nodulation appears to be a characteristic specific for a particular plant species and depends on the right genetic information being carried by the infecting bacterium. Perhaps these specificities could be modified, or new bacterial strains found in nature. There has been some success in transferring genetic information between bacteria but not so far into *Agrobacterium*, the main organism which transfers genetic material into plants.

2. Transfer the genetic information directly into plants, by-passing the bacteria-plant interaction. There are many difficulties:

   - overcoming the differences in genetic organization between plants and bacteria (see Chapter 2);
   - one of the critical enzymes involved in nitrogen fixation is extremely sensitive to the presence of oxygen. Any engineered system would have to provide a suitable low-oxygen environment as already obtains in nitrogen-fixing root nodules;
   - that same enzyme actually functions rather slowly so large amounts (several percent of the bacterial cell protein) are needed. That might not suit the plant;
   - as the Haber process illustrates, fixing atmospheric nitrogen is very energy-demanding. (Fundamental laws of chemistry tell us that the amount of energy needed to promote a particular chemical change — such as converting hydrogen plus nitrogen into ammonia — is independent of how it is done and what catalysts are employed. Industry may use high pressure and high temperature, neither available to bacteria but, for every unit of product actually made, each route requires the same energy input.) The recipient plants may not be energetically capable of supporting their own nitrogen-fixing requirements — they might become independent of fertilizer at low growth rates but still require it for rapid growth and economic yields.

Although research has been going on for decades, very much more still needs to be done before endogenous nitrogen-fixation becomes feasible for major crop plants like wheat or maize. There will be problems of funding: is such R & D to be categorized as "curiosity" and supported from the public purse, or is it to be commercially driven and paid for by industry and the agribusiness? Who actually wants it — the farmers, the seed producers or the existing suppliers of fertilizers? Who would be prepared to pay? Bear in mind that nitrogen may not be the only factor limiting yields: solving the nitrogen-fixing problem will do nothing to satisfy requirements for potassium, phosphorus and other elements. If the farmer has to go to the trouble and expense of spreading potassium and phosphorus fertilizer, how much more will applying nitrogen fertilizer cost in the same operation compared with the likely price of seed for genetically-engineered nitrogen-fixing crops?

Exploiting biotechnology

Other alternatives?

Some people have proposed at least a partial way out by increasing the proportion of legumes in the human diet but their digestibility by people is limited; some contain toxic materials and their amino acid composition is often poor for human needs. The production of legumes has indeed been increasing but it remains far below that of cereals: at the beginning of the 1980s, the US (the major soya-producing country) produced 20 times more cereals than soybean. Furthermore, while soybean production is in theory independent of nitrogen fertilizer, in practice some nitrogen is needed for good yields and high returns from agricultural land.

A traditional form of biological nitrogen fixation in rice-growing areas of the Far East uses nitrogen-fixing blue-green photosynthetic bacteria grown between the rows of rice plants in the flooded paddy fields. The bacteria grow and die in a few weeks, releasing their nitrogen to the developing rice plants; some 40-75% of the nitrogen is provided in this way, a most important economic factor for poor farmers.

Algal Culture

Most algae are microscopic green plants which can be grown in culture just like bacteria and other microbes. Some are very efficient at converting the light captured in photosynthesis into chemicals and have been perceived as possible sources of food, animal feed, pigments, chemicals and fuels. Because of their dependence on light, it has long been technically difficult to grow them in photobioreactors under factory conditions but recent designs now permit cost-effective culture, particularly in sunlight rather than artificial illumination. Even in notoriously dull regions like Britain, algal cultivation in photobioreactors is much more effective than in outdoor ponds because the systems are closed and conditions can readily be controlled to optimize growth.

Growing algae outdoors in shallow open channels is an alternative method and obviously more of an option in climates with lots of reliable sunshine to provide light and warmth: California, Arizona and Israel are regions that spring to mind. With low capital and running costs, the system need not be as efficient as factory production, and proposals speak of channels several metres wide and 20-30 cm deep. Harvesting and drying are relatively expensive (or time-consuming if the harvested algae are sun-dried), and contamination of the open, unprotected cultures by other organisms can reduce yields or increase processing costs. Nevertheless there are certain advantages over conventional agriculture; estimates have indicated very much smaller inputs of land, water, energy, labour and capital per unit weight of useful organic matter produced. Algae can be used to treat raw sewage to render it suitable for agricultural irrigation. The algal mass may be suitable as fish food, while certain types are of interest because of their content of specific chemicals including pigments, vitamins and polyunsaturated fatty acids.

Soil Inoculants

Microorganisms have many important functions in agriculture:
- the earlier account of nitrogen fixation pointed to the role of the bacteria which colonize the roots of legumes;
- winter feed for cattle is prepared from vegetable matter by a fermentative preservation process called *silage* or *ensilage*. The grass, sorghum, maize, etc. is stored in plastic bags, concrete drums or metal silos for the fermentation to be carried out mainly by lactic acid bacteria naturally present in the plant material;
- the roots of many plants and trees are associated in the soil with *mycorrhizas*, fungi better than roots at absorbing certain nutrients from the soil which are then passed on to the plants. Mycorrhizas are believed also to protect tree roots from unfavourable soil conditions such as acidity, drought and high temperature; the fungus, not itself capable of photosynthesis, presumably derives benefit from photosynthetic products released from the plant roots.

In all these cases the processes can often be helped by inoculating the plants or silage with the requisite microorganism. Major commercial activities have grown up to supply the inoculants; the seeds of legumes, for instance, are commonly dipped in a liquid containing nitrogen-fixing bacteria before they are sown (Figure 9.5). Silage inoculants ensure that proper fermentation takes place without having to rely solely on the bacteria already present in the feedstock. Some types of mycorrhizal inocula are also now produced on a large scale for use in reafforestation and land reclamation projects.

Animal Healthcare

Farm livestock are all vertebrates (animals with backbones) and all have the same type of immune systems as humans. The details vary, of course, but in principle everything relevant to human healthcare described in Chapter 7 holds equally for farm animals, and indeed for vertebrate pets as well. Improved diagnostic methods will permit early recognition of disease while more advanced prophylaxis and therapy offer better prospects of control.

The rules for testing, manufacturing and using animal drugs may be a little less stringent than for their human counterparts, the technical procedures not quite the same, and clearly there are ethical differences and the lengths to which one might go to save life, but the biology is virtually identical. However, the economics are very different. People of various political persuasions argue heatedly about what, if anything, should limit the availability of human healthcare. Cost inevitably enters into consideration (no person nor any government has unlimited resources for medical expenditure) but there is a widespread sense that, as far as possible, whatever help is needed should readily be made available to prevent and cure human disease. The morality of disease management is necessarily different in agriculture (though not, perhaps, with domestic pets): farm animals are kept for profit, not for sentiment, and the benefits of maintaining good health has to be set against their cost.

Exploiting biotechnology

Figure 9.5 The first Midwest application of genetically engineered microorganisms was made by Biotechnica Agriculture, Inc. at the Chippewa Agricultural Station, near Arkansaw, Wisconsin. The company was testing genetically engineered microbes from increased nitrogen fixation in alfalfa.

Animal vaccine development is accordingly directed mostly towards improved safety and efficacy in preventing economically important diseases. Ideally, vaccines should have a long shelf life and be effective with a single administration. Labour-saving delivery systems are very important because, unless there is an impending threat of a particular disease, farmers may not be willing to spend the time and effort to vaccinate a large herd or flock. Some of the new delivery possibilities are:

- vaccination of chick embryos while still in their shells;
- intra-nasal administration;
- use of aerosols for poultry and pig vaccinations;
- ballistic systems in which a biodegradable bullet coated with freeze-dried vaccine is fired at the animal from an air gun. Vaccination of animals in the wild can be done like this;
- it may be possible to protect animals from insect pests by spraying them with a biopesticide analogous to *Bacillus thuringiensis* toxin used for protecting plants.

What we eat: agriculture and food

Important veterinary diseases include:

- viral
    - diarrhoea in pigs;
    - foot-and-mouth disease. In economic terms, the world market for anti-foot-and-mouth vaccine is said to be larger than for any other disease, human or animal;
    - rabies, a particularly nasty disease when transmitted to people. The most common wild carrier in Europe is the fox and a novel oral antirabies vaccine contained in chicken heads has been distributed by helicopter in high-density fox habitats;
    - fowl plague;
    - bovine rhinotracheitis;
- bacterial
    - scours (*colibacillosis*), which causes diarrhoea, dehydration and eventually death among young pigs and calves;
    - *pasteurellosis* (a respiratory infection in cattle);
    - swine dysentery;
- other infections
    - anaplasmosis giving rise to anaemia and death;
    - tropical diseases such as *trypanosomiasis* (nagana) and rinderpest. The vaccine currently in use for rinderpest needs refrigerated storage, not always available in infected areas. A more robust version and a new type of vaccine are now under test;
    - *coccidiosis* in poultry;
    - *chlamydial abortion* in sheep.

Control and treatment of many animal infections are reminiscent of those for humans but, because constructing transgenic animals is more acceptable than making transgenic people, one could also conceive of protecting agricultural livestock from disease by genetic manipulation of the animals themselves. For example, scientists in Australia have come up with a most intriguing idea for dealing with insect larvae pests on the skin. The hard exoskeleton (or cuticle) of insects is made mostly of a chemical called *chitin*. Some organisms contain an enzyme (*chitinase*) able to break down chitin and the notion is to transfer the chitinase gene to sheep where it would be expressed in the glands in the skin which make lanolin for lubrication. The chitinase would thereby be spread over the skin and attack the insect cuticles.

Diagnostic tests based on monoclonal antibodies, potentially as important for animals as they are for people, are early targets or already available for such animal diseases as blue tongue, bovine leucosis virus, canine heartworm and parvovirus, equine herpes and infectious anaemia, feline infectious peritonitis and leukaemia, furunculosis of salmon, toxoplasmosis, tuberculosis and others.

Diagnosis can reveal useful information other than the presence of disease. Gene probes can be used to test for sex-linked characteristics and hence for the sex of an individual. Since mammalian males have the XY chromosome pattern while females have XX, the presence of

Exploiting biotechnology

one or more traits associated with the Y chromosome as revealed by gene probing would prove the individual to be a male. Sex control is enormously important for breeders, for instance for replacing female calves with males; the market value of steers is higher than that of heifers because they are heavier at weaning and gain weight more rapidly.

The Use of Bovine Somatotrophin (BST) to Improve Milk Yields

This protein, also called *bovine growth hormone*, is a naturally-occurring hormone which increases milk yields if additional doses are given to lactating cows. While too expensive for commercial use when originally discovered, the advent of genetic engineering offered production at much lower cost in microorganisms. Administered to mature lactating dairy cows, it results in a 10–25% increase in milk yield for an increased food intake of about 15%. The hormone, normally present in small amounts in milk, appears to have no effect on humans and the sale of milk and meat from experimentally treated cows has been approved for human consumption.

A good deal of debate has centred on the use of BST. There has been some concern about the health of the animals: in pigs made transgenic for somatotrophin (perhaps not quite the same thing as administering the hormone via a syringe) there have been reports of skeletal and joint problems. Then there are economic and social arguments: that BST will favour large, sophisticated farmers at the expense of small ones, that farmers will produce the same amount of milk but with smaller herds and, in the view of milk lakes and butter mountains, who wants more milk production anyway?

Transgenic Animals

Improvement of animal stocks by selective breeding from parents with the most desirable characteristics is a long, slow business. With large farm animals, every step in the process requires several generations of breeding, each with a long gestation period; it must take many years before a marked improvement is obtained. Genetic manipulation offers a revolutionary prospect — an opportunity both to widen the range of alterations and enormously to accelerate the whole process. The major technical problem is to know exactly what to do: animals are very complicated creatures so deciding how to alter them in some specific manner is not at all easy.

In principle, the objectives for constructing transgenic animals fall into two major categories:
- making foreign proteins. Somebody has described doing this with animals as *biopharming*; the name could be applied just as well to the production of foreign proteins in plants;
- improving the conventional performance of the animals themselves.

Genetic techniques

Modifying animals for making foreign proteins has to take account of how that protein is to be obtained. Just as in the bacterial case, in which there is a processing advantage if the cells can be genetically instructed to export the protein to the medium (Chapter 2), so with animals, a readily accessible fluid or tissue is the best source of the protein. The obvious options are blood, milk, saliva, urine and hens' eggs. In the laboratory, therefore, many researchers have sought to integrate the foreign protein gene with one already producing a milk protein. They hope that when the recombined genetic material is introduced to the animal, the foreign protein, too, will be secreted in the milk from which it may be purified relatively easily. This is already known to work in mice, convenient, small experimental animals useful for working out basic methods. Secretion into milk is nevertheless not entirely without its problems both because it does require the secreting females to be reproductively mature and diseases of the mammary glands are not uncommon.

There are various ways for getting the recombinant DNA into the animal. Viruses can be used to carry foreign genes into animal cells. In *pronuclear injection* use is made of the fact that when a sperm fertilizes an egg, some time elapses before the two haploid nuclei fuse into the diploid; during that period the sperm nucleus is present as a male pronucleus. The eggs are fertilized in the laboratory, the new DNA injected directly into the pronucleus and the genetically modified zygote grown into a complete animal. (Attempts to incorporate genetic material directly into sperm before fertilization have proven difficult, although recent experiments suggest that just immersing them in a solution of DNA may work.) In some systems it may be possible to replace the nucleus of the egg with one from an embryonic cell which has earlier received novel DNA. Another interesting possibility, especially when expression is to be female-linked (e.g. in milk), is to insert the DNA into mitochondria since they (and their genome) are inherited solely from the mother. However, these micromanipulations and microinjections are experimentally tricky and success rates are so low that many eggs have to be injected. Nor can they be used with birds' eggs because, by the time the eggs are laid, the embryos already have tens of thousands of cells.

Obtaining genetically improved animals is not wholly dependent on gene transfer. An alternative is *embryo cloning*: i.e. very productive individual cows and bulls are mated by artificial insemination. When the zygote has grown to a 16-cell embryo, the nuclei of the individual cells are transplanted into 16 enucleated eggs from other cows. The constructs are cultured *in vitro* until large enough to be transplanted into the uteri of surrogate mother cows and grown into complete animals, all of them identical siblings.

Making foreign proteins in animals

This use of animals is always likely to be fairly expensive. Thus, the intended products need to be sufficiently valuable to be worth producing even when enhanced supplies resulting from costly animal production greatly increases their availability. Efficiency will probably depend on the use of at least medium-sized animals: sheep, goats and pigs are preferred, while mice and rabbits are probably too small. Low rates of transgenic success, years to reproductive maturity, seasonal breeding and long pregnancies (characteristic of large animals) would make transgenic cows less practical.

Human antitrypsin and haemophilia Factors VIII and IX made in sheep (Factor IX transferred to sheep is active and heritable from one generation to the next), together with lactoferrin (for treating certain forms of bacterial infant enteritis) produced in cattle, are

Exploiting biotechnology

expected to be high on the list of desirable foreign proteins. Further down may come large volume, medium value products such as polyhaemoglobin (a blood replacer), human and bovine serum albumins, and calf foetal serum (important constituents of growth media for mammalian cell tissue culture), antithrombins from leeches (effective, unlike streptokinase, against existing and reformed clots), and tissue plasminogen activator. To be on the safe side, animals used for the production of pharmaceuticals should never be used as human food. This would prevent inadvertent access to the pharmacological products themselves and avoid possible distaste at the idea of consuming human genes and gene products even when made within and as part of another animal.

Improving animal quality

Some of the desirable improvements in animals are:

- better, more tender but leaner meat;
- larger muscles: more meat;
- more milk, and of higher quality;
- better resistance to disease;
- more wool and/or of higher quality;
- faster race horses.

These are pretty difficult targets; not enough is yet known of animal physiology, biochemistry and genetics to be able to predict just what to do to achieve them. But there are a few good leads and some work has already been done. An obvious example is growth hormone; the gene has been transferred to mice and the mice do actually grow larger. Growth hormone gene transfers to larger animals are less successful — although the growth hormone gene is expressed, there has so far been no growth response.

Genetics appears to have no bounds. In recent years scientists have reported on a gene which may allow animals to grow indefinitely without showing the normal signs of ageing. The phenomenon appears to be based on a substance (appropriately called *longevin*) released by certain bacteria in the gut of some animals interacting with a host protein named *tithonin* (after the mythical hero Tithonus upon whom the gods are said to have bestowed eternal life without eternal youth). The discovery could lead to important benefits to human sufferers of premature ageing syndromes, while farm animals manipulated to contain the appropriate genes might never die from old age but always from an accident or in an abattoir. Not everybody seems to welcome the possibility.

All this activity at the sharp end of science notwithstanding, conventional breeding, too, continues to have its successes. Pollution drove native salmon out of the River Thames in the middle of the last century but the river has now been cleaned up and a new salmon strain bred. Juveniles from other rivers were released into the Thames, the adults later returning to breed after several years out at sea. Traps were set at one of the weirs and the best performers selected for breeding. Things seem to be going well: a 17-pound fish has been hooked, the largest in the river for 150 years.

What we eat: agriculture and food

FOOD

In all its ramifications the food industry is so huge that, if a description is not to fill an encyclopaedia, it must either be very selective or read like a catalogue. This account does not aim to be exhaustive and it is certainly not the authors' intention simply to produce a listing. It will therefore try to combine enough breadth with sufficient depth to avoid drowning the reader in information yet give him a sense of what is going on, particularly where there is movement away from traditional practice. The perspective will be mainly technical, but with commercial indications where appropriate.

Beer, Cheese and Bread — Ancient Biotechnological Traditions

Taking a long view, one could say that, through fermented foods and beverages, biotechnology has been a human activity since before the dawn of recorded history. The production of yoghurt, cheese and other sour milk products, and of soy sauce, tempeh, pickled cucumbers and sauerkraut, the leavening of dough, brewing of beer and fermenting of fruit and cereals to make wines and other alcoholic drinks are all microbial activities of exactly the sort we now associate with biotechnology. At the time, of course, people had not the faintest idea of how (chemically) these processes worked — they just knew that they did, not altogether reliably but often enough to be very desirable and useful. Only in the last 120 years or so have scientists and technologists come to understand what is going on.

Now that the underlying science is well understood, the traditional technologies intended to give very reliable and reproducible products appear sometimes to be remarkably elaborate. As a starter culture for a fermentation, the manufacturer's own strain of the microorganism will be added to the fermentation mixture in sufficient quantities to out-compete the feedstocks' natural microflora which it is usually not practical to eliminate completely by sterilization. Product flavours are chemically rarely simple and the natural "contaminants" may very well be essential contributors to the final flavour pattern via their own fermentation products.

All foods, of course, have to be wholesome and meet safety requirements; even though there is unavoidable variation in the raw materials, the finished products must also satisfy the customers who expect a specific product always to have the same taste and texture. Decades, if not centuries, of experience have gone into achieving those aims so tradition is very well established, with mystique often an essential element. While improved efficiency is always welcome, producers are careful about changing their methods for fear of altering their products and putting off the consumers; many of the products are sold into highly competitive markets and if word began to get round that somebody's brand of beer sometimes had an odd taste, drinkers would soon shift their allegiances. Genetic engineering of the yeast strains used for manufacture would thus be undertaken only with very great caution.

In practice, modern methods have probably done most for the constancy of starter cultures and the design of equipment. In the old days each new fermentation batch was started by using some of the last one as the inoculant. Nowadays this will not do because of the risk of contamination with spoilage organisms or pathogens. Pure strains must be isolated, maintained unaltered and provided to the production staff in the form of starter cultures which will ensure

Exploiting biotechnology

a constant product and avoid variation between batches. Strain maintenance, though labour intensive and therefore expensive, is critical to all parts the modern fermentation industry: food, beverages, antibiotics, industrial products and so on.

Equipment design has focused on moving away from batch towards continuous culture — the downtime for cleaning, sterilizing and recharging the vessel at the end of each batch run is thereby eliminated with a resultant increase in overall efficiency. But the traditional experience was all based on batch culture, so a good deal of technical development was necessary to convert to the continuous mode and success has not yet always been forthcoming. Nevertheless, continuous operation can offer much greater efficiency, so manufacturers have a strong incentive to try to nudge market preferences towards products made on a continuous basis; they may therefore decide to charge premium prices for ones made by traditional batch methods.

Genetic engineering is a possible option for altering certain microbes, as in the case of yeast intended for low alcohol beers and wines. Such organisms and their products will, of course, be used only if they are absolutely safe and customer opinion may need careful nurturing.

Food Diagnostics

Maintenance of quality is an essential requirement for marketing food in a modern industrial society. The difficulties of doing so arise not so much in making sure that the right things are present but that the wrong things are not — "wrong things" in a food context often mean microorganisms and their products. Nasty consequences may follow from microbial contamination; every so often there are reports of canned meat and fish not being properly sterilized during cooking and canning, allowing anaerobic microbial growth to continue in storage, exacerbated because canned food is not usually stored under refrigeration. One of the most serious contaminants is the strictly anaerobic bacterium *Clostridium botulinum* which grows only in the absence of oxygen (and is thus very happy inside a sealed can stuffed with meat or fish protein); it produces an extremely poisonous neurotoxin destroyed only by thorough cooking. Canned meats and fish are normally consumed without further cooking and people sometimes contract botulism and die from eating such contaminated food.

Immunoassays

Monoclonal antibodies are valuable for diagnosing food contaminants just as they are for recognizing disease (Chapter 7). The procedures are fast, easy to use, capable of being automated and extremely sensitive and specific. The materials to be recognized must, of course, be immunogenic; that requirement is met by proteins and some other macromolecules while a number of small molecules, including hormones and various plant toxins, can be made immunogenic by chemical linkage to a protein or other large compound.

The technologies are again similar to those described for healthcare in Chapter 7: radioimmune assay (RIA), enzyme-linked immunoassay (EIA) and enzyme-linked immunosorbent assay (ELISA). The targets include:

- the presence of proteins from particular sources (bacteria, blood, eggs, fungi, legumes), both as indicators of contamination and to ensure absence of specific types of protein for consumers who wish to avoid them;
- insecticide, herbicide, and plant and animal hormone residues left over from agriculture;
- antibiotics sometimes given to livestock to keep them free from infection;
- bacterial and fungal toxins (there are a number of dangerous fungal toxins [such as *aflatoxin*] which can accumulate during storage if the food is allowed to become mouldy);
- anabolic steroids.

Microbial Contamination

Traditional tests for microbial contamination in food are time-consuming and labour intensive, just as they are in healthcare: the test material has to be spread onto a suitable nutrient jelly or inoculated into special test media which are incubated for one to several days to allow microorganisms to grow. Each dish of jelly must be spread individually, and each tube separately and carefully inoculated to avoid entry by microbes always floating in the air, and there is no way of shortening the incubation time in conventional test procedures.

Just as an infected patient needs to be diagnosed rapidly so that treatment can start without delay, so fresh foods often cannot be held in storage until a battery of traditional microbiological tests is complete. The manufacturer has several options:

- risk of staling by holding the food until testing is complete;
- if contamination is normally rare and the food really does have a very short shelf life, take samples and start testing but ship the food out anyway. Should the tests a day or two later show serious contamination, there may be time to recall the food or at least warn the public and the medical profession;
- adopt novel, faster test procedures.

Rapid methods have already been outlined in Chapter 7; some of the details for food diagnosis will, of course, be different from healthcare methods but the principles are just the same.

Modifying Raw Materials and Processing

The way the raw materials are used in the kitchen to prepare meals, and in the food factory to make products for sale, has evolved from long periods of experimentation and experience. In the same way that a good cook knows how much of each ingredient to add to the dish, and knows, furthermore, when to add it, so the food manufacturer is able to choose from among a wide range of available raw materials to produce for the market he has in mind.

Genetic engineering of plants and animals to improve the quality of the raw materials, and the use of enzymes to change their chemical and physical properties, are some of the new ways of generating novel and improved products and processes. Examples of the modern

Exploiting biotechnology

methods are to be found in the development of new emulsifiers, improved crystallization of sugar and the modifications of oils and fats, and of polysaccharides, to give better or more valuable properties.

Emulsifiers

The phenomenon of emulsification allows two liquids which refuse to mix (usually oil and water) to form a continuous single-phase fluid. Oil droplets become coated with molecules of the emulsifier so that they become "water-wettable" and no longer repelled by water. Or it can be other way round: a chemical "skin" envelopes water droplets and lets them snuggle up to oil. Milk is a familiar example of an emulsion: in low-fat milk, the fat droplets are dispersed in water (giving the milk its white, opaque appearance) but high-fat milk has so much that the water cannot hold it all and a cream layer floats on the surface. However, high-fat milk can be emulsified (homogenized) by breaking the fat up into extremely fine droplets, natural emulsifiers then maintaining the single phase.

There are many uses for emulsifiers in preventing the separation of food components. One recent improvement uses an enzyme found in the pancreas to modify a natural emulsifier from egg yolk to a form which can be heat-sterilized and needs less thickening agent in low-oil products.

Sugar crystallization

Table sugar (sucrose) is a disaccharide made up of the two monosaccharide units, glucose and fructose. Much of it comes from sugar beet but beet syrup also contains a trisaccharide (i.e. a sugar molecule three units long) called *raffinose* which slows the crystallization of sucrose in the manufacturing process. An enzyme derived from a microorganism can split that third unit off and allow a higher yield of product.

Improving oils and fats

Among the natural oils and fats, cocoa butter, a major constituent of chocolate, is one of the most valuable because of its consistency and melting properties. Chocolate needs not only to have a desirable taste, it also has to melt at the right temperature: in the mouth but not in the packet. Rather a sharp melting point is needed because the mouth is at 37°C while the temperature of the environment in which people are eating the chocolate can easily exceed 30°C on a hot day; a melting temperature of 35°C is about right.

Cocoa butter is expensive compared with palm oil and others. So an enzyme process has been developed which makes certain modifications to some of the palm oil components in such a way as to change the balance of its chemical components to something much closer to cocoa butter. Because palm oil does not dissolve in water, and the enzyme will not dissolve in oil, the reaction is carried out in a special non-aqueous liquid which will hold enough water to dissolve the necessary amount of enzyme.

What we eat: agriculture and food

Improving polysaccharides

These long-chain sugar molecules are used in food for their "mouth feel" properties, to thicken and gel products (e.g. chocolate and fruit-flavoured mousses made by whipping raw materials or a pre-prepared powder in milk or cream), and to stabilize aerated foods (whipped cream, mousse again, and some chocolate products). Among the natural polysaccharides used in food are *carrageenan* and *alginic acid* from seaweeds, plant gums from guar and locust bean (carob), and *xanthan* from a bacterium.

The properties of each type of polysaccharide depend on the rigidity of its long chain, the way in which individual molecules group together to form blocks, how they interact chemically with other polymeric molecules and the changes caused by heat, salt, acidity and like factors which may occur in the food itself or during preparation or cooking.

More understanding of their structure and behaviour, plus modern enzymological methods for specific modifications of their chemical composition, may increasingly allow their qualities to be improved in specifically desirable ways.

Enzymes

The food industry already uses enzymes for:

- bleaching in baking;
- clarifying and decolorizing fruit juices and wines;
- conditioning dough;
- controlling volume in baked goods;
- curing sausages;
- eliminating undesirable bitterness, off-flavours and -odours;
- improving milk digestibility;
- liquefying vegetable purees and soups;
- macerating dehydrated vegetables and fruits;
- maintaining colour and flavour in eggs;
- making breakfast cereal, cheese, ice cream, soft centres for chocolates, syrups and whole milk concentrates;
- mashing (converting barley malt into fermentable sugars in brewing);
- reducing viscosity in fruit processing;
- removing beer hazes;
- stabilizing evaporated milk;
- tenderizing meat.

Exploiting biotechnology

Rennin

This preparation, obtained from the lining of calves' stomachs and containing the enzyme *rennet*, is used to promote clotting in the manufacture of cheese and other products of milk souring. Consumer demand for veal seems to be falling (changing tastes? compassion for young animals reared by intensive farming?) and a fall in the number of calves slaughtered has led to the gene being transferred to a microbial host for manufacturing the enzyme in a fermenter. Meanwhile, a search for other sources has shown that a related enzyme from a species of fungus similar to bread mould can replace calf rennet with no significant changes either in processing technology or taste of the product. (This development makes it possible to produce cheeses according to Jewish Orthodox law which prohibits mixing milk with meat products, including rennin from calves. Such cheeses are often labelled "vegetarian".)

Flavours and Fragrances

This is another enormous subject about which we can do little more than point to interesting new developments.

The perceptions of tastes and odours are sensitive and discriminating, highly subjective and physiologically complex. Humans can distinguish only eight different tastes (sweet, sour, salt, bitter, metallic, astringent, "hot" [as in peppers] and cold [menthol]) but large numbers of odours. Many "tastes" actually have strong odour components as becomes all too clear when one contracts a heavy cold; a blocked nose can make tastes indistinguishable. The sensitivity to odours is often much greater than to primary flavours and some important fragrance components can be detected in minute concentration. (Other animals are, of course, a lot better than we are: everybody knows about dogs, but how about male moths being able to detect the pheromones [odours in the moth world] borne on the breeze for hundreds of metres?)

The perceived sensations of taste and smell are the result of the combination of all the taste and odour signals impacting upon an individual person at any one time. Natural flavours and fragrances are well recognized and it is the skill of the perfumer, as well as of the cook and food manufacturer, to concoct new ones for the delectation of their clients and to make sure that the old familiar ones expected in a well-known product are always reliably present. As with other technologies, understanding brings with it the power to manipulate; increasingly, artificial food flavours can be produced as single chemicals later blended to give the complex effect, or made directly as multi-component flavour "blocks" which contribute more individuality than single chemicals.

One aspect of flavour and fragrance science which has come a long way in recent years is the chemical identification of the major flavour components in a range mainly of fruits and vegetables: peas and beans, cucumbers and tomatoes, oranges and lemons, apples and pears, olive oil, garlic, cloves and watercress. Some special flavours are expensive enough to make it worthwhile looking for cheaper sources by chemical, biochemical or microbiological synthesis. *Vanillin*, the flavour component of vanilla beans is one of them. The annual world market for vanillin from the beans is approaching 1,500 tons worth more than $150 million, compared with 6,500 tons of the synthetic product selling for $120 million.

It is likely that, as skills develop, some of the more expensive flavour and fragrance compounds will be synthesized in genetically modified microorganisms or by other biotechnological procedures, such as enzymic synthesis, from lower cost raw materials. Vanillin is again an example: production by plant tissue culture gives a product at one-fifth the cost of the natural vanilla extract and with better flavour than the synthetic material. A second example is *ambergris*, a natural material made by male sperm whales when they suffer from certain forms of stomach irritation: the whale eventually vomits the stuff into the sea — what a way to get a perfume component! Vomiting whales are rare in most places and, not surprisingly, ambergris is expensive so attempts are currently being made to find an enzyme route to one of the main odour compounds using a material extracted from a species of sage as the feedstock. Yet another endeavour now in hand is to find a way of making one of the natural musk odours using cheap palm oil instead of more expensive starting materials.

Thus, decisions about production methods and whether or not to develop a biotech. route are primarily economic: what are the relative production costs, how large are the perceived markets, what price can the product command and hence is it worth doing the genetics or other scientific development and later setting up a production plant?

Novel Sweeteners

While people retain their liking for sweet foods, they know nowadays that too much sugar is bad for them and they feel guilty about eating it. "Traditional" artifical sweeteners like saccharin (very cheap indeed, and 500 times sweeter than sucrose, but said to have a metallic or bitter aftertaste) and cyclamate (pretty cheap and 30 times sweeter, but accused of causing cancer in rats) having failed to capture the market for sugar substitutes, a large and lively industry has grown up to make low-calorie and non-metabolizable sweeteners.

High fructose syrup

This product has acquired major importance in the food industry. Its manufacture is based on the enzyme *glucose isomerase* which catalyses the conversion of one sugar (glucose) into another (fructose). Weight for weight, fructose is perceived as sweeter than glucose but has a lower calorie content because it is less well metabolized by humans. Nothing is added or subtracted in the conversion reaction, the atoms making glucose simply being rearranged a little to make fructose. Intramolecular rearrangements are called *isomerizations*, hence the name of this enzyme.

The glucose for the process comes from maize starch, a prolific agricultural product in the United States where high fructose syrup originated primarily for the dietetic soft drinks industry. Indeed, the development of the syrup in the US was prompted at least partly by a cessation of cane sugar imports from Cuba, coupled with the ready availability of maize starch. Maize processing was already a large industry supplying oil for cooking, steep liquor as

Exploiting biotechnology

a feedstock for the fermentation industry, with the solid residues (hulls, fibres, etc.) being sold for cattle and chicken feed. In Europe, by contrast, protection for sucrose made from domestic sugar beet, or imported from former colonies with whom special trading arrangements were in force, kept high fructose syrup at bay for a long time.

The starch is broken down by amylase enzymes to yield glucose. At this point glucose isomerase enters the picture to convert part of the glucose into fructose; most commercial sources of the enzyme are microbial and it is used in an immobilized form. The nature of the chemistry ensures that glucose is not wholly converted to fructose but a mixture results, the actual proportion of the two sugars depending on the conditions under which the process is carried out; the resultant mixture is the high fructose syrup. A typical composition after enzyme catalysis might be 42% fructose and 51% glucose, together with some polysaccharide remaining from the original plant material. On a weight basis this mixture has about the same sweetness as cane sugar but fewer usable calories. It might be further processed to about 55% fructose but such enrichment adds significantly to the production costs.

The proportion of fructose in the final mixture is determined in part by the temperature at which the isomerization is carried out: the higher the temperature the greater the proportion of fructose to glucose, but pushing the actual temperature much above the present operating levels of 60–65°C produces dark colours from the breakdown of fructose. Although the enzyme preparations in current use are stable at the operating temperatures, it might be possible to locate more heat-resistant versions which would last longer, not need replacing so often and hence might be cheaper; bacteria living in hot springs or in the neighbourhood of marine thermal vents are possible sources, while developments in protein engineering might show the way to the specific changes in the protein molecule required to confer greater heat stability. Such changes in protein structure could be encoded in a synthetic gene for insertion into a suitable microbial host by the usual methods of genetic manipulation.

High intensity sweeteners

Two other products materials have been successful in this market. *Aspartame*® is a simple chemical derivative of two amino acids joined together enzymatically to make a dipeptide. Weight for weight it is 200 × sweeter than sucrose but one of its amino acids is phenylalanine which has to be avoided by people suffering from phenylketonuria (see Chapter 7). Sachets of the material carry a suitable warning. A second example is *Sucralose*® (600 × sweeter than sugar), an artificial derivative of sucrose. Manufacture involves fermentation and enzymic conversion followed by a chemical modification using non-biological methods, a good example of how biotechnological procedures can successfully be integrated into quite different process technologies.

Amino Acids

Glutamic acid, in the form of *monosodium glutamate*, is familiar as a flavour enhancer while *aspartic acid*, a near relative, has both pharmaceutical uses and is one of the substrates for the production of Aspartame®. *Lysine*, made microbiologically with an annual production of some 40,000 tons worth around $250 million, is important as a feed supplement both for humans and animals. It is one of a category of amino acids which are essential for humans: a required component to enable people to make their own body proteins, yet one which cannot itself be made biochemically by human beings because we lack the enzymic machinery for doing so. It must therefore be consumed in the diet, no problem for people well supplied with a mixed diet but possibly deficient in the diets of those unable or unwilling to consume enough animal protein. Microbial lysine can be added as a supplement to correct the dietary deficiency. Another such amino acid is *tryptophan* but the technology for producing adequate yields from inexpensive substrates has not been developed and its production from specialist feedstocks makes it generally too expensive to be used as a feed supplement.

Single-Cell Protein

During the 1960s, with apparently unlimited cheap oil available and a growing concern for the adequacy of world food supplies, especially proteins, the fascinating concept arose of solving global food shortages by growing vast amounts of microbial biomass from oil in huge fermenters. Technology was developed using natural hydrocarbons as feedstocks to produce protein-rich microbial biomass as a food supplement; this was biotechnology on an enormous scale. Of course, baker's and brewer's yeasts have for long been produced for the food and brewing industries but they are grown on agriculturally-derived substrates and it was the desire to move towards hydrocarbon feedstocks that prompted the developments described here.

Some processes were based on crude oil or its derivatives; because of the immiscibility of oil and water, such techniques required special measures and a considerable input of energy to mix the two fluids thoroughly. An alternative water-soluble feedstock was methanol produced from methane, the major component of natural gas (see Chapter 4). Technically the development was a major success; economically almost all the initiatives have failed. As it turned out, its significance was probably greater in terms of engineering developments in the fermentation industry than it was as a source of food.

A number of production facilities were built in western countries, the Soviet Union and elsewhere. Perhaps the most novel and adventurous was the one in the north of England, built by ICI, to make a product called *Pruteen*®. This enormous fermenter, with a capacity of some 1,300 tons of culture medium, was designed to grow a bacterium rejoicing in the name of *Methylophilus methylotrophus*[*] on methanol as the carbon source, ammonia to supply nitrogen and other minerals salts as necessary. The bacterium was chosen carefully: it had to be non-pathogenic, able to grow in high concentrations to ensure efficient use of the plant, be robust enough to withstand the vigorous physical agitation within the culture fluid, show a high conversion of feedstock to protein, be readily separable from the medium and, of course, have a high nutritional value. All these demands were satisfied. The very design and construction of the plant represented major advances in fermenter technology and, when it went into production in the early 1980s, it performed pretty much as expected, with output

---

[*] From the Greek, meaning "methanol liking, methanol eating". Not a gourmet palate!

Exploiting biotechnology

rising to about 80% of the target. The product was to be sold as granules and powder to blenders of animal feed and milk replacers. Annual production (requiring more than one plant) was predicted in 1978 to reach 1 million tons by 1990. In 1981, 6,000 tons a month were being made with production runs lasting 100 days.

There can be no doubt that *Pruteen*® was a remarkable technical success; Figure 9.6 shows a view of the impressive plant in its heyday. Economically, however, neither Pruteen® nor most of the other initiatives were successful in the west and have mostly been abandoned, some before any production took place; *Provesteen*, a yeast product made by Phillips Petroleum is one of the few survivors. Why did they fail? The reasons include:

- oil prices went up much more than most people expected;
- the production of alternative and cheaper forms of supplementary protein (mainly fish meal and soya flour) expanded to meet demand;
- food shortages did not develop to the extent and in the manner anticipated. The prevailing view in the 1960s wrongly ascribed lack of food to insufficient production and natural disasters; they were more often due to poor organization, armed conflict and unworkable political and economic philosophies;
- few people predicted the "green revolution" and worldwide rise in agricultural productivity, with Europe and North America in enormous surplus and rapid increases in productivity in China, India and other countries in South-east Asia. In effect, severe, chronic food shortages were largely confined to regions torn apart by war or suffering from inept government — for them single-cell protein would have been of little help.

Only in the USSR did large-scale single-cell protein production continue, presumably because of that country's plentiful supplies of hydrocarbons and perennial shortage of foreign currency with which to purchase American soya products. However, there are some other microbial protein products now on the human food market; one called *Quorn* is based on an edible filamentous fungus grown in culture and sold as a meat replacer and extender. Capable of being spun into a meat-like texture and with little or no taste of its own, it can be flavoured to suit the dish or product in which it is to be used. As a niche market product its selling price is comparable with other high protein foods; it is perhaps more likely to appeal to vegetarians and people concerned about animal products than serve as a major alternative to meat in its traditional form.

Except in a few specialized instances, single-cell protein as an additional source of food has probably not been worth the money and effort invested in its development; rather than as contributions to human nutrition the main benefits of its development might ultimately emerge in the form of technological advances in fermenter design and biochemical engineering.

Regulatory Approvals

The regulation of human medicines (see Chapters 5 and 7) is also the model for animal drugs and the time, effort and expense is not likely to be very different. Although such approval may not be needed to begin clinical trials, it will be necessary for the use of vaccines and for edible products derived from animals receiving any drug.

What we eat: agriculture and food

Figure 9.6 The Pruteen plant at Billingham in the north of England. The fermenter itself was housed in the slender tower with the white cap seen in the photograph just below and to the left of the large circular cooling tower. (*Reproduced by permission of ICI, London, UK*)

Exploiting biotechnology

All new pesticides, whether chemical or biological, must receive regulatory approval before they can be used generally. Several years of testing may be required, first in greenhouses and then in field plots up to 10 acres (4 hectares) in area. Microbiological pesticides, even those using naturally-occurring strains, as well as chemical preparations require larger-scale field tests, while the release of genetically-engineered microbes for whatever purpose must undergo prior individual risk assessment.

The safety of proposed foods and food additives must be established before approval is granted. Transgenic plants have initially to be tested under confined conditions which prevent flowering and pollen dissemination, although these restraints will later be removed; food products obtained from them will, in turn, need their own approval.

Economic and Social Effects of Biotechnology in Agriculture

Like all new technical developments, advances in agriculture and the food industry will have many social and economic consequences, some of them becoming clear only with the passage of time.

The shapes of the likely problems are familiar enough to followers of current events. There is a limit to the amount of food which can be consumed, a limit which has more or less been reached in the industrial countries. Supplying food to deficient areas is generally regarded as feasible only on a temporary emergency basis, to tide a population over an acutely difficult period. Care must be taken not to send so much free or cheap food aid into a country that domestic agriculture is undermined because local farmers, unable compete on price, are driven off the land.

Some biotechnological initiatives in agriculture and related activities are directed to novelty or specialist products (exotic flowers and fruits) or improved quality (better meat, better storage characteristics for fruits, etc.), but most are intended to enhance efficiency. That can only mean more food or the same food with less effort. As we do not need more food it must mean less effort — less land under cultivation and fewer farmers employed, although there may be more demand for fertilizer and hence more jobs and investment in the chemicals industry. What is to happen to land taken out of production? Are fewer farms and more factories desirable? Would changes in employment from rural agricultural to urban industrial be welcome — and to whom? Yet if nitrogen fixation can successfully be conferred on crop plants, even the possibility of increased employment in the fertilizer industry might not materialize. The outcome could instead be both less agricultural employment and fewer industrial jobs.

This is neither the place to argue one way rather than another nor to seek to balance the pros and cons of technical advances. Our purpose is solely to define some of the more obvious problems; the choice of answers must be the subject of public debate.

10 Wealth from the earth: metals and hydrocarbons

INTRODUCTION

Apart from agricultural and forest products, almost all raw materials for non-food industries are extracted from the beneath the ground: oil, metals and other minerals, and coal. In every case, and from every site and deposit, extraction is incomplete either because the cost of physical recovery from the source becomes progressively more expensive as the most accessible material is removed, or because the processing costs of extracting the desired material from tailings or poor ores are too high compared with the present or anticipated market price of the product.

As high-grade ores and deposits around the world are gradually worked out, attention turns to poorer or more remote sources from which recovery is inevitably more expensive and investment costs often greater. It becomes more urgent to improve the percentage of material recovered and, of course, it is always desirable to increase the rate of extraction and secure a more rapid return on invested capital.

This is the context in which biotechnology will have a growing part to play in the coming years by lowering extraction and processing costs and so bringing lower-grade sources within the bounds of economic operation. Its value comes in the generation of low-cost chemicals which, in various ways, can form the basis of recovery procedures. The chemicals do not need to be purified; mixtures containing other substances are acceptable so long as they do not interfere with the workings of the active chemical(s). In theory it might be possible to use isolated enzymes to generate the desired compounds but in practice living microbes are probably necessary because:

- isolated enzymes are unlikely to be sufficiently stable for these purposes in the environments in which they would have to operate;
- enzymes are relatively expensive;
- it is very difficult to conceive of simple reactions, requiring no more than one or two enzymes working together, which would be capable of producing useful chemicals from cheap feedstocks. Living microbes have the ability to do just that, starting with molasses or other inexpensive agricultural products (Chapter 4).

Biotechnology for mineral extraction has a long and respectable history: unwittingly, it was used by the Romans to help extract copper and there are graphic accounts of what it was like to work in a "copper stew" in Wales early in the 19th century.

Exploiting biotechnology

BIOLEACHING OF METALS

Conventional recovery of metals from mineral ores requires first that the ores are mined and brought to the surface; heating and treatment with chemicals later extracts and purifies the metals. The remaining crushed spoil is discarded, usually as a heap or dump, the residual quantities of metal it contains being too low and too dilute to be worth the additional expense of securing a comparatively low incremental yield. Indeed, any mineral must contain an economically extractable concentration of metal to be worth mining in the first place. In copper mining, over half the energy cost and more than half of the capital investment is directed to conventional ore removal and beneficiation (separation of as much as possible of the unwanted mineral material before final purification and refining, which accounts for the rest of the cost). For conventional metal recovery to be economically viable, the value of the metal extracted must exceed these combined costs of energy and capital as well as the various other operating expenses. But if and when that relationship no longer holds, less expensive processing techniques become essential or the operation as a whole will not be worth continuing, even though potentially valuable product still awaits recovery.

Discarding the spoil in a heap is not necessarily the end of the story. Chemicals in the ore provide convenient sources of energy for certain types of ubiquitous bacteria which colonize the dumps and some of the residual metal from the crushed tailings is dissolved by acid produced in the course of bacterial metabolism. Such processes have been in use for centuries: three millennia ago, Mediterranean civilizations were recovering copper from waters which had percolated through tailings dumps. In the 17th and 18th centuries, miners in Wales and Spain used leaching to extract copper from minerals, although they were no more able than the Romans and Phoenicians to explain the processes. Bacteria were not discovered until the seventeenth century and, only within the last hundred years, have scientists begun to appreciate the detailed mechanisms of natural leaching processes.

As the industrial age has progressed, the demands on mineral resources have increased relentlessly and more and more of the known high-grade reserves have either already been exhausted or will become so in the foreseeable future. Future mining operations will thus be forced to make do with lower-grade deposits, inevitably more expensive to operate because of the amount of rock which it will be necessary to mine and process for each ton of metal recovered.

Bioleaching of Copper

In the US, for example, the average ore grade for copper has fallen from about 2% early in the 20th century to near 0.6% at the present time and parallel decline has taken place elsewhere. Yet vast quantities of metal remain available for extraction from existing waste; copper dumps in Chile containing 2–4% copper and, valued at $35 a ton, are actually better sources of copper than newly mined ores in Arizona at 0.15–0.35% copper, worth not more than $4 a ton. (The commodity price of copper metal is about $1,900 a ton.)

A slow process, the bioleaching of copper nevertheless requires little capital expenditure and operating costs are particularly low when, as in leaching from dumps, mining of ore is not

Wealth from the earth: metals and hydrocarbons

involved. The operations themselves are simple, a factor important for developing countries like Zambia, Chile, Brazil, India and China which have large copper resources but are short of capital to build the more modern extraction and processing plants. Bacterial leaching also has important environmental benefits since it eliminates the emission of sulphur dioxide from the smelters; control of sulphur dioxide release is estimated to add more than 10% to the production cost of copper produced conventionally.

How Bioleaching Works

The process depends on:

- the presence of the right bacteria;
- the existence in the rock of certain minerals, mostly compounds containing iron and sulphur;
- a source in that rock also of compounds of copper or other metals, the recovery of which is the objective of the operation.

Many of the largest commercial leaching operations use rock previously excavated from open cast mines. The source material, either crushed low-grade ore or waste from the metallurgical extraction process, is heaped on an impermeable, sloping base in such a way as to provide natural run-off for the aqueous solution which will later drain from the base of the dump; this solution is pumped to a neighbouring copper recovery plant. Some dumps are very large indeed, rising several hundred feet from their bases; the largest, US, contains more than three billion tons of material (Figure 10.1).

Figure 10.1 A large-scale copper leaching operation in the southwestern part of the United States.

Exploiting biotechnology

Leaching occurs naturally in wet climates as rain falls on the dump but in arid regions acidified water is sprayed onto the dump surface or injected into the interior through vertical pipes. As the rainwater or acid solution percolates downwards through the rock, the acid environment encourages the growth of the bacteria which are found naturally in *sulphide*-containing rock and which require the presence of acid in order to thrive. While sufficient acidity would probably eventually develop anyway, it might take a long time for the whole dump to become acid. Incorporating acid into the sprayed or injected water facilitates the bacterial attack and ensures a more uniform reaction throughout the dump.

The nutritional requirements of the bacteria are easily satisfied. They acquire the inorganic compounds they need from the rock itself and obtain their carbon from carbon dioxide, some in the air which diffuses between the rock particles and the rest dissolved in the percolating water. The rock also satisfies their need for energy which they derive from the oxidation of chemicals, mainly from a form of iron sulphate called *ferrous sulphate*, from sulphur itself and from iron pyrite ("fools' gold"). Sometimes the interior of the dump becomes hot as the result of bacterial activity and the waste heat the bacteria generate; when that happens, varieties tolerant to the high temperature become the dominant microbial population rather than the species more common at lower temperatures. The chemical products of bacterial activity include a more highly oxidized variety of the iron sulphate (now called *ferric sulphate*) together with *sulphuric acid*. These are able in turn to oxidize and dissolve various insoluble *copper sulphides* in the rock to give water-soluble *copper sulphate* which flows as a dilute solution from the bottom of the dump (Figure 10.2).

At the recovery plant the metal is extracted by passing the copper sulphate solution over scrap iron: a further chemical reaction interchanges the copper in solution with the solid iron to give iron sulphate in solution and a solid powder of copper metal which is collected and purified (Figure 10.3). For certain high-grade electrolytic applications the copper salts are

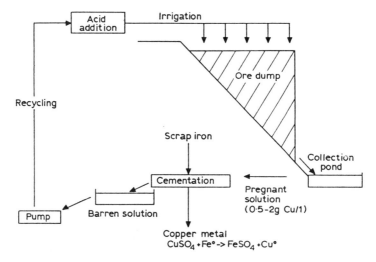

Figure 10.2 Schematic representation of a sophisticated leaching operation. (*Reproduced by permission of Prof. A E Torma*)

Wealth from the earth: metals and hydrocarbons

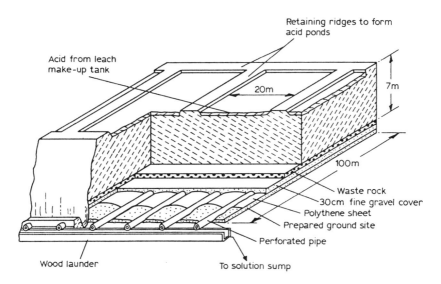

Figure 10.3 Diagram of a dump or heap leaching process. (*Reproduced by permission of Prof. A E Torma*)

then concentrated using solvents. Acid and scrap iron are the most expensive components in copper leaching, between them accounting for more than twice the sum of all the other costs put together, including labour, electricity and equipment.

Other Metals, Especially Uranium and Gold

Many metals other than copper may be solubilized by bacterial action. In addition to uranium and gold they include antimony, arsenic, bismuth, cadmium, cobalt, gallium, indium, manganese, molybdenum, nickel, selenium, tellurium, thallium and tin; while bioleaching is probably not yet commercially viable for any of these, in the near-term the greatest interest is likely to be taken in the so-called "strategic metals": uranium, molybdenum, cobalt and nickel.

Certain types of uranium-containing rock are suitable for bioleaching in a way generally similar to that for copper. This is done in much smaller quantities than for copper, the crushed ore being contained in a large vessel (*vat leaching*) rather than heaped in piles in the the open (*dump leaching*). Bioleaching can also be applied to gold recovery but in that case the technique is employed in the reverse way to that used for copper. Gold often occurs naturally in the form of "nuggets" (*particles*) of the pure metal. Metallic gold is very resistant to chemical attack (that is one of its attractions and why it does not tarnish) but, if present in sufficient quantities, it can economically be dissolved out of rock with cyanide. However, gold does sometimes occur in chemical combination with iron, arsenic and sulphur, chemical forms of gold not susceptible to solution with cyanide; bacterial leaching can then be used to dissolve

201

Exploiting biotechnology

out not the gold itself but some of the other materials, thereby concentrating the residual gold and making it easier and cheaper to recover with cyanide. Residual gold is now being recovered this way in Australia, Canada, South Africa and probably other countries.

There are problems with this approach. The tailings from which the residual gold is to be recovered will usually have been through a *cyanide* extraction stage and certain cyanide compounds are detrimental to bioleaching; *thiocyanate*, for example, is very toxic to the leaching bacteria. This can be a difficulty because of the need to recycle cyanide waste back to the oxidation plant to keep the costs down; various approaches have been designed to overcome it.

In situ Leaching

Even more cost-effective would be recovery of the target metal by leaching not from tailings, dumps or crushed mined ore, but directly from the ore body still in the ground.

One application of *in situ* leaching has been the recovery of uranium from the roofs and walls of worked-out mines, so depleted that conventional mining at that site was no longer economic. The tunnel surfaces were sprayed with water or acid solution and left for several months for bacterial growth to take place; just as in copper dumps, the bacteria are naturally ubiquitous and will begin to grow when the conditions are right for them. The walls and ceilings were subsequently hosed down to yield a solution of uranium compounds from which the uranium metal could be recovered. Costs are said to be broadly in line with working the mine conventionally before it had become exhausted.

Another intriguing possibility might be to introduce the bacteria and acid via injection wells directly into an ore body in its original location, with recovery either from production wells at an appropriate distance or from drainage tunnels drilled beneath the deposit for collecting the run-off. The mechanism for metal recovery depends on the bacteria generating chemicals to solubilize the target metal as they percolate downwards — or are driven by injection — through the ore body to the collecting wells. When the ore-containing rock is not naturally sufficiently permeable to let the liquid and bacteria penetrate, explosives might fracture it before the operation starts. Such *in situ* leaching appears to have been confined to small-scale operations because of uncertainties with respect both to the pattern of liquid flow underground and the behaviour of bacteria when injected in this way into an ore body. As we shall see later in this chapter, the analogies with the use of bacteria downhole for enhanced oil recovery are rather obvious and no doubt synergy will develop between the two technologies just as they might also with *bioremediation* (see Chapter 11).

Efficiency, Economics and the Future of Bioleaching

Prediction is always uncertain but trends can be discerned and opinions sought from those who ought to know. Albeit originally serendipitous, the fact that bioleaching has been in use

Wealth from the earth: metals and hydrocarbons

for thousands or years, that it continues in some locations profitably on a huge scale and that average ore quality is on the decline, make it virtually certain that bioleaching will steadily become more rather than less important in the future. As well as developing new operations, companies are always interested in maximizing returns from investments already made in resource recovery while governments encourage in various ways the exploitation of mineral resources within their own control. As the technologies themselves become more adaptable and experience with some of the newer applications accumulates, increased deployment of these methods seems inevitable wherever they appear appropriate and the right skills are available.

The industry view is interesting. In a survey carried out two or three years ago, the chief executive officers in 200 North American mining companies were asked to identify their key economic and technical problems. Responsibility for environmental damage, together with labour costs and shortages, were among the problems posing the greatest financial and organizational difficulties — while improvements in smelting, refining and ore extraction were seen by various respondents as the most important technical questions. It is noteworthy that the bioleaching of copper has indeed addressed these very technical problems, thereby reducing production costs by half.

Many of the same executives identified *in situ* mining and advanced rockbreaking as the technology developments most likely to improve their own competitiveness. Not far behind came advanced biorecovery methods, seen as the most desirable technical improvements by more than half of the companies dealing particularly with precious metals. While the possibility of new, genetically-engineered leaching organisms was considered, it was recognized that there would be many problems associated with their deployment, ranging from regulatory limitations to competition in the rock from wild strains of the same organisms which might well be more robust than their engineered cousins.

In situ leaching would clearly benefit from better methods of rockbreaking. Blasting with explosives continues to be the most common way of fracturing rock and is likely to remain so. The use of explosives offers by far the least expensive way of introducing enough energy into the rock to break it up into small enough particles to allow good access by the bacteria. New advanced blasting technology might turn out to be the most important factor in facilitating the wider application of *in situ* bioleaching.

MICROBIAL METHODS FOR IMPROVING THE RECOVERY OF CRUDE OIL

Crude oil is probably the major feedstock both for industrialized and developing countries. Although the total quantity of reserves in all the world's oil reservoirs is enormous and is certainly not going to run out in the near future, enormous, too, is the rate of usage. Even if the global supply is in no danger of exhaustion in the short-term, one day it will presumably all be gone. Of more immediate urgency to a number of countries and to many oil companies, particularly small independents, is that the local oil reserves over which they have jurisdiction or possess extraction rights may soon become uneconomic to operate, or might already have done so.

Exploiting biotechnology

And yet, in no case of actual or approaching abandonment of an oil-producing lease or concession has all the original oil-in-place been extracted. Quite the contrary: as a worldwide average, albeit with wide variations, the quantity recovered from the average reservoir is little more than a third. Huge quantities remain in the ground when the costs of conventional production dictate cessation of activity. Can anything be done to enhance the total yield? Does biotechnology have something to contribute? Strange though it may seem, microbiological intervention may prove to be a major factor for increasing the proportion of recovered oil. But first we must identify the problems.

What Oil Reservoirs Are Like

Crude oil comes from ancient plant and animal remains, covered by sediments and buried before they had a chance to decay. Often deep underground, under pressure and at high temperature for millions of years, the chemicals in the carcasses become converted to the complex mixtures of substances, perhaps thousands in number, that make up natural crudes. Mostly those substances are *hydrocarbons*, containing just carbon and hydrogen. Large quantities of *methane, propane, butane* and other hydrocarbon gases may be dissolved in the oil, with the excess forming a gas cap (bubble) at the highest point of the reservoir. Other elements such as oxygen, nitrogen and sulphur are also present in small quantities. They complicate the refining process: the presence of comparatively large amounts of sulphur, in particular, lowers the value of the crude because of higher processing costs.

Oil floats on water. Aquifers often exist beneath oil reservoirs and in many locations the crude oil has tended over geological time to migrate upwards through porous rock, floating on rising waters. Occasionally the oil reaches the surface; when it does so the more volatile components evaporate and the residue is left as tar pits like those in Trinidad and Southern California. Conventional oil reservoirs are formed when the crude is prevented from breaking surface by becoming trapped under a layer of impervious rock which prevents further upward travel (Figure 10.4). The oil in such reservoirs is never present as a lake but always dispersed through the fine pores and cracks of the rock which appears to the naked eye to be quite solid. Most oil is probably present in sandstone, but limestone and chalk reservoirs are also common, some of them granular like sandstone and others cracked and fractured. Oil reservoirs are found at all depths to 12,000 or 13,000 feet and beyond; the deeper the reservoir the higher the temperature and pressure. Those within a few hundred feet of the surface may be close to air temperature, or at about blood heat, and the pressure may not be much greater than that in a soda syphon. But thousands of feet down the oil will be hotter than boiling water and the pressure thousands of pounds per square inch.

Producing the Oil

Typically a series of production stages characterizes the working life of an oil reservoir. First the oil must be found. In regions with the right sort of geology, possible reservoirs are located

Wealth from the earth: metals and hydrocarbons

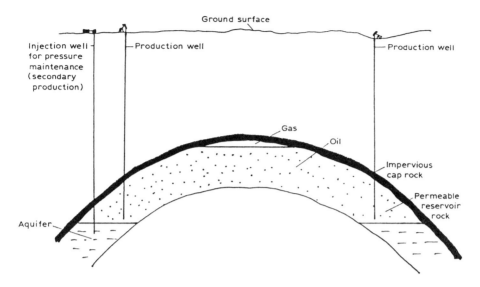

Figure 10.4 Structure of a typical oil reservoir. The porous oil-bearing rock is often sited above an aquifer and oil is prevented from migrating any further upward by an impermeable cap rock. The oil usually contains dissolved gas, some of which may be present as a bubble (gas cap). Reservoir pressure results from the presence of the dissolved gas, the upward force of the aquifer (which, via ultimate contact with the surface, will be under a head of pressure corresponding to its depth), and the weight of the overburden.

by prospecting for domes of impervious rock or other structures under which oil might have accumulated: taking account, of course, of general geological features, this is done by exploding small charges on the surface and listening for the sound echoes from reflecting rock layers. But proof of oil always depends on drilling and literally showing its presence in the rock; the history of the industry is littered with well-researched prospects yielding only "dry holes".

Primary production

Wells are drilled into promising locations in order to produce the oil. Just how many wells are to be drilled, and where they should be located, are decisions for the operators and reservoir engineers — they will be influenced by the geology and economics of the system. Reservoirs are not homogeneous structures and geological discontinuities (faults) are often present which prevent the underground migration of oil in certain directions. Technically, the more wells that are drilled the faster the oil is produced, but cost and the effect of rapid depletion on the ultimate level of recovery must be major determinants. A comparatively cheap operation in an uninhabited wasteland, drilling wells is very expensive offshore, in other inhospitable environments or where local amenities must be protected. So the minimum number of wells to be drilled is governed by cost, the geology of the system, the anticipated return in the

Exploiting biotechnology

form of the rate and quantity of oil expected, and special local considerations. Wells can be drilled at an angle deviating from the vertical so that surface facilities clustered together at a convenient site can tap a much greater area below ground.

Matrix acidizing with and without bacteria

Drilling wells is an elaborate procedure and well completion includes lining the hole with metal pipe (casing) both to prevent the sides from caving in and to confine fluid production to the oil-bearing zones and avoid any incursion of water. At the bottom of the well, in the oil-bearing zone(s), explosive shot is used to make perforations in the casing in order to allow oil to drain from the reservoir matrix into the wellbore and thus to the surface. Whilst most sandstone reservoirs tend to be fairly permeable to the flow of oil, and have rather a uniform rock structure at a microscopic level, chalk and limestone reservoirs are often not very permeable, with insufficient natural drainage channels to allow the oil to flow freely into the wells.

Chalk and limestone rocks are composed mainly of *calcium carbonate* which, unlike sandstone, can readily be dissolved by acid; it is common practice to inject *hydrochloric acid* into the well to dissolve some of the rock and improve the flow of oil. The trouble with acid treatment is that hydrochloric acid is so powerful and reacts so rapidly with carbonate rock that all the injected acid tends to be spent creating a cavern close to the wellbore rather than enlarging flow channels; extending its beneficial effects very far out becomes difficult. Sometimes the action of acid is enhanced by injecting it under such pressure that the rock is cracked or fractured, enabling the acid to travel a good deal further. This technique of *acid fracturing* is much more expensive than simple *matrix acidizing* because of the equipment and expertise needed.

A biotechnological alternative to hydrochloric acid is to employ one of a number of bacteria which, in the absence of air (missing, of course, from an underground reservoir), produce acid as part of their normal metabolism. Chapter 2 discussed the process of fermentation in which microbes, for example, those that turn milk sour during cheese and yoghurt manufacture, generate organic acids as waste products for the same reason that yeasts excrete alcohol in the brewing process. These properties can be applied to the anaerobic environment of an oil reservoir. Given the right nutrient feedstock, appropriate bacteria can produce acids within the flow channels of the reservoir rock, acids which dissolve the very carbonate from which the rock is made. Using bacteria to make the acid actually in the reservoir has certain important advantages compared with injecting acid directly from the surface:

- because hydrochloric acid is very corrosive, precautions must be taken to protect metalwork (piping, tubing, pumps, etc.) in the well itself, in the tanks used to store the acid and in the vehicles which transport it to the site. In the bacterial method the fluid injected is not corrosive;
- a population of bacteria plus nutrients (such as molasses — see Chapter 4), injected into a reservoir, produces acid slowly to begin with and then more and more rapidly in line with its pattern of growth (described in Chapter 2): as the population grows in size so

does its rate of acid production. Thus, if the injection were to take place over a protracted period, perhaps several hours or even days, the bacteria injected first would have been pushed far from the well by the time their foodstuff was exhausted and acid production had ceased. Since most of the acid production, like most of the growth, occurs late in this period (because half the final bacterial population is produced by the very last cell division), much of the rock dissolved would be at a distance from the well rather than close to it. Thus the effect is spread over a comparatively wide area rather than being concentrated close to the wellbore;

- under some circumstances it might be possible to move the injected bacteria far enough from the injection well to use bacterial matrix acidizing instead of the much more expensive fracturing procedures;
- bacterial oilfield technologies are low cost.

There is an important potential extension of bacterial acidizing to a new set of techniques. In conventional practice, wells are drilled either vertically or at no more than a relatively modest angle to the vertical in order to enlarge the area which can be tapped from a single surface facility. Such vertical wells penetrate through the underground oil-bearing zones which are often no more than a few tens of feet or metres thick. Thus, each comparatively short length of wellbore passing into an oil zone has to drain oil from a wide surrounding region. More effective drainage of a reservoir can be obtained by drilling more wells but that obviously increases costs. To overcome this problem the oil industry in recent years has made increasing use of a relatively new technology called *horizontal drilling* as a way of allowing wellbores to pass through more of the actual oil reservoirs without drilling more wells. Drilling from the surface starts vertically downwards as normal but at an appropriate point the drill is steered gradually through a right angle to continue horizontally through the oil zone, perhaps for thousands of feet. Drainage from the formation into the well is far more effective; it is as if a long drainpipe has been laid in the oil-bearing zone. But for horizontal wells in chalk and limestone, treatment with hydrochloric acid becomes essentially impossible because the acid cannot be driven to the far end of the well without reacting with rock on the way. The delayed-action biological acidizing process is thus ideally suited to making sure that long horizontal wells can be acidized throughout their length because they can be filled with injection fluid before that fluid itself becomes acid.

Field testing of bacterial matrix acidizing has already begun in southern England and in Texas. A large proportion of the world's oil reservoirs are located in carbonate rocks so the bacterial method, once successfully tested, is expected to have wide application. Acid treatments have a further application which we shall meet later on in this chapter.

The problem of coning

The high pressure in the reservoir, originating from the dissolved gas or the underlying aquifer, or both, forces oil into the well and, if the pressure is great enough, up to the surface. Indeed, unless proper precautions are taken, oil may migrate upwards through a newly drilled well

Exploiting biotechnology

with such force that it gushes like a fountain into the air, a wasteful and dangerous sideshow. As oil is progressively removed from the reservoir, the gas cap expands and the underground pressure tends to drop so that the flow into the wells becomes more sluggish and the oil may now fail to reach the surface under its own pressure. The oil is then lifted by a pump installed downhole, usually driven mechanically from the surface via a long rod operated by an oscillating beam, the familiar *nodding donkey* (Figure 10.5).

Pumping sets up a suction pressure around the well. This encourages the flow of oil but it can also promote the establishment of a cone of water drawn upwards from the aquifer, especially if the well bottom is far down into the oil-bearing zone. The *coning* problem is exacerbated if the oil in the reservoir is rather viscous, as many oils are: then the more mobile water is sucked preferentially into the well and, as time passes, the produced oil contains more and more water. In many fields coning is a major production problem, difficult to control; one method of dealing with it is to stop pumping until the cone subsides and then start again, but that obviously reduces the overall rate of oil recovery from the well and lowers the return on investment. When production is started up again, the cone in time reappears and the cycle has to be repeated.

Figure 10.5 A typical wellhead pump. The oscillating boom, operated by an electric motor, actuates the pump via a string of rods extending to the bottom of the well. (*Photographed in Romania*).

Wealth from the earth: metals and hydrocarbons

Ideally, coning could be prevented by introducing between the aquifer and the well some sort of waterproof or water-retardant barrier such as a cement or gel. However, because most cements and gels are viscous it is very difficult to ensure that, when injected, they are properly sited to stop water flow and do not block the oil drainage channels instead. Furthermore, while they must be injected as at least fairly freely flowing liquids, the injected fluids must eventually solidify if they are to prevent water flow. Predicting the time to solidification of a time-setting gel or cement, and hence the distance penetrated before gelling occurs, is not always easy. Faced with such uncertainties many, perhaps most, oilfield operators appear dissatisfied with injecting cement or polymeric blocking agents to control coning problems and rely instead on intermittently turning the pump off for a period, a procedure called *stopcocking*.

Even in the absence of air, some types of bacteria are able to produce viscous or solid polymers from simple sugar nutrients. As they grow from a small starting population, the bacteria generate not only more cells but also large amounts of polymer which may simply be very viscous or actually sticky, welding the mass of bacterial cells into something like a stiff jelly. The injection fluid, an aqueous medium containing a few bacterial cells plus the nutrients from which the polymer is eventually produced, is barely more viscous than water itself; injected via the well into the top of the cone, the medium can be expected to migrate through the water flow channels of the cone itself. Just as in acidizing, most of the jelly-like polymer production occurs at the end of the bacterial growth period, thereby largely or entirely blocking the flow channels and slowing down or completely stopping further water flow.

Coning is a widespread problem in oil production and its control would significantly improve the economics of production in many fields. Bacterial coning control methods have been developed in the laboratory and are likely to be tested in the field in the near future.

Secondary production

Inevitably the rate of oil flow progressively declines as time passes, the proportion of water increases and eventually this phase of primary production is no longer profitable. The actual proportion of the original oil-in-place recovered during primary production varies widely from one reservoir to another, with a worldwide average value of about 20-25%.

Further recovery requires the injection into the reservoir of a fluid to help push out more of the oil. One candidate might be the gas that escapes from the oil as it reaches the surface, but gas is itself a useful commodity and it might be best to install a pipeline and sell it directly to customers. An alternative is to inject water, easy with offshore fields and onshore operations in well-watered areas, but possibly difficult in the middle of a desert (Figure 10.6); however, the water normally produced together with the oil may satisfy much of the need for fluid reinjection into the reservoir.

There are two general patterns of water injection: it can be introduced directly into the aquifer simply to increase reservoir pressure, although coning difficulties might thereby be magnified. Alternatively, a pattern of injection and recovery wells can be drilled so that water is injected through a set of injector wells and pushed through the reservoir to a matching set of production wells some distance away, driving the oil ahead of it. This technique is

Exploiting biotechnology

Figure 10.6 Injection equipment at the wellhead. There is not much to be seen: just the valving of the wellhead itself together with the water line, gate valves and gauges. This injection facility was actually being used for a microbial flood. (*Photographed in Oklahoma*).

called *waterflooding*. (It is in this procedure of water injection that the second use of matrix acidizing is to be found. A certain pressure is needed to pump water into a reservoir at some specified rate. For a number of technical reasons, mostly deposition of scale and impaction of fine particles in the rock pores, this pressure tends in some locations to rise over time, thereby reducing the rate at which water can be injected. Raising the injection pressure risks fracturing the rock and establishing non-uniformity of fluid flow. What is needed is a way of cleaning up the rock in the neighbourhood of the wellbore, so reducing the impediments to water entry. The susceptibility of carbonate rocks to acid treatment suggests in their case the use of the biological system which again has the advantage, compared with hydrochloric acid, of offering a wider radius of activity. Acidizing injector wells is currently under test in the US.)

On average, approximately another 15% of the original oil-in-place might be mobilized by water injection but, by then, so much water is likely also to be produced that the process becomes uneconomic. One reason for such a high proportion of the oil being left behind is that the more mobile water tends to travel through the easiest and most accessible channels in the heterogeneous rock and fails to sweep out much of the oil. Another reason for less than complete recovery is that water does not push oil smoothly and uniformly: the two fluids repel one another and the water by-passes much of the oil immobilized onto the rock faces or stuck in narrow pore throats.

Between them, the primary and secondary phases of production yield on average about 35% of the original oil-in-place, with large variations between individual reservoirs. The 65% or so which is left represents an enormous residual resource, notionally worth more than $20,000,000,000,000 (twenty thousand billion dollars, or $20 trillion) at a crude price of $18 a barrel, and proportionately more at higher prices. (A barrel contains 159 litres, equal to 35 Imperial gallons or 42 US gallons.) As a comparative benchmark, $20 trillion is several times the gross national product of the US and more than 20 times that of the UK. For a whole host of reasons not all of this oil is ever likely to be recoverable, but obviously even a proportion of it would be a very valuable resource.

Tertiary (enhanced) recovery

Recovering more than the typical 35% of the original oil can be achieved by infill drilling of more wells, which is expensive, or by adding costly chemicals to the water either to make it better at shifting the immobilized oil or by making the flow pattern more uniform, or both. Per barrel of oil recovered, all tertiary recovery methods are significantly more expensive than primary and secondary technology.

Profile improvement/selective plugging

Different reservoir flow pattern problems need different sorts of chemicals to correct them. Irregular water flow may be the result of major cracks or other high permeability zones running through the rock; for good reason these are called *thief zones*. Plugging thief zones would force the water to flow through the oil-bearing channels and so push out more oil but, just as it is difficult to correct coning, so it is also difficult to use injected cements and polymers to block the thief zones without also plugging the other (oil-bearing) channels.

Making use of bacteria to plug thief zones is in principle not so very different from using them to control coning. An aqueous injection fluid containing anaerobic polymer-producing bacteria, together with their appropriate nutrient feedstock, is injected into the well. Where necessary, downhole packers are set in place to ensure that the injected fluid enters just the right zones in the rock. Enough fluid is injected to penetrate far into the thief zone, so providing a deep plug. Just how far that will be depends on the macro- and micro-structure of the particular reservoir. If fluid flows easily between adjacent vertical zones it will be no help to insert a shallow plug because the subsequent waterflood will simply by-pass it. But, if vertical communication between zones is poor, by-passing will not be possible and the plug then needs to be deep enough only to ensure that the injected water does not enter the thief zone.

Exploiting biotechnology

Some field tests of bacterial plugging procedures have already taken place, mostly in Canada, and others are planned there. One field pilot undertaken in 1988 worked well from a microbiological point of view: the bacteria produced the materials required of them but the relevant water channels were not blocked. Subsequent analysis showed that the extent of water flow in the reservoir had not properly been understood when the tests were designed and, as a consequence, the bacterial system had not been injected in the right place. Failure of a test, however, is not a tragedy: one learns from experience. In this case the limitation was probably one of reservoir engineering rather than microbiological design, and no doubt the methods will be applied again in better understood circumstances to yield a more favourable outcome.

Microbial polymer- and surfactant-flooding

Even in the absence of major cracks, water flows irregularly because of microinhomogeneities in the rock and by-passes oil because of the mismatch in viscosities; the effect is rather like water spilled on a gently sloping paving stone not running downhill evenly but in a series of rivulets. While the oil may in principle be susceptible to mobilization, in practice much is left behind because the water finds it "easier") to flow through the rock without displacing oil. The reservoir thus behaves as if it has large numbers of small thief zones, far too many to be blocked individually. This problem has to be addressed differently.

Dissolving a polymer in the injection water (*polymer flooding*) increases its viscosity and improves oil recovery by suppressing the irregularities of water flow resulting from reservoir microheterogeneity. Some of the polymers already in use for this purpose are derived from petrochemical feedstocks, while others are polysaccharides of microbiological origin. Polymer flooding is too expensive for most production, but using bacteria to produce the polymer in the reservoir may bring the feedstock price down to as little as one-tenth of the cost of preprepared polymer; with little difference in the other operating costs, the economics of the operation become far more favourable.

An indication of market opportunity and sensitivity for polymer flooding is provided by US data published in the mid-1980s. Using the market price then current for polymer intended for direct injection (polymer manufactured conventionally, not generated downhole by a bacterial population), it was estimated that, at a crude price of $20 a barrel, polymer flooding could recover economically an additional 250 million barrels of crude in the US alone, with a total value of $5 billion. At a price per barrel of $40, the market size would expand to 550 million barrels, then worth $22 billion. The effect on the market size of reducing the cost of polymer is the same as increasing the price of the oil. The system described below for the downhole generation of microbial polymer is that sort of low-cost procedure, but precise calculations have not yet been made of the economic benefits to be expected from its deployment in the field.

The type of polymer needed for polymer flooding is viscous, but not so viscous as to form a solid or semi-solid jelly of the sort needed to control coning or plug thief zones. Again, the appropriate types of anaerobic polymer-producing bacteria are used; together with their nutrients, the cells are introduced into the reservoir via the injection wells. The injected fluid spreads out from the wells in all directions as permitted by the local conditions and flow patterns. When the calculated quantities of bacteria and nutrients have been introduced, further injection is suspended to give the cells time to grow and develop a working population around each injection well. Once developed, that population is used to convert into soluble polymer a modified feedstock which allows polymer production, with little or no further cellular growth, so as to avoid random plugging of the rock pores by accumulations of bacterial mass.

The modified feedstock is injected for a period of time which, in reservoirs with wells spaced far apart, might be months or even years. The polymer generated continuously is washed free from the stationary producing cells and driven through the reservoir by the ongoing waterflood. The water is now more viscous ("thickened") and better able to push oil more uniformly as it flows through the reservoir rock towards the production wells.

Bacterial polymer flooding is another biotechnological aid for oil production which has been laboratory tested and is expected to undergo field trials, probably in the US, during the coming two or three years. Reservoir engineers anticipate that, depending on how much oil has already been recovered from a field when it is started, a successful polymer flood in a susceptible reservoir might increase recovery by 5–15% of the original oil-in-place.

Polymer flooding helps to recover more mobile (or freely flowing) oil. But a good part of the oil remaining after secondary production is stuck onto the rock or impacted in the narrow constrictions in the rock (pore throats). Detaching such oil and making it flow with the water stream requires *surfactants*, chemicals reminiscent in their action to that of domestic detergents: they help the injected water to get the oil moving just as grease and fat are much more easily removed from dishes with the aid of detergent than by water alone. Mobilizing oil in this way is called *surfactant flooding* and a number of pilot trials have shown it to be an effective way of recovering more oil under appropriate conditions. However, using current industrial chemicals, the cost of surfactant flooding is generally too high; to be cost-effective either the cost of surfactant must be reduced or the price of oil increased.

A bacterial surfactant flood design conceptually resembles the polymer flood which has just been discussed but makes use of different chemicals. Some anaerobic bacteria produce surfactant substances which are effective in helping to detach oil stuck to rocks and impacted in pore throats (Figure 10.7). A system design similar to that of microbial polymer flooding needs also to be developed for biosurfactants; a number of candidate microbes have already been identified, but the protocols for their use have not yet been perfected in the laboratory to the point of being ready for testing in the field.

Several microbial field pilot trials have been run in recent years in which surfactants have probably played a significant role, but insufficient information about them has been published to assess whether or not other factors were of equal or greater importance. One such test carried out in Oklahoma over a four-year period has proved encouraging; however, because other technical operations going on nearby in the same field eventually began to interfere with the test activities, the trial had to be started again at some distance from the original site and new data have only recently begun to emerge.

The US study on the economics of polymer flooding referred to earlier also provided market estimates for (non-biological) surfactant flooding and showed the benefits to be very sensitive

Exploiting biotechnology

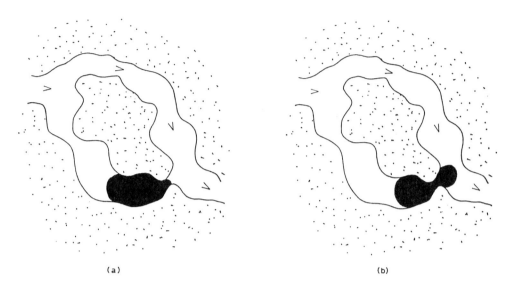

Figure 10.7 Entrapment of oil droplets. (a) oil droplet trapped in a pore throat and by-passed by the water flow; (b) expulsion of the droplet entails increasing the oil/water contact area against the mutual repulsion of the two fluids.

to the market price for crude. With oil at $20 a barrel, the surfactant market was estimated at 700 million barrels of incremental oil recoverable (value $14 billion); however, at a crude price of $40 a barrel, 14,800 million barrels (valued at $592 billion) would be worth recovering. As with polymer, the anticipated markedly lower cost of microbial surfactant flooding compared with non-biological surfactants is equivalent in its market effect to raising the oil price; as yet no detailed calculations have been published.

We might return for a moment to quantities and make use of a specific example. Data in Table 10.1 for the Permian Basin in Texas and New Mexico show the progress of production and the problems which remain.

Table 10.1
Oil reserves in the Permian Basin of West Texas

Category	Billions of barrels
Total original oil-in-place	105.7
Oil produced to date	25.3
Estimated recoverable reserves (producible by current technology)	5.9
Mobile oil (in principle recoverable with polymer flooding)	30.0
Residual oil (non-mobile, requiring surfactants or other agents to impart mobility)	44.5

Single well stimulation

Just as matrix and fracture acidizing is used in carbonate reservoirs to assist in the drainage of oil into producing wells, so existing wells in both carbonate and sandstone may need periodic cleanout to maintain their flow. With some types of oil, drainage into the well becomes impeded by the deposition in the pump or riser tubing of wax which can be removed by circulating hot oil, though care must be taken not to flush the molten wax from the well out into the reservoir matrix. Metallic scale, rather like the "fur" which deposits in kettles in hard water areas, sometimes clogs pumps and tubing; acid, or even mechanical scraping and reaming, is needed for its removal. Asphaltic material may become deposited in the reservoir around the well, while fine sand or clay particles can be carried along by the flowing liquids and impacted in fine pore spaces. Some of these factors contribute to a "skin" around the well which slows the ready flow of oil.

Most microbiological field operations, and there have been hundreds of them in the US and a number of Eastern European countries, have been carried out on single wells. The bacteria are usually chosen primarily for their ability to generate copious amounts of gas (mainly carbon dioxide), although they also produce a mixed cocktail of various solvents, surfactants and acids. After their injection into the well and surrounding rock, production is shut down for a few days to allow the fermentation to proceed. The gas production raises the downhole pressure, while the other chemicals are believed to help in removing some of the substances preventing liquid flow. When the well is reopened, the offending materials are dissolved or washed out of the rock, dislodged from the well bottom and driven up the riser tubing to the surface where they can be discarded. This treatment is not usually tailored to the particular production problem of each well; indeed, those problems are rarely specifically identified. But the cost is low and, for marginal operations yielding no more than a few barrels of oil a day, a cheap workover treatment like this might mean the difference between closing a well down and keeping it going, especially when crude oil prices are as low as they were in the period 1986–90. Even if the treatment has little beneficial effect, not much money is lost and many operators are prepared to take the risk.

A Comment on Bacterial Survival in Oil Reservoirs

The conditions for microbial life in an oil reservoir are not necessarily harsh. Air and oxygen are absent but many microbes exist quite happily without them. High pressure is not too serious: the pressures at the bottom of the deepest ocean trenches are higher than in most reservoirs, yet microbes are normally found there and, as it turns out, bacteria are not very sensitive to modest increases in pressure. High temperature is more of a problem and so is very saline water. A few varieties of bacteria will grow at temperatures up to that of boiling water but most types prefer a cooler environment. While some reservoirs are very hot, others are not and it is to the cooler oilfields that, for the moment, the bacterial recovery systems are targeted. Finally, some reservoir waters are so salty that many forms of microbial life would not be able to grow in them. But all the methods outlined in this chapter are based on the

Exploiting biotechnology

injection of water from the surface; because of the labyrinthine nature of the reservoir matrix, injected water, the salinity of which can be controlled, mixes slowly with the water already present in the rock.

Another problem facing petroleum microbiologists is that some (perhaps all) oil reservoirs already have resident populations of bacteria. They may have gained entry via the wells. either during drilling or in subsequent injection operations, although there is a growing body of microbiological opinion that microbes exist naturally even in deep virgin reservoirs (how they got there and what they live on are fascinating questions for microbiologists and others). Any system design has to take possible resident microbes into account and consider how they will interact with enhanced recovery procedures introduced downhole.

Applications and Prospects for Microbial Oil Recovery Technologies

While the clearest benefits of using advanced microbial methods for the improvement of oil production will be economic, significant technical advantages are also to be expected, particularly in the control of water flow and deep matrix acidizing. Once proven in a series of well-conceived and well-documented field trials, they will offer the promise of a set of low-cost technologies readily applicable to a wide range of oilfield problems throughout the world. Few scientific advances remain to be made — most of the microbiological parameters are well understood and the targets still to be reached are almost all developmental: ensuring compatibility and integration of the microbial technology with existing oilfield practice, while acquiring experience of reliability and effectiveness under the real working conditions of oil recovery operations.

Since the major oil companies have the largest operations and produce most of the crude, the greatest potential value of these microbiological techniques must ultimately accrue to them. But it is quite likely that smaller independent operators will be the first to put them to use, partly because independents may have a greater immediate need and partly because large, well established businesses are often less willing than small ones to experiment. Thus:

- the majors are not always the most innovative organizations and some of them suffer from the "not invented here" syndrome;
- their need for low-cost technology may not be so acute as that of the independents: they operate the most profitable fields and are more likely to be able to weather the ups and downs of the market;
- because of their normally higher overhead costs, majors also tend to sell fields which are in serious decline to independents;
- as an alternative to maximizing the yield from their existing operations, the majors are more likely than independents to have the resources with which to explore and develop new fields even if, as in offshore locations, the up-front capital costs are very high;
- independents, on the other hand, may operate with smaller profit margins and so are frequently reliant on cheap technology;

- however, in some oil provinces, and at some stages of market cycles, even small independents risk going out of business altogether: unless cheap technology is forthcoming they may be totally unable to survive.

Inexpensive production technologies, including the microbiological methods described here, have both economic and political implications:

- existing fields will yield more oil recovered economically and at faster recovery rates. Hence, the return on the original capital investment will increase;
- the reliance on imported oil, a major problem for countries like the US, will decrease. American production has tended for years to fall behind domestic consumption, yet enormous quantities of technically recoverable oil remain within the boundaries of the US. Nevertheless, with the current expensive enhanced recovery technology, the exploitation of that residual oil cannot compete with imported crude from cheap primary production, even if it does originate in politically unstable or potentially hostile parts of the world.
- in poor countries, short of foreign exchange, improving domestic oil production with affordable low-cost technology will enhance export earnings and/or diminish the cost of imports.

Prediction in the oil industry is notoriously unreliable. Even so, the signs point to a rapidly increasing use of microbial methods, once a series of successful field tests has built confidence in their cost-effectiveness and utility. The next five years should tell.

11 Keeping the place clean

Pollution And Waste Management

Unwanted materials in the wrong place are the essence of pollution but unfortunately virtually all human activity, domestic and industrial, individual and collective, generates chemical and physical waste products which it would be best not to have around. Equally unfortunately, more and more of the waste products once thought of as innocuous are now considered undesirable, if not actually sinister: carbon dioxide (the most important "greenhouse gas") is perhaps the best example.

When wastes escape from our control, actually or potentially degrading our environment by presenting health hazards, reducing amenity value and diminishing our quality of life, or by imposing economic or other social costs, we term them *pollutants*. Since wastes in general are unavoidable, we seek, with growing enthusiasm as the years pass, to minimize pollution of the environment by effective *waste management* procedures, designed either to contain the undesirable materials or to convert them into acceptable alternatives. This may be done voluntarily by enlightened individuals and corporations, but the stimulus of official regulations is frequently necessary. Sometimes waste management can be made directly cost-effective because valuable materials are recovered which would otherwise be lost; often, however, this is not so and the expense of undertaking the procedures is simply an additional cost in making a product or providing a service. In a nutshell: preventing the release of undesirable materials constitutes waste management but, if they do escape, they give rise to pollution.

Someone who has not hitherto given the matter much thought may be surprised to find how much waste management and pollution control is already effected through biological means; for convenience, we may consider all of them as falling within the general area of biotechnology. Before taking a closer look at these activities it might be helpful to consider briefly but broadly what actually constitutes pollution.

Pollution, Mainly Chemical

Although they might well be local eyesores, dumps of old bedsteads, bicycles and car tyres are a very minor part of waste disposal and are probably furthest from any involvement with biotechnology. In industrial countries, most waste substances are released in water streams (channelled into sewers or dumped into rivers), collected and disposed of as solid or semi-solid garbage, or discharged into the air.

Much of this material, particularly from domestic sources, is biodegradable — much is, indeed, routinely biodegraded. For example, municipal solid wastes, generated at the rate of 3–4 pounds (1.50–2.00 kilograms) per person per day in Western Europe and North America,

Exploiting biotechnology

typically have a high decomposable paper content. But some of the waste is not biodegradable and must be dealt with in other ways. Industrial waste is both more varied and often more concentrated; some forms of it are discharged without treatment while others are converted to less harmful materials or retained in long-term storage.

Location of Pollution

Pollution can be found anywhere: in the air, in the rivers, streams, lakes and oceans, and on land.

Pollution in the air

Aerial pollutants are usually thought of as arising mainly from industrial activities: among the most important polluting gases are the oxides of sulphur and nitrogen formed during the burning of fossil fuels in power stations and car engines. These gases dissolve in atmospheric moisture to be washed back to earth as acid rain, with particularly devastating effects on the forests and wild life areas downwind from industrial regions burning high-sulphur fuels. Biotechnology can do little to ameliorate these problems once the noxious gases have been formed, but this chapter will include a discussion of the proposed use of microbiological systems for removing some or all of the sulphur from coal before it is burned. That might be one way of reducing the acid rain problem.

Other gases are also serious pollutants. One of them is methane, a naturally-occurring hydrocarbon, the major component of natural gas and thus an important fuel. From geological sources it seeps into the tunnels and airways of mines from which it is rapidly vented because of its potentially explosive properties. It is produced biologically by the decomposition of vegetation in lake bottom sediments and muds and, by a similar process, is also one of the main products of the biological degradation of wastes in landfills. At many modern landfill sites the methane generated is collected and used locally for fuel; however, in some older installations the gas is not channelled in this way and sooner or later finds its way into the atmosphere. If there is an impervious layer preventing its escape directly into the air from the landfill, it may travel surprisingly long distances through the soil and reach the surface under or near buildings where, if mixed in the right proportion with air and ignited, it may cause an explosion. All manner of underground voids and caverns accumulate methane, particularly when located in the vicinity of abandoned mine workings which are no longer ventilated. As well as being potentially explosive, methane is also one of those greenhouse gases which, when they accumulate in the atmosphere, are believed to contribute to global warming. Interestingly enough, certain microbes use atmospheric oxygen to oxidize methane, producing carbon dioxide and water: biotechnological systems based on such microorganisms, requiring little or no maintenance, might be used to prevent methane accumulating in poorly ventilated voids.

Keeping the place clean

But carbon dioxide produced by most forms of biodegradation is not a desirable end-product either. It is the main greenhouse gas, released in vast quantities not only by the burning of fossil fuels (and hence, directly or indirectly, by most industrial and automotive activity) but also by the respiration of living organisms. Until a century or two ago, the global production of carbon dioxide was roughly balanced by its uptake by green plants in photosynthesis, but this balance may now be badly upset both by the greater production of carbon dioxide resulting from the large amounts of fossil fuels which are burned for energy and the felling of many of the forest trees which would have used carbon dioxide for their growth. Biotechnology has no immediate answers to these problems but it should be noted that, even when biodegradation is used to get rid of an undesirable pollutant, one of the products may be carbon dioxide, itself a pollutant when present in excessive concentrations.

Pollution in water

Because water is a ubiquitous and convenient liquid, it is used to transport waste materials to treatment facilities where they are rendered harmless and safe for discharge: sewage obviously falls in this category. But all over the world many domestic, industrial and agricultural waste streams are not processed through treatment works but are instead discharged directly into waterways, in some areas polluting them so badly that virtually all life is extinguished and they become stinking and unsightly lakes and rivers of chemicals. Further sources of freshwater pollution include acid mine drainage from abandoned workings and oil spilled from tankers and barges at sea. Marine pollution also arises from production water, still containing some hydrocarbon from oil recovery procedures, which may be discharged overboard from oil platforms, while nuclear reprocessing facilities release radioactive elements in small amounts but on a continuous basis.

Pollution of land

Wastes and residues dumped or accumulating on land also contribute their share to the total picture of environmental degradation. Crops are sprayed with certain insecticides, and with fertilizers containing nitrates which, not being degraded by natural agents, accumulate in the soil and are subsequently leached into rivers. All sorts of chemicals continue to be dumped into pits and elsewhere: in *sludge farming*, now banned in a number of countries, oil wastes are ploughed into the soil in the hope of achieving natural breakdown. Solids, including the non-liquid components of domestic wastes (as well as those bedsteads, bicycles and car tyres) are buried in landfill pits which, when full, are covered with soil and allowed to biodegrade by natural microbial action.

Exploiting biotechnology

The Chemistry of Pollutants

With so much variety among them, there is no simple way to categorize the many different types of pollutants except perhaps on the basis of their susceptibility to enzymic attack. Like all biochemical reactions, biodegradation depends fundamentally and critically on the relationship between the chemistry of the pollutants and that of the enzymes; if those integrate appropriately, biotechnology might be able to render the operation practically feasible and commercially viable.

Pollutants readily susceptible to biodegradation

These include most substances in the liquid discharges from households which contain largely human and food wastes, all of it biological in origin and hence all of it biodegradable. In industrial countries, some 60% or more of the solid and semi-solid domestic wastes, comprising animal and plant remains, together with paper and paper products, is also readily broken down biologically by microbial degradation when deposited in a landfill.

Unless during their manufacture chemicals are employed to inhibit subsequent biological breakdown, the food, natural fibre, wood and paper industries all generate product and processing wastes composed mostly of biodegradable substances because they all make use of natural raw materials. Some natural substances are much more resistant to degradation than others: wood, generally, is slow to degrade compared with softer vegetable tissues, particularly if it contains natural plant chemicals which have probably evolved specifically to protect the plant from microbial attack.

Many hydrocarbons are degraded by various microorganisms. Destruction and utilization of the short-chain, "light" hydrocarbon molecules in petrol is more rapid that that of the longer-chain, "heavy" compounds characteristic of viscous oils, tars and waxes; but even crude oil, a very complex mixture, may degrade slowly in nature. The lighter components in oil often evaporate before the natural microbial population can attack them, while degradation of the heavier components, if it occurs at all, takes a very long time. Microbial degradation of mineral oils appears to be totally dependent on the presence of oxygen, so it is extremely slow — and perhaps totally absent — if air is excluded.

Pollutants biodegradable with some difficulty

Constituting the next group, this category includes the intermediate fractions of crude oil, a host of industrial chemicals and a number of pesticides. Many of these compounds are totally artificial but some have sufficient resemblance to natural biological counterparts to exhibit a limited susceptibility to enzyme attack. In those cases, prolonged contact between the pollutants and the microbes possessing the relevant enzyme(s) may, by mutation and selection, adapt and improve the microbe's ability to attack the compound(s).

Some pesticides fall into this category. Synthesized to resemble natural compounds, they are constructed deliberately to resist degradation by the insects and other pests which they are designed to control; the compounds act by interfering with some vital metabolic activity and so kill the target organism. However, those very molecular modifications sometimes also make it difficult for natural microorganisms to metabolize such pesticides and that is why they often persist in the soil long after application.

Pollutants essentially resistant to biodegradation

This is the last group. Once more, crude oil is prominent, represented now by the "heavy ends". Also very important here are many of the products made from petrochemicals — most plastics cannot be attacked by microbes and concern is growing all over the world at the enormous quantities of polyethylene, polystyrene and other plastics which are thrown away to remain almost indefinitely where they lie. Some plastics exposed to sunlight do slowly break down spontaneously and some of the breakdown products are probably used as nutrients by soil and other microbes. But the process is slow and plastics in landfill, shielded from the decomposing effects of sunlight, will remain there for tens or perhaps hundreds of years. However, biodegradable plastics are now available for a number of uses and are gradually becoming more common as packaging materials.

Among the virtually non-biodegradable chemicals are certain chlorinated compounds like the *polychlorinated biphenyls (PCBs)* used until recently in the manufacture of electrical transformers. Living systems have never encountered such compounds in their evolutionary history so they have not developed enzyme mechanisms to attack and make use of them. Prolonged natural contact will possibly allow them to do so but that might take many decades, as happened with detergents. The earliest synthetic detergents, introduced after World War II to replace soap powders, were in use for about 20 years without good biodegradation systems developing naturally. The formulations of the detergents were therefore changed to facilitate their degradation in the natural environment. Thus, rather than wait for natural biodegraders to develop an ability to destroy refractory chemicals, we might intervene by using our growing understanding of protein structure to design new enzymes specifically for the task of degrading these pollutants. Genetic engineering will then allow genes for the new proteins to be constructed in the laboratory and inserted into appropriate microorganisms for inoculation into the polluted locations. Alternatively, as happened with PCBs and the early detergents, another and more likely eventuality would be to ban them from use and find more acceptable substitutes.

Other pollutants are chemically entirely different. They are the heavy metals, elements such as cadmium and mercury which are generally toxic to animals as well as to many plants. They are not naturally present in large amounts in soil and water but are brought to the surface through mining, either for their own industrial value or as by-products in other metallurgical processes. Unless they are deliberately buried again they are going to remain on the surface and in surface waters indefinitely. Cadmium and its compounds are employed in photocells, rechargeable batteries, alloys, stabilizers in plastics and as protective coatings. Mercury is used

Exploiting biotechnology

in switches, thermometers, vacuum pumps and other instruments, as a pesticide and in the metallurgical and chemical processing industries. While it is probably not feasible to recover the greater part of these elements for recycling or burial because they are gradually dispersed in use, the waste quantities produced in their manufacture could be recovered for recycling or permanent disposal.

It is not necessarily the metals themselves, but certain of their compounds which are so poisonous; in some instances these toxic derivatives arise by microbial or other chemical action in the environment after discharge. The problem is aggravated because some organisms living in aqueous environments have the ability to concentrate within their own bodies heavy metals from their surroundings. One notorious case in the 1950s was the release into the sea of a mercury compound from a plastics factory in Japan. Marine microbes converted the original compound into a toxic derivative which accumulated in fish later eaten by the local human population. An outbreak of mercury poisoning resulted, known from its geographical location as *Minimata Disease*, and a number of deaths ensued.

It has also happened that toxic compounds have accumulated in the vicinity of smelting plants or, as the result of industrial discharges into the sewers, in sewage farms, the sludge from which is often sold as agricultural fertilizer. By such routes, cadmium has on occasion contaminated soil used for growing lettuce plants, which are particularly good at accumulating the metal, and poisoned people who ate the plants.

There is really no way in which these materials can be made safe except by preventing their release in the first place. They could be vitrified into glass blocks for deep burial but any chemical form in which they might be disposed of to the environment, however safe at the time of discharge, cannot be relied upon not to change into a toxic derivative. For the toxic heavy metals, removal from industrial effluents is essential to prevent them from exerting their poisonous effects downstream from the point of discharge. Biotechnology is beginning to be rather important in helping to remove such materials from liquid waste effluents by concentrating them into small volumes for easy handling, storage or entombment.

Radioactive products resulting from the reprocessing of nuclear fuel elements are a special case of dangerous substances which cannot be converted to safe forms for discharge and which therefore must be confined and stored. Although there are very few reprocessing plants in the world, the public is increasingly concerned about the release of radioactive elements from them; even if concentrations are low, over time the quantities discharged become significant. While biotechnological techniques have not been used for such control, they have been under consideration for a number of years and will be discussed briefly later in this chapter.

POLLUTION CONTROL AND WASTE MANAGEMENT USING BIOTECHNOLOGY

Since the atoms in chemical compounds cannot be made literally to disappear, but can only be rearranged into different molecular structures, there are two ways in which biotechnology can help both in the eradication of pollution which has already occurred and in its prevention in the first place:

- by the chemical degradation of polluting compound(s) to (more) acceptable products;
- by the sequestering and concentration of particular toxic elements into materials which can be permanently disposed of in a harmless way.

Biochemical degradation is applicable primarily to the "organics", compounds composed mainly of carbon, hydrogen and oxygen but which may also have an appreciable content of nitrogen, sulphur, phosphorus and some other elements. Aside from water (which always comprises the bulk of living organisms) and some salts, all biochemicals are organic. Indeed, the very term derives from the fact that they were first recognized in and isolated from living organisms; all the chemicals referred to in Chapter 2 are of this type, as are many industrial chemicals. Sometimes additional elements like chlorine are present but, while this may greatly influence the chemistry of the compound in its intact state, once it is degraded the presence of chlorine atoms among the degradation products is usually of little pollutional significance. "Inorganic" compounds, by contrast, are those (apart from carbon dioxide and carbon monoxide) which contain no carbon but are made up of selections from all the other elements, including the metals; they occur naturally in the air, in rocks, minerals and in natural waters.

Except for coal and oil, which have been modified by prolonged burial, all natural organic compounds are susceptible to biochemical degradation: food goes bad, leather rots and wood decays, all the consequences of the same general process of microbial attack. The most common end-products resulting from the biodegradation of organic compounds are carbon dioxide, water and a variety of other simple substances containing nitrogen, sulphur, phosphorus and the non-toxic metals common in biochemistry. Most are harmless in moderate concentrations and, for the microbes that undertake the degradation, the rewards come in the nutrient value of the breakdown products.

Sequestration is quite different. As we noted earlier, there are no really safe compounds for the discharge of such toxic metals as cadmium and mercury, or metalloids like arsenic; there is always a risk that apparently acceptable compounds will be changed in the environment to toxic derivatives. The prudent course is to remove them altogether from waste streams and lock them into some sort of solid matrix from which they cannot readily be leached by ground waters after burial. The problem is that these toxic metals may be present in aqueous waste streams in low or very low concentrations, accompanied by much larger quantities of other, harmless metals which they resemble fairly closely chemically. Simple, non-biological methods of extraction are rarely selective enough to remove the small amounts of toxic metal without being swamped by the much larger quantities of the harmless ones.

Yet again, biochemical specificity can come to the rescue because certain living beings (and even the dead remains of some of them) have very considerable powers of specificity and selectivity which enable them to bind to low concentrations of one metal in the presence of much higher concentrations of a near relative. By flowing a waste stream over a biological sequestering agent, immobilized as described in Chapter 3, the unwanted metals are removed from the water and concentrated into a small bulk of bioabsorbent. Plants may help, too — a range of shrubs, flowers and trees from Pacific islands and alpine regions turn out to be very good at absorbing heavy metals from soil; indeed, they seem to thrive on them. These plants may be able not only to clean up soil for normal agricultural use but may be usable as an

Exploiting biotechnology

economic source of some of the metals (nickel and chromium) they accumulate. So effective are they that, when burnt, their ash might have as high a concentration of the heavy metals as a good ore.

Both biodegradation and *biosequestration* work within the usual biochemical rules of specificity: the enzymes responsible for the degradation, or the receptor molecules binding the metals, will act only on a single or very limited range of substrates. As always, the more fully the chemistry of the offending substances has been identified and understood, the more effective will be the means of dealing with them.

The rest of this chapter will explore a number of existing or prospective biotechnological contributions to the resolution of some of these problems, grouped into two broad categories.

Pollution control, or environmental clean-up activities include:

- bioremediation of soil, involving the removal of pollutants;
- dealing with oil spills and leaks: oil released at sea, as well as petrol which has leaked from storage tanks into lakes, streams, underground water and aquifers.

Waste management operations encompass:

- disposal of hydrocarbon wastes: in oilfield operations, the removal of hydrocarbons from production waters before discharge and, in refining, the disposal of oily wastes and sludges by sludge farming;
- sewage treatment;
- anaerobic digestion, including landfilling and the production of biogas;
- disposal of industrial effluents, the treatment of factory wastes both for degradation of toxic organics and for removal of heavy metals;
- microbiological desulphurization of coal as a way of reducing sulphur dioxide emissions.

BIOREMEDIATION OF SOIL

Microbiological and Biochemical Problems

The soils in some industrial sites are contaminated with hazardous and toxic chemicals: examples are to be found among coking ovens, chemical factories, plants refining and processing oils and other hydrocarbons, and those using wood-preserving chemicals and other toxic substances. If the contaminants are potential targets for biodegradation, the question must be posed as to why, after perhaps prolonged exposure out in the open to natural microbial populations, they have not already decomposed. Sometimes degradation will in fact have started but, because of the chemistry of the pollutants, will be proceeding very slowly. In other situations, the chemicals may be difficult ones for microbes to begin to attack or, perhaps because they are present in low concentrations, there is not enough of them to trigger the microbial attack mechanisms.

Further reasons might relate directly to the local microbiology: too little nutrient for the attacking microbes to build up enough of a working population, perhaps because other microbial species which do not attack the pollutants are more competitive in capturing those nutrients that are available. Local environmental factors (too little moisture, wrong soil texture, too much acidity or alkalinity, too many metallic salts or other interfering substances) may prevent a potentially active microflora from actually getting started. Identification of which factors are limiting biodegradation will enable such appropriate corrections to be made as increasing the water content, breaking up the soil or adding alkali or acid as necessary to bring the local environment close to neutrality.

In some cases the microbiology of degradation may be complex because the pollutants themselves are complicated. No single microbial species may be able to do the whole job and cooperation between several types might be necessary. Quite often a bottleneck arises at an early stage of degradation and, if it can be properly identified, the introduction of the right microbes in sufficient quantities to start the degradative process moving might overcome it; such microbes are sometimes called *vanguards*.

The Mechanics of Decontamination

Microbial decontamination of the soil can sometimes be carried out *in situ*, while at other times it is more effective (though probably also more expensive) first to excavate the soil and then to treat it either in specially engineered beds or heaps, or in equipment designed for the purpose.

The engineering details of *in situ* decontamination are adjusted to fit the local circumstances. If the contamination is shallow, microbes, nutrients and other necessary materials may be directly ploughed into the soil. If the contamination is deeper and the soil at least reasonably permeable, a pattern of injection and recovery wells might be drilled in order to irrigate the contaminated ground with injected water containing the microbes and their nutrient support. The latter procedure is, of course, highly reminiscent of microbial polymer- and surfactant-flooding for enhanced oil recovery (Chapter 10) and experience with either activity helps to perfect procedures for the other.

Excavated soil is usually treated on top of drained sand beds, using plastic or metal foil to isolate the contaminated material from the natural subsoil. Aerobic degradation may be encouraged by circulating air through the mounds or heaps. Heat may be produced as a result of the chemical action and this can help to speed up the process. In some circumstances the soil is treated in rotating drums, brought to the contaminated site for the purpose.

The costs of biotreatment will obviously depend on the local set of problems and will inevitably be compared with other possible means of decontamination. Because it uses "natural" mechanisms, biotreatment might be preferred as being environmentally friendly. Just as with enhanced oil recovery and bioleaching for metals, both of them engineering applications of biotechnology, the biodegradation processes must be properly integrated and compatible with other aspects of on-site operation.

Exploiting biotechnology

DEALING WITH OIL SPILLS

More than a decade ago, nearly 50,000 gallons (200,000 litres) of petrol spilled from a ruptured underground pipeline in Alaska and leaked into a nearby lake. To facilitate clean-up, a dam was quickly constructed to restrict the spill to a shallow part of the lake and wells were drilled into the gravel lake bed to collect as much of the spilled material as possible; about half was recovered. Environmental factors spontaneously took care of most of the rest: as the natural microbial population and the rate of oxygen consumption in the lake both increased, biological processes removed more petrol, over 80% of the original amount in three weeks, and 90% by five weeks after the spill. Nutrients to encourage microbial growth were then added, resulting in a total of 96% removed after a further five weeks. Finally, two oil-utilizing microorganisms were inoculated together with more nutrients: degradation eventually reached 97% of the original spilled hydrocarbon. Natural processes are often effective and can be made even more so by supplementation with additional microorganisms and their nutrients.

Oil Spills at Sea

The occurrence in recent years of major oil spills at sea and in coastal waters has provided some of the most widely publicized ecological disasters, none of them more so than the *Exxon Valdez* episode in Alaska. In major incidents, the oil, often unrefined crude, spills from tankers as a result of collision damage or other accidents, although deliberate discharges when shipboard tanks are cleaned, usually further from land, are not unknown. The worst ever spill close inshore was the deliberate release of crude during the Gulf War.

Because initially it is less dense than water, the crude oil floats on the ocean surface, originally as a coherent slick. If the water is calm, and remedial measures are begun very quickly, much of the spilled oil may be contained within booms and recovered by skimming. However, with the passage of time, the oil begins to weather: the lighter components evaporate to leave a progressively denser and heavier sticky, tarry sludge of oil and water (the so-called "chocolate mousse") which tends to sink below the surface, beyond any hope of further easy mechanical recovery.

Under the influence of onshore winds and tides, slicks originating from spills not too far out to sea may be swept onto the shore where the oil coats beaches and rocks, causing enormous damage to wildlife as well as to inshore shellfish cultivation beds or other fish farming. (In the January 1993 incident in the Orkney Islands, the main concern was for salmon farming. Fortunately the stormy weather dispersed the oil so rapidly that little lasting damage seems to have resulted.) The longer the spilled oil has weathered without actually sinking, the stickier the mess that comes ashore. It is therefore very important to recover and remove the oil quickly from the shoreline in order to minimize economic losses resulting from the costs of cleaning, the interference with local fisheries and a possible fall in tourist revenues. With widespread ecological awareness, the instantly televised damage to wildlife tends to elicit a vigorous vocal public response.

Keeping the place clean

Non-biological treatments for oil spills

Apart from mechanical skimming, industrial detergents (surfactants) have often been sprayed onto the slicks as dispersants to break them up into small globules and promote more rapid natural degradation. This procedure has now fallen from favour, partly because it tends to encourage the oil to sink without actually having been degraded, and partly because, even when the detergents themselves are not toxic to marine life, they spread the oil slick over a such wide area that fish are unable to swim clear and die from a shortage of oxygen. Some people feel that detergents do more damage than oil and they are now banned for offshore use in the US and most European countries.

Once the oil has come ashore, various methods are used to try to clean it off. They include high pressure water hoses, with detergents where permitted, to wash the oil down the beach where it can be collected and removed. Many of the available techniques were eventually used in the *Exxon Valdez* episode, with varying reports of their effectiveness but, because of a total ban on the use of dispersants on that occasion, much of the work had to be done manually. A number of ecologists and others concerned with the environmental damage caused by oil spills on beaches remain unconvinced that all the remedial measures in Alaska were effective.

Microbiological treatment of oil spills

Several attempts have been made to assist natural microbial degradation of oil both by encouraging the growth of indigenous marine oil-degrading organisms and by inoculating the slicks with microbial cultures prepared and held in stock specially for the purpose. Stimulating microbial growth might entail applying soluble inorganic nitrogen and phosphorus compounds; these elements are necessary for growth and are not present in sea water in sufficiently high concentrations to support the comparatively luxuriant microbial growth preferred for oil degradation. To serve as a binder, straw is sometimes also dumped onto the oil; there are no deleterious long-term effects as straw is fully biodegradable. In seawater these nutrients tend very easily to be washed away and diluted so new *oleophilic fertilizers* have been developed. They comprise the nitrogen, phosphorus and other supplements for microbial growth compounded together with an organic material which makes the preparation attach to oil droplets and not dissolve in seawater. The nutrients are not released until the bacteria themselves attack the fertilizer but the criticism has been levelled that the microbes then go on to use as food the organic part of the fertilizer and no longer bother with the spilled oil.

An extension of these microbiological procedures has been used in attempts at beach clean-up. In Alaska, for example, some rocks were sprayed with microbial cultures and nutrients which, it was hoped, would help to degrade oil. The reports suggest that they did so, but that it was rather a slow process; some people wonder whether there is really any need to do anything at all as in time the natural microbial population will clean things up. At most, they suggest, sprinkle some fertilizer.

One of the reasons for the relatively long time it takes for microorganisms to remove oil is their need first to make the chemicals which allow them to mount an enzymatic attack on

Exploiting biotechnology

the oil itself. Since water and oil repel one another, oil and oil components are, of course, very insoluble in water. The microbes, like all other living beings, are based on water which constitutes 80% of their body weight. The problem for them is how to begin an attack on the oil from their base in water, a problem they commonly solve by making their own *biosurfactants (biodetergents)*. This enables them to emulsify small globules in water, just like the industrial detergents mentioned earlier, but the microbes do this comparatively slowly, which partly accounts for the length of the whole clean-up process.

Note again the relationship with the proposed use of microbial surfactants for certain enhanced oil recovery procedures (Chapter 10). A new development, which grew out of work on oil recovery and is still at an experimental stage, is to imitate the microbes and use biosurfactants instead of industrial detergents to wash the oil off the beach. This would have several advantages:

- like industrial detergents, they facilitate rinsing the oil off the sand and rocks into collection equipment;
- they rapidly emulsify any remaining oil and so speed up microbial action;
- biosurfactants are not toxic to the degrading microbes, as are many industrial detergents;
- biosurfactants, as biological products, are biodegradable and, in marked contrast to the persistence of the industrial detergents, rapidly disappear from the natural environment;
- at the present time, however, they are relatively expensive.

Much of the initial thick layer of oil deposited on a beach is probably best removed by pressure hosing; the deployment of dispersants is next used to mobilize the remaining oil by emulsifying the hard surface layer that develops. In recent trials, biosurfactants effectively washed oil from an artificial model beach. However, to be useful in real oil spills, all the necessary materials must be readily available in adequate quantities to deal with problems rapidly when they arise. The production of biosurfactants for beach cleaning has not yet been scaled up to provide the stocks which should be held in reserve. So far, no more than expensive experimental quantities have been prepared but scale-up will markedly reduce production costs and, if the market is there, the products will surely be developed.

Leaks From Fuel Tanks

A not uncommon problem is the leakage of petrol or diesel fuel from storage tanks, either as the result of corrosion or because of some operational mishap. The fluid may drain into the soil and eventually find its way into a water course or underground aquifer.

Microorganisms will degrade petrol and most components in diesel oil but they need the oxygen of the air to do so. Depending on the local circumstances and where the escaped fluid is judged to have gone, it might be possible, as in soil bioremediation, to degrade the hydrocarbons by irrigating *in situ* with aerated water containing the appropriate microbial inoculum together with such additional inorganic nutrients as the microorganisms may require. Degradation can take weeks or months to complete, depending as it does on a host of factors, including the ease with which oxygen and bacteria reach the hydrocarbons. If the pollution has extended beyond local soil contamination, prolonged flushing with a bacterial system able to accomplish the degradation may be attempted through the same route as the fluid leaks. A major difficulty with all deep underground remediation is to know exactly what is happening. Sampling is difficult and limited, especially when there are buildings in place, so a knowledge of activity below the ground often relies on modelling the events on the basis of such data as might be available.

DISPOSAL OF HYDROCARBON WASTES

Wastes which must be rendered innocuous before disposal arise from two sources in oil production and processing:

- reservoir water coproduced with crude oil, from which it is separated on the surface;
- refinery wastes and sludges of various types.

Disposal of Coproduced Water

As production from an oilfield gradually matures, more and more water is produced together with the oil; some of the reasons are discussed in Chapter 10. The combined fluids, which may also contained dissolved gas, are passed into separators, first to remove gas and then to separate oil from water. A simple separator tank is not sufficient and the water to be disposed of still contains appreciable quantities of dissolved and suspended hydrocarbons. The two common alternative methods are then:

- disposal on site by reinjection into the reservoir, either for waterflooding or via special disposal wells into regions of the field no longer used for production. Such disposal procedures might require no further treatment of the water other than that needed for operational reasons;
- removal from the site by discharge into waterways onshore or into the sea offshore.

We have already observed that hydrocarbons in the environment are eventually dispersed and degraded. However, the primary wellhead separation of the two fluids may leave enough hydrocarbon in the water close to the point of discharge to give local concentrations which are toxic to aquatic life and might enter the human food chain via plankton and fish. Regulations for off-site disposal usually require extensive scavenging of the remaining hydrocarbon before discharge of the water from the production site.

As an example of regulatory requirements, produced waters discharged into the North Sea must not contain more than 40 parts per million of hydrocarbon, no simple task when a single offshore platform might be producing more than 100,000 barrels (15,000 tons) of water a day, all of which has to go somewhere. The platforms have elaborate separation equipment which can achieve this low level of hydrocarbon in the discharge water and, incidentally,

Exploiting biotechnology

recover significant additional quantities of oil in the process. Nevertheless, environmentalists continually demand from legislators ever lower levels of hydrocarbon discharge into the sea, and so the industry is alert to the possibility of a reduction of the discharge limits to 25 or even 20 parts per million. Reducing the residual hydrocarbon by extending the conventional types of separators will in many cases impose severe logistic engineering problems on the limited platform areas; removal of residual hydrocarbon by microbiological methods might prove to be more cost-effective.

Microbial remediation of oily wastes usually starts with one or more naturally-occurring strains of microbes which are modified or improved in the laboratory, either to yield more product per unit of feedstock supplied, or to make them more amenable for incorporation into an industrial process. The obvious place to look for oil-degrading microbes able to dispose of oily wastes is in environments (like those noted in the case of oil spills) which already harbour microorganisms capable of decomposing crude oil and refined petroleum products.

The best microbes for hydrocarbon oxidation will be candidates for further improvement by finding the optimum conditions for them to function. This will be done by mutation and selection to more effective forms, and perhaps eventually by genetic modification to alter specific properties. In a suitably dense suspension, contained in a vigorously aerated treatment tank, they will convert most of the residual oil to carbon dioxide and water. The ability of microbes to catalyse chemical change at very low substrate concentrations probably makes biological treatment plants practical and cost-effective even for the very rapid rates of waste water through-put on oil platforms.

Disposal of Refinery Wastes and Sludges

Mention was made early in this chapter of the practice of sludge farming, until recently the way of disposing of more than 80% of refinery wastes. The bulk of such wastes are now classified as hazardous and sludge farming is increasingly frowned upon because of the possibility of incomplete decomposition, with toxic chemicals being leached into aquifers and other sensitive environmental structures. In sludge farming a natural population is relied upon to degrade the waste material and, just as in the removal of hydrocarbons from discharged coproduced water, those natural populations can be isolated, cultured under controlled conditions, and used to attack the sludges in a confined tank or *bioreactor*.

The biotreatments of refinery process streams, and of soil or water already contaminated with oil, both represent opportunities for biotechnology:

- the treatment of process streams is the more urgent in order to eliminate as soon as possible the need for disposing of any more oily wastes on land;
- bioremediation of existing contaminated sites, while potentially a great opportunity for dealing with many years of disposal into landfills, soils and lagoons, is perhaps less pressing since those locations have already been despoiled.

Both *in situ* treatment and decontamination of the excavated material in an appropriate facility can be used to deal with these problems; in general, the techniques are similar to those for the bioremediation of soil, using microorganisms specifically chosen and adapted for the degradation of those types of wastes. Treatment times tend to be long, with some components of the wastes more susceptible to degradation than others; a small intractable residue may remain at the end. Another disadvantage is that, particularly with *in situ* degradations, it is difficult to prove effective decontamination without extensive monitoring for the presence of specific toxic chemicals. However, operating costs are usually very much lower for some biotreatments than for such alternative procedures as chemical stabilization, solvent extraction or incineration on- or off-site, and the necessary equipment can often be brought to the treatment location in an easily portable form.

Sewage Treatment

In some places raw sewage is still discharged directly into the sea but the practice is declining in developed countries where sewage disposal is a major industry. A decade ago estimates for the UK showed that 100 tons of water were purified per year for each member of the population, with a capital investment per head of £100–200. In the US, spending on pollution control in the ten-year period 1976–86 is believed to have been some $400 billion.

Sewage comprises domestic as well as industrial liquid wastes; most of the materials in it will inevitably degrade in the natural environment but it is pleasanter, more effective and less risky to do so under contained and controlled conditions. In some localities rainwater also flows into the sewers and provision will then be made to divert large quantities of storm water (which in itself does not warrant the treatment accorded to sewage) in order not to overload the processing facilities.

The treatment of sewage, aside from its obvious amenity value, has four objectives:

- the removal of readily oxidizable organic materials contributing to biological oxygen demand. If these substances are discharged into rivers and streams they will encourage microbial growth which will deplete dissolved oxygen and kill the fish;
- the removal of solids in suspension which would otherwise result in the silting of waterways;
- the removal of toxic materials, such as ammonia (which arises from the biological degradation of many natural wastes), as well as heavy metals and other pollutants of industrial origin;
- the elimination of pathogenic bacteria and viruses, some of which originate from human disease victims and carriers and which might infect others downstream.

In coastal regions, treated sewage is discharged at sea; inland, the water is returned to the rivers usually in so clean a condition that it is virtually or actually suitable for people to drink with no further treatment.

The presence of industrial wastes may seriously complicate the treatment process. Domestic sewage is rather constant in composition and contains few really toxic materials. Industrial waste may be much more variable depending on schedules and other factors related to the

Exploiting biotechnology

producing facilities: some of the chemicals released might even be directly deleterious to the processes of sewage degradation. Various regulations control such discharges and, if the public authorities agree to their release into the sewers, the specific costs for dealing with them may be levied on the polluters.

The Treatment Processes

After extracting gross particulate matter (cloth, wood, residual metal, etc.) by coarse screening, grit is removed by being allowed to settle and the sewage is then passed to sedimentation tanks for the removal of most of the suspended solids as a sludge, simultaneously eliminating much of the biological oxygen demand.

The sludge is further digested using aerobic or anaerobic conditions, or both. A final refractory sludge emerges from this stage of digestion which is no longer objectionable or harmful; it may be buried, dumped at sea or, if the heavy metal content is acceptable, used as fertilizer.

The liquid waste undergoes further microbial treatment with the objective of oxidizing the organic wastes. There are a number of configurations for doing so. The *biological* or *percolating filtration* process sprinkles the water from a slowly rotating dispenser over pieces of stone or clinker, often arranged in circular beds, a familiar suburban scene. The material of the filter bed spontaneously becomes populated with microorganisms, including bacteria, fungi and protozoa. Together they form a community supported by the organic nutrients in the sewage. Most of the oxidation is carried out by the bacteria and the oxygen they need diffuses into the filter bed from the air; the bed must therefore be sufficiently permeable to allow air to enter, and wetting the surface with a water spray encourages the dissolution of oxygen. Bacterial growth sometimes forms so much bulk that the filter bed becomes blocked. However, the protozoa in the beds graze on the bacterial film and liquid flow rates are controlled to avoid excessive build up. Shaped plastic particles have recently been used instead of stone or clinker to facilitate the penetration of water and air through the bed.

Alternatives to biological filters include the *rotary biological contactor*, a specially designed piece of equipment in which a series of discs with honeycomb surfaces are rotated several times a minute, partly submerged in the liquid being treated. Microbes colonize the disc surfaces and the efficiency of the degradation process is improved by their alternate access to nutrients in the sewage and oxygen in the air.

In the *activated sludge process* the microbes are not attached to a solid support but form flocs suspended in the aerated liquid. As the incoming organic material is degraded, the sludge increases in bulk; some of it is drawn off to be degraded anaerobically while the remainder is recycled through the system. This process is much faster than a biological filter of similar volume, so it is less demanding of land area and there is less investment in equipment. However, the running costs are higher and technically it is more difficult to operate; pure oxygen used instead of air accelerates the process but at an additional cost.

Keeping the place clean

A variant of the activated sludge configuration is the *deep shaft process* in which the treatment takes place in an upright cylinder, 50–150 metres (165–500 feet) deep and up to 10 metres (33 feet) in diameter, wholly or partly buried in the ground. The shaft is so designed that the injected air keeps the liquid circulating and sedimentation tanks are provided to remove the excess sludge. Although the biological processes are basically similar in all the aerobic treatment processes, other engineering designs have been developed; an important one is the fluidized bed process which seeks to combine the benefits of biological filtration and activated sludge. In this system the microbes, instead of forming flocs which take a long time to settle, are allowed to attach to particles small enough to be recirculated but large enough to settle quickly. This enables denser microbial populations to be used and reduces the size of the facility accordingly.

ANAEROBIC DIGESTION AND ITS USES

Reference was earlier made to the ability of microbes to degrade sewage sludges anaerobically, that is, in the absence of air and oxygen. Although the process can be slow, and speed will depend both on the nature of the material being degraded and on the conditions under which it is taking place, this form of degradation has a number of benefits including low cost and the generation, under some circumstances, of a useful product.

Landfilling

Much of the solid and semi-solid waste produced in developed countries is buried in landfills. Rough estimates for the past few years suggest a daily *per capita* generation of about 4 pounds (2 kilograms) of such material in the US with about two-thirds of that amount in Europe. The composition of the waste varies from country to country. In the most industrialized regions, decomposable vegetable matter, paper and card may comprise more than 60% of the total dry matter, the rest of the bulk being made up of glass, textiles, metals, plastic and other material.

In a landfill operation, the waste is dumped in layers, each up to 8 feet (2.5 metres) thick, the layers being covered with a foot or two (30–60 centimetres) of earth. The wet material decays under attack from the microorganisms always in the environment and also present in the waste material itself. Once the material is compacted in place, and particularly after being covered by an earth layer, fresh air is barred from further access; oxygen buried with the garbage is rapidly used up and the system becomes anaerobic. As the refuse decays, the mass settles and compacts further, often allowing additional dumping at the same site.

Under ideal conditions the main products of this decomposition are ammonia, almost all of which remains in solution in the water, and a gas composed roughly of 70% methane and 30% carbon dioxide; the gas is called *landfill gas* or, more generally, *biogas* because of its biological origin. The gas mixture is combustible in air and might be useful as a source of fuel. However, problems often arise with poorly managed landfill operations resulting in the formation of

Exploiting biotechnology

acid *leachates*, liquids which may cause extensive environmental damage if they are allowed to seep from the base of the landfill. The amount of water in the landfill is important: too much encourages the formation of leachate but too little inhibits the production of methane. Efficient management uses optimal quantities of water for the particular circumstances. Most of the gas is generated in the first 15 years or so but some production continues long after that.

The Importance of Biogas

The methane/carbon dioxide mixture is both beneficial and dangerous. As we observed early in this chapter, if methane is not properly vented from a landfill it might find its way out of the underground site, seeping along soil layers until it is able escape into the atmosphere. At certain concentrations, methane/air mixtures are explosive and methane originating from a neighbouring landfill or from a mineral source does occasionally accumulate in the basement of a building or in some other sensitive location; if detonated, it may cause a dangerous explosion.

Some digester plants are specifically designed to make use of the methane they generate for their own power needs, or for some other use. It is not easy to predict the net amount of useful fuel gas to be produced because, as well as intrinsic fluctuations in the rate of gas generation, the microbiological degradation is sensitive to variations in the composition of the feedstock and to the presence of inhibitors like ammonia or antibiotics from farm wastes; there may also be energy costs associated with heating the fermenter and operating ancillary equipment.

The relative importance of anaerobic digestion for waste control, versus the generation of methane as a fuel, varies in different locations and is dependent on the price and accessibility of alternative fuels, the ability of a population to buy that fuel and the availability of feedstock. In developed countries, disposal of waste may be the more important, with gas generation a useful by-product. In poor countries with a high proportion of their population in agriculture, as well as large numbers of farm animals, heavy use is made of simple anaerobic fermenters in rural areas to generate biogas as fuel for cooking, heating and low-wattage electricity generators. There are said to be more than 7 million such small-scale facilities in China in addition to more elaborate installations for power generation on a larger scale. They are also used in other Far Eastern countries, in Africa and in India, mostly powered by human and animal faeces, and waste vegetable matter; in all those countries wood for fuel is becoming increasingly difficult to find, and petroleum fuels are too expensive for much of the population, so that biogas makes a significant contribution to the total fuel supply.

With more capital investment, however, farm wastes can be made to do better. A UK company has pioneered an almost completely automatic plant for turning poultry litter — bird droppings, woodshavings and straw — into electricity. The material, brought in from intensive chicken and turkey farms, is burned at around 900°C to make superheated steam for driving a turbogenerator. To satisfy the power needs of one person requires the litter from 1,000 chickens. The burn is said to be so clean that the exhaust gas is virtually all nitrogen and steam, neither an environmental hazard. The ash, high in potassium and phosphorus, can be used as fertilizer for crops and grassland; it contains no undesirable nitrate to leach into water courses, a major benefit because poultry litter used directly on the land does release a lot of nitrate.

While waste disposal is usually the main reason for operating the digesters in developed countries, and the importance of cheap fuel is less significant, the generation of landfill gas is not trivial. Estimates suggest that the material discarded from an average household in an industrialized country would generate enough biogas to boil a few litres of water a day. While not making much of a dent in the electricity bills, arguments can be made that, on a national scale, biogas could offer major savings on often imported petroleum products although considerable capital investment would be required. For the US it has been calculated that the digestion of all its crop and animal residues might produce 1-10% of total energy requirements. The European Community expects to satisfy a small percentage of its total energy needs in this way.

In the UK at the present time, about 30 million tons of refuse containing organic matter are disposed of each year, 90% of it in landfills. Most of the gas produced is employed to generate electricity, with subsidiary uses for firing kilns, boilers and furnaces, for refining as a chemical feedstock, for transmission by pipeline or as a vehicle fuel. After the US, Britain is the world's second largest commercial user of landfill gas. In 1990 the country had 36 landfill gas utilization schemes, producing power equivalent to 150,000 tons of coal annually; an additional 60-70 projects are expected to be developed within the next decade, giving a total annual coal equivalent of about 1 million tons.

DISPOSAL OF INDUSTRIAL EFFLUENTS: TOXIC ORGANICS AND HEAVY METALS

Hazardous Wastes

A wide range of manufacturing and other industrial processes produce solid and liquid wastes which may be described as "hazardous", and special precautions must be taken for dealing with them. Some 3.5 million tons of such waste were generated in England and Wales in 1985; about 80% was deposited in landfills, the remainder being disposed of by incineration, solidification, chemical treatment, or disposal at sea or in mineshafts.

In 1981 the Land Wastes Division of the UK Department of the Environment defined hazardous or special wastes as follows:

> *If a substance contains prescription only medicines or substances which have flash points of 21° C or less, or contains known or probable human carcinogens at a concentration of 1% or more or is likely to cause serious tissue damage on exposure for a period up to 15 minutes or if ingestion of 5 cubic centimetres* (about one teaspoonful) *is likely to cause death or serious tissue damage to a 20 kilogram* (44 pound) *child, then it is defined as a special waste.*

Exploiting biotechnology

Such a definition covers an enormous multitude of compounds, divisible generally into a number of classes:

- some 20 different metal and metal-containing compounds, including arsenic, cadmium, copper, and mercury;
- asbestos;
- biocides;
- pharmaceutical compounds for treating the diseases of man, animals and plants;
- phosphorus and its compounds;
- complex organic compounds, tarry wastes and residues;
- inorganic compounds containing *halogens*, sulphur and cyanides (halogen, meaning "salt-forming", refers to elements like chlorine and iodine; the most common of their salts is table salt, or sodium chloride);
- organic halogen compounds.

Biological Disposal of Organic Wastes

In landfill, degradation depends on mixed microbiological populations, culminating in the action of the methane-producing bacteria. These populations may be unable to attack many of the hazardous waste materials or the materials may be present in concentrations too low by themselves to support microbial growth. But, when the microbes are also well supplied with easily biodegradable materials — such as the municipal solid waste which is also usually disposed of in landfill — a process called *co-disposal* takes place: the municipal waste supports the bulk of the microbiological activity in the course of which the organic components of the hazardous wastes are also degraded.

Co-disposal requires careful management of the landfills. It is important to ensure that the proper proportions of municipal and industrial wastes are present. The nature of the industrial wastes themselves is very important because their solubility in water, acidity or alkalinity and chemical properties are all factors which can greatly influence the speed of biodegradation. Conditions which promote a rapid production of methane from the municipal wastes are likely also to encourage degradation of the industrial effluents while, conversely, anything seriously harming the microbial populations will correspondingly inhibit toxic waste decomposition.

Removing Heavy (Toxic) Metals

Mention was made earlier of the special problems of toxic metal disposal. In many cases it is difficult to render these elements really safe for general environmental disposal and they need to be collected, concentrated and entombed (for example, in a disused mineshaft) right out of harm's way.

Keeping the place clean

Contemporary chemical methods are satisfactory for absorbing the bulk of the metals from solution onto comparatively small volumes of absorbent material appropriate for the specific metals present in particular waste streams. It is then possible either to put the absorbent together with its absorbed metal into permanent storage or to displace the metal into a small volume of liquid and re-use the absorbent.

Two problems may arise:

- the chemical absorbents may not scavenge the toxic metals down to sufficiently low concentration for the safe discharge of the liquids from which they were removed;
- the presence of other, non-toxic, chemical elements and compounds in much higher concentrations than those of the toxic materials may overload the absorbent and cause it to become saturated before all the toxic substances have been immobilized.

Both difficulties lend themselves to biotechnological resolution.

As well as the Pacific and alpine plants discussed earlier, many other living organisms, including microbes, can absorb metals from their environment, often from solutions of very low concentration and sometimes with quite amazing specificities: a number of bacteria accumulate lead, cadmium or zinc up to 60% of their (dry) body weight. Certain aquatic algae (tiny plants) do even better and accumulate 90% of their weight as uranium. Some marine animals are able to concentrate within their own bodies vanadium from sea water even when the amount of metal present is less than about 1 gram in 3,000 tons of water! No doubt some would do the same for silver or gold. (Incidentally, the economics of extracting gold from sea water is limited by its very low concentration [1 gram for every 125,000–250,000 tons of water] and the consequent burden of sending huge quantities of water through an extraction system. The cost of recovering one ounce of gold, worth a little less than $400, would have to include moving 3.75–7.5 million tons of water, plus other costs of extraction. That is hardly a promising investment unless perhaps, by some happy chance, a location for performing the extraction could be found where a rapid natural current eliminated the water pumping costs.)

The range of metals absorbed, the ability to discriminate between them, the weight of metal each microbial cell can accumulate, and its ability to extract metals in very low concentrations, vary for different metals and different microbes. In some cases the absorbing biomass (*biosorbent*) can be reused after removal of the bound metals under controlled conditions. Several sorts of biochemical metal-absorbing mechanisms exist in nature:

- certain chemical complexes can form between the metals and the biochemical components of cell surfaces, usually without much discrimination between different metals. This is a passive absorption phenomenon, sometimes shown by dead as well as living biomass, when clearly it cannot involve active metabolism. While often rapid, this process may be influenced by the local chemical environment, including the degree of acidity. The presence of certain sugar polymers on the cell surface can be an important factor in the absorption process;
- the use of one or more enzymes to change a soluble form of a metal present in the cell's environment into an insoluble compound which precipitates on the cell surface or in its external environment. Outside the cell, insoluble precipitates are likely to sink to the bottom of a natural body of water to become part of the bottom sediment, or become

Exploiting biotechnology

trapped in an industrial precipitation filter as we shall shortly see. Precipitated inside the cell, they remain there;

- attachment to *metallothioneins*, proteins which very specifically recognize and bind certain metals. Metallothioneins are characteristic of higher organisms, including man, but — using the techniques of genetic engineering — the genes specifying them can be inserted into microbial cells in which the proteins they code for can subsequently be expressed and used as binding sites for particular metals.

An additional practical advantage which can be important for eradicating certain types of pollution is the capacity of some microbial species to absorb organic compounds together with heavy metals.

Biotechnological processes for removing heavy metals

There are several engineering configurations for biological systems binding and extracting metals from effluent waste streams. One is to use either an aerobic or anaerobic sludge system in a suitable vessel, rather like the use of sludges in sewage treatment, except that here the purpose is to absorb metals rather than to degrade organic chemicals. Another is to immobilize the cells onto a support matrix, enclosed in a cartridge on the end of a waste pipe, through which is passed the contaminated effluent stream; the metals accumulate in or on the immobilized cells, while precipitates are trapped within the matrix. A third way is for the waste stream to be made to flow slowly through a winding open channel (a *meander stream*) in which is grown an alga chosen for its ability to absorb the offending metal(s) in the waste. This use of algae is well suited to regions with bright sunlight and a warm climate where they grow well.

Once the metal has been absorbed by the biomass a variety of choices exist for its ultimate disposal. In some biosystems, appropriate chemicals are used to detach the bound metal from the biomass and concentrate it into a small volume of solution for easy disposal: by evaporation to leave a quantity of dry solid residue, by absorption of the liquid into a solid or by retention of the concentrated liquid in special storage facilities. When regeneration of the biomass is not feasible, the cells together with their bound metal might be incinerated, reducing the bound metals and a little of the biomass to ash.

Metal Separations

Although no processes have yet been developed for commercial use, there may be opportunities for employing specific biosorbents to separate from one another industrially valuable metals which are difficult and expensive to purify by conventional chemical means. There has been mention in the technical literature of the desirability of new methods for separating and purifying rare earth metals as well as better ways of separating hafnium and zirconium, which are used in the nuclear industry.

The Special Case of Radioactive Wastes

Nuclear reprocessing generates waste streams containing radioactive elements. Not only is their radiation hazardous to all forms of life because of the damage it causes to biological chemicals, some are also acutely poisonous.

As they release the energy used industrially to generate electricity, the radioactive atoms of the uranium and plutonium in nuclear fuel break up into a series of *decay products*, some of which are themselves radioactive; they are recovered for long-term storage when the exhausted fuel is reprocessed. The procedures generate large volumes of spent reaction fluids which contain some of the radioactive atoms originating from the fuel rods. Because of their hazardous nature, strict regulations govern their release into the environment; public concern maintains continual political pressure for ever lower permissible levels of discharge.

The bulk of the radioactivity is stripped out of the waste streams by chemical technologies which concentrate them into small volumes of solution either for permanent disposal in special containers or for vitrification into solid blocks which are buried in chambers located deep underground in stable geological formations. After removing most of the radioactive elements, the difficulty — as with other metal scavenging procedures — is to eliminate the last traces of radioactivity, particularly as the residual concentrations are at or below the limit for further chemical or physical treatment at reasonable cost. The difficulties may be compounded by the simultaneous presence of comparatively large amounts of related harmless chemicals which occupy so many of the binding sites on most absorbents that their ability to bind the radioactive elements is impaired. For example, of the four major radioactive elements in nuclear reprocessing waste streams (uranium, plutonium, strontium and caesium), strontium is chemically similar to calcium and caesium to sodium, both present in large amounts compared with their radioactive relatives. Because of their specificity properties, the ability of biological systems to capture very low concentrations of certain metals in the presence of high concentrations of interfering substances has clear attractions for helping to deal with the radioactivity problem; although their value for this purpose has been under public discussion for several years they are, however, not yet used in reprocessing technologies.

MICROBIOLOGICAL DESULPHURIZATION OF COAL

Coal Burning and Acid Rain

Coal remains one of the major fuels of industrial countries, most of it burned in electricity generating stations. Like crude oil, it originates from the remains of ancient living organisms, decayed and transformed under high pressure and temperature over long periods of geological time.

The transformations have been so extensive that, while traces of the original biological molecules can still be identified, the bulk of the material bears little resemblance to the biochemicals of the organisms from which it once came. Compared with lignite (brown/soft

Exploiting biotechnology

coal), hard (black/bituminous) coal — composed largely of carbon atoms linked in complex arrangements — is the more completely transformed type. The matrix has less hydrogen per average carbon than does crude oil and includes small quantities of other elements, in particular oxygen, nitrogen and sulphur. Coal also contains minerals which, while physically embedded within the matrix, are chemically quite separate from it. Coal is therefore a mixture, with the minerals forming distinct crystalline particles lying randomly within the bulk of the material. Prominent among those minerals is iron pyrite, that same sulphur-containing substance which plays such an important part in metal leaching.

Burning the coal, a process of oxidation which gives carbon dioxide and some water as the main products, also results in the oxidation of much of the sulphur, both that which forms an integral part of the matrix and is described as "organic" and the separate "inorganic" mineral pyrite inclusions. The oxidized sulphur is released as a gas called *sulphur dioxide*; in the atmosphere some of this gas reacts further with oxygen to give the even more oxidized product *sulphur trioxide* which, when dissolved in water droplets in the air, becomes sulphuric acid and returns to earth as acid rain. Sulphuric acid is not the only undesirable component of acid rain but it probably does the most damage.

Social pressures and government legislation, particularly in developed countries, are aimed at reducing emissions of sulphur dioxide, above all that major contribution emanating from power stations because it is quantitatively the most significant and technically the easiest to deal with. While all coal contains some sulphur, some coals contain much more than others; those with a sulphur content below 1% are regarded as being *low-sulphur coals*, producing quantities of sulphur dioxide per unit of coal burned and energy generated which, for the moment at any rate, are regarded as acceptable. The problems arise mainly from *high-sulphur coal* which may contain 5% sulphur or more.

The Problem of Dealing with the Sulphur

Coal is a variable material. Deposits are very diverse in their sulphur contents and even coal from a particular seam will show quite distinct differences in sulphur content from place to place. The variations within a seam present no great problems because whatever methods are used to control or prevent sulphur dioxide emissions can be designed for the average value. Much greater problems arise when major deposits, or the whole of a country's available coal, is of the high sulphur type. The choices might then lie between:

- importing an alternative supply of coal with a lower sulphur content;
- installing equipment for scrubbing the sulphur dioxide from the flue gases;
- removing some or all of the sulphur from the coal before it is burned.

Each has its problems, both social and economic.

Compared with coal having a high sulphur content, low-sulphur coal not surprisingly attracts a price premium: its use avoids the additional costs of pollution control. As legislation becomes more widespread and more restrictive, the price of low-sulphur coal will tend to

Keeping the place clean

rise (Figure 11.1). Higher prices for low-sulphur coal will make the costs of emission control or prior sulphur removal from lower cost, high-sulphur resources increasingly attractive as cost-effective options. Furthermore, the progressive elimination of high-sulphur coal from power station use would have a devastating effect on coal-mining areas in many countries (including the UK and the eastern half, in particular, of the US), with high consequential social and economic costs. When framing their legislation, governments must be expected to take those costs into account. Hence, in making such decisions, the direct and social costs of using highly priced, low-sulphur coal must be set against that of low cost, high-sulphur coal — plus the expense of pollution control.

The Non-Biological Options

Before outlining how biotechnology might contribute to the resolution of these problems it is important to survey briefly other ways in which it can or might be done. It may seem a little churlish to lump these under the apparently dismissive heading of "The non-biological

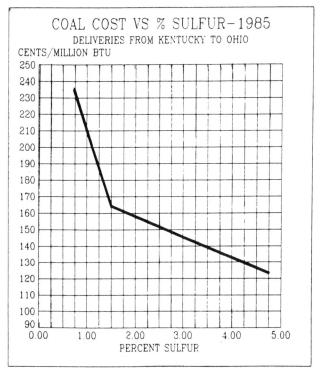

Figure 11.1 Average cost of coal versus sulphur content for Kentucky and West Virginia coal delivered to Ohio utilities during 1985. (*Reproduced from Mining Engineers 39, p806, August 1987, with permission of the Society for Mining, Metallurgy and Exploration*)

Exploiting biotechnology

options" but the obvious alternative of calling them "conventional" is equally misleading because some of them are quite as innovative as the biotechnological proposals.

Coal washing

This procedure is carried out before burning in many power stations. Usually the finest particles are first removed because of the high cost of cleaning them, the coarser material washed and then blended again with the unwashed fines (Figure 11.2). Although this does not necessarily reduce the sulphur content on a weight basis — sometimes it is actually increased — sulphur is reduced per unit of heat generated because the calorific value of the fuel rises with washing.

Fluidized bed combustion

Keeping a mass of finely-divided coal particles suspended by a powerful upward draught of air which also supplies the oxygen for combustion ensures that the coal is completely burned. The temperature and other conditions in the chamber may combine to retain the sulphur in the ash rather than releasing it as sulphur dioxide. One modern design uses coal straight from the mine without washing, and limestone added to the powdered coal traps the released sulphur in the molten slag which is later discharged, cooled and solidified. Such new combustor designs might prove to be important elements in solving the sulphur emission problem.

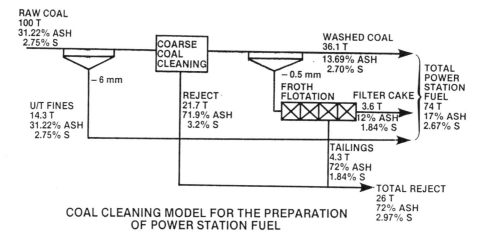

Figure 11.2 Flowchart for current preparation of coals for power stations. (*Reproduced by permission of British Coal*)

Keeping the place clean

Flue gas desulphurization

This is a further option. Equipment fitted to existing plant scrubs the exhaust gases free from sulphurous components which are converted into gypsum (used for making plaster and wallboard for buildings) or otherwise disposed of. Capital costs are high, estimated at £160 million for a UK coal-burning plant generating 2,000 MW, and problems arise in the disposal of toxic waste discharged from the process. But removing each ton of sulphur this way costs only one-fifth as much as a ton removed by coal washing. This technique, for the moment, is the preferred method in the UK, although the original £2 billion programme has been reduced to £1.2 billion by fitting fewer units.

Other treatments

All more-or-less experimental, they include:

- grinding the coal into particles fine enough to allow the pyrite to become free might allow the two solids to be separated physically by virtue of the greater density of the pyrite;
- using a magnetic field to extract the weakly magnetic pyrite from powdered coal;
- chemical treatments, which can remove much of the organic as well as the pyritic sulphur. However, they are expensive and some of the calorific value of the coal may be lost;
- flotation procedures, in which coal particles are suspended in water containing a frothing agent while air is pumped vigorously through the liquid. The less dense coal particles attach to the air bubbles and are skimmed off the surface.

None of these many technologies is entirely satisfactory and their development continues. This leaves room for the possible use of microbial methods for removing sulphur; like the other techniques, they are not perfect either but offer a different range of choices.

Microbial Systems for Removing Sulphur

Organic sulphur

As we have seen in Chapter 2, biochemical reactions depend on the action of specific enzyme catalysts which underpin the whole of biochemical organization. Coal generally does not participate in the chemistry of contemporary organisms and its matrix does not serve as a source of nutrients; probably no enzymes have evolved specifically to catalyse coal chemistry. It is true that the chemicals in coal were once part of living organisms but they have undergone such extensive change that they are now far removed from present-day biochemicals. Nevertheless, because enzyme specificity is not always quite perfect, microbes do exist which fortuitously

Exploiting biotechnology

will weakly attack the organic sulphur atoms forming an integral part of the coal structure, but it is indeed a weak effect.

Some microbes have been discovered which, given enough time, are able to remove about half the organic sulphur of some coals, and a pilot plant has been constructed in Sardinia for testing at least one process on an industrial scale; the results are awaited with interest. Conventional strain improvement or genetic manipulation might improve the microbial performance but, for most high-sulphur coals, simply removing half the organic sulphur would not be sufficient to bring the total sulphur content down to acceptable levels.

Pyritic sulphur

Sulphur in this form is another matter. Bacteria can use the oxidation of pyrite sulphide to sulphate as a source of energy, just as they do in bioleaching (Chapter 10). Some do so at ambient temperatures up to about blood heat, while others work more rapidly and in conditions as hot as 80°C (nearly 180°F). Bacterial action depends upon individual cells literally attaching themselves to the particles of pyrite. They have no mechanism for deliberately looking for the pyrite; they simply adhere randomly all over all the solid material so that cells finding themselves not attached to pyrite have no source of energy and die. In order to be attacked and its sulphur oxidized, each pyrite particle must be at least partly exposed because the bacteria cannot sense the presence of pyrite from a distance and are anyway unable to erode overlying coal. The coal particle size thus becomes critical for bacterial action: the more finely the coal is ground the greater the chance that pyrite particles will be at least partly exposed and thus susceptible to attack.

Like the bioleaching systems, bacteria for coal desulphurization need an acid environment; they must have air (oxygen) for converting the sulphide to sulphate, and carbon dioxide plus small quantities of various mineral salts for their own growth. In principle, bacteria can remove all the inorganic sulphur if the coal is well pulverized and that reduction in sulphur content might be enough to convert high- to low-sulphur coal.

Pros and Cons of Microbial Desulphurization

The complexities both of coal mining and the use of coal as a power station fuel make it difficult to identify the single most cost-effective method for preventing/controlling sulphur dioxide release. Even if microbial desulphurization were the preferred method, there remains the option of whether to establish facilities at the mine or at the power plant. Each situation poses its own set of industrial, logistic, economic and social problems and opportunities, some of them briefly discussed above.

Depending on the precise nature of the coal, the following factors generally support the possibility of employing a microbial system to remove some or all of the sulphur:

- microbial desulphurization at the mine would reduce the weight of coal to be transported;

- the coal slurry needed for desulphurization could also be the means of transporting the coal by pipeline to a power station;
- there would be no loss of fines, which in some conventional procedures can cause a significant reduction in the calorific value of the coal;
- the microbial process is good at removing pyritic sulphur;
- it has low operating costs and is therefore likely to be competitive with other methods of removing sulphur before combustion and with procedures for controlling emissions after burning;
- because of its low cost, microbial desulphurization might allow existing coal extraction and processing industries to remain viable;
- it might also avert the need for the possibly expensive importation of foreign low-sulphur coal;
- hence, it might offer a realistic means of pollution control in poor countries dependent for electricity on coal-fired power stations but without their own domestic supplies of low-sulphur coal.

Other considerations, less encouraging for industrial biotechnologists, are:

- common though coal pulverization already is in many generating plants, it is not universal. If it became an additional requirement for microbial desulphurization it would impose an incremental energy cost;
- the coal particles must be slurried in water, requiring subsequent expensive dewatering;
- the microbial process is slow, requiring as long as 10–20 days for completion depending on temperature;
- the long retention times lead to a requirement for large holding tanks and a demand for land; this is likely to be expensive in congested urban regions near power stations, but not necessarily in rural mining areas. High land prices will favour treatment at the mine rather than at a power station;
- the acidity of the microbial process will increase problems of corrosion damage and incur a cost for corrosion protection;
- at present the microbial process is poor at removing organic sulphur, but that might be improved with further research;
- decision makers in the coal-mining and power-generating industries are generally not familiar with microbiological concepts and methods. They may not be willing to invest in microbial desulphurization facilities, preferring more familiar solutions.

MARKET PROSPECTS FOR ENVIRONMENTAL BIOTECHNOLOGY

The largest commercial opportunities for environmental biotechnology might not reside in the very well established practices of sewage treatment and landfill; in both of these

Exploiting biotechnology

cases improvements in engineering design and operating procedures will probably be more significant than scientific advances.

However, for two main reasons the situation will be different with industrial wastes from certain manufacturing and processing procedures:

- at the present time the involvement of biotechnology in their disposal is comparatively modest and not extensively tested but new and improved treatment technologies using biological methods are undergoing rapid development in several industrial countries;
- the variety of industrial wastes and effluents requiring their own specific forms of treatment lends itself well to the employment of tailor-made biotechnological methodologies.

Over the coming years, as these considerations become increasingly familiar to companies and public bodies burdened with waste disposal problems, the wide range of specificities available for the design and construction of biotechnology-based disposal procedures seems likely to result in their being chosen for a growing number of treatment facilities. Of course, the great advantage of living microbes for environmental treatments is that they reproduce and maintain themselves on site as effective agents as long as biodegradable material is present. No non-living chemical compounds will do that whatever their origin.

12 Bioelectronics: a courtship between technologies

Electronics and biotechnology are two of the most rapidly advancing technologies and it is very much to be expected that synergy will develop between them. It is beginning to do so in two quite unrelated ways:

- in the form of *biosensors*, elegant and effective tools for the detection and measurement of chemicals;
- as *biochips*, a possible new technology for replacing silicon-based electronic data processors with much smaller ones made of biological molecules.

BIOSENSORS

Biochemical (and hence biotechnological) systems are characterized above all by their remarkable specificity, their ability to distinguish between closely-related chemical entities. Thus, catalysis with a particular enzyme might enormously speed up a reaction involving one sugar and simply do nothing at all with another sugar which is almost, but not quite, identical. Or a receptor on the surface of a cell might respond to a specific stimulating molecule such as a hormone, while remaining totally indifferent to one which differs apparently only trivially in structure. The importance both of specificity and catalysis in biological organization have repeatedly been stressed in earlier chapters.

Biochemical procedures offer two very great advantages for the detection and measurement of different individual chemical compounds in complex mixtures: specificity and speed. A common difficulty in chemical analysis is not easily being able to measure a particular chemical in the presence of a host of others, some of them perhaps in much higher concentrations than the sought-after species; those other chemicals might cross-react in a conventional (that is, non-biological) analytical procedure and give spurious results. Avoiding such pitfalls while using conventional methods may enormously complicate the task of analysis: using the right enzyme as an analytical tool introduces specificity and speed from the start. Results which once took hours or days to achieve can be obtained in minutes.

What are Biosensors?

They might be very simply described as devices which make use of biochemical specificity (like that exhibited by an enzyme) to detect, and often to measure the concentration of, a particular chemical or group of closely related chemicals. The interaction of the substances

Exploiting biotechnology

under test with the biochemical agent produces a response, usually but not always electrical, which can be used to move a needle on a dial, generate a digital response on a display panel or perform some other function such as activating a pump to introduce an external component into the system.

In practice, a common biosensor configuration has the enzyme or other sensing agent immobilized on a membrane in intimate contact with a means of converting a product of the biochemical interaction into an electrical signal. The latter is then used either to provide numerical information about chemical quantities or to generate a response, like operating a switch. Depending on their purpose, many different types of sensing entities are available. They include:

- individual enzymes, each of which specifically catalyses the conversion of one or more substrates into defined products;
- multi-enzyme systems (several enzymes working in concert in an organized way), more elaborate than using simple, individual enzymes. Such systems may be useful for conversions not possible with individual enzymes;
- whole cells, which are also specific in various ways with respect to chemicals able to bind onto their surfaces. They may catalyse chemical conversions on their surfaces by virtue of individual enzymes or multi-enzyme systems located there, perhaps enzymes which will not readily work if they are removed from their native environment. Or a substance might gain entry into the interior of the cell, there to undergo a more elaborate series of reactions than is technically achievable at the present time using isolated individual or multi-enzyme systems;
- subfractions isolated from cells. All living cells have an elaborate architecture with complex internal structures, each of which accomplishes important tasks in the cellular economy. Those internal structures, more complicated than multi-enzyme systems, are large enough to be seen with microscopes. Their effectiveness with respect to complex functions normally depends on their retaining structural integrity. They represent a level of complexity midway between whole cells and multi-enzyme systems;
- antibodies and antigens constitute another aspect of specific recognition and offer a further set of opportunities for detecting particular components.

Biosensor Design

The purpose in each case is to mount the biochemical sensing element in such a way that a useful instrument results. That instrument might be a sensor which is dipped into the liquid in a fermentation vessel or one to which a drop of liquid (blood, perhaps, or urine) is applied. Biosensors are usually made for measuring substances in solution since that is where the majority of biochemical reactions take place; in most cases the sensing element itself will be in some sort of aqueous environment. So designing a biosensor to measure the properties of a solid is difficult; gases are much easier — as long as the gas is allowed to dissolve in the water associated with the sensor. In some cases, as we shall see, it might be possible to detect and measure gaseous components without their first being dissolved in water.

Bioelectronics: a courtship between technologies

For liquid use, the sensor molecules must be protected from being washed away in the liquid being analysed. This is done by immobilizing the sensor elements: either they are attached chemically to a supporting membrane or other structure, or else they are physically entrapped in a suitable matrix. These techniques were described in Chapter 8 with respect to enzyme immobilization for the manufacture of biochemical products. Even when immobilized, the sensors must of course be protected from abrasion and mechanical damage, so they are often covered by a porous membrane which allows the molecules to be detected to pass through the pores and reach the sensors. Many natural and artificial membranes are available with a range of different pore diameters and other properties. The ability of membranes to allow the passage of chemicals selectively can be very important. For some biosensors it is desirable to keep interfering species away from the sensor elements and, if the protective membrane can admit those chemicals it is desired to detect while excluding others which might interfere with the measurements, a more effective instrument results.

There are many ways in which a biosensor might be packaged for use; Figure 12.1 shows one design.

How Do They Work?

Biosensors are based on chemistry: their ability to detect specific compounds relies on chemical interaction between the compound in question and the sensor element. When the interaction takes place, energy is released because the total amount of chemical energy associated with the

Figure 12.1 Advanced prototype of a hand-held glucose biosensor for diabetic monitoring. (*Reproduced by permission of Cambridge Life Sciences – ACE Technology*)

251

Exploiting biotechnology

reaction products is less than that of the original reactants. The law of the conservation of energy states that energy can neither be created nor destroyed; hence, if the products of a chemical reaction contain less energy than the starting materials, the excess must have been liberated. It is this liberated energy which is used to trigger the sensing system. (Do note that by no means all chemical reactions release energy; on the contrary, many consume it — the biological implications of energy demands were discussed in Chapter 2. But energy-demanding reactions, while not totally excluded, are less suitable for use in biosensors precisely because the operation of the sensing device usually requires the release of chemical energy in order to activate the detector.)

There are many ways of using chemical and other forms of energy to drive electrical recording mechanisms. All biosensors depend on energy transducers, devices which convert one form of energy into another. In some cases the chemical reaction produces atoms or molecules which carry electrical charges; these are made to move in a certain direction under the influence of a voltage gradient and result in the flow of a current, measured by a meter or other device. In some recent developments a light-emitting *diode* is used to create a photoresponse.

Certain chemical compounds form so-called *piezoelectric crystals* whose electric properties, especially their oscillating frequency, are sensitive to surface distortion resulting, for example, from the mass change which follows the attachment of particular molecular species. Such crystals might be usable in biosensors. The specific detector molecule is first attached to the crystal surface where it acts as a receptor; the crystal then responds electrically when this receptor binds selectively to its corresponding activator molecule, the response being amplified to drive a readout. Piezoelectric crystals do not work well immersed in liquid but might prove very useful for detecting compounds in air, including combustion products which cause pollution, nerve gases and other materials. Such systems have even been equated to selective electronic noses with possible forensic uses, for example, in detecting drugs; the compounds to be detected would, of course, either themselves have to be gases or else be sufficiently volatile partly to vaporize so that the detector could sense them.

The energy derived from chemical reactions is often released as heat, resulting in a rise in temperature in the immediate vicinity of the reactants; if this forms the basis of sensing, the temperature change is detected by a sensitive *thermistor* linked to an electrical recorder. The actual amount of heat generated is minute and it is important to insulate the sensing thermistor very well in order to stop the heat being dissipated and so invalidating the response. Because they require good thermal insulation, detection devices relying on temperature changes are rather cumbersome and difficult to miniaturize.

Bioaffinity Sensors

Yet another type of sensor depends on displacement mechanisms. Suppose that the specific detector, immobilized in some way on a membrane, already has fairly loosely attached to it a molecule which can be used to evoke a measurable response — it might be a substance which *fluoresces* and emits light of a certain colour and intensity whenever it is irradiated by light of another colour. The light emitted by fluorescence is then picked up by a sensitive *photocell* and recorded electrically. The system is so designed that the molecular species to be detected is one which, while not fluorescent itself, will displace the original fluorescent material from its binding with the detector, the displaced material being washed away. The extent to which the fluorescent response is diminished when the system is irradiated in the presence of the test compound will be related to the concentration of that compound, and hence to the extent to which it has displaced the original fluorescent material from its binding with the detector.

Immunosensors

A whole host of optical devices have been and are being developed for use with biosensing. Generally, the way they work is to alter the optical properties of the system when their bound detector molecules react with specific counterparts in solution. This is proving very useful for antigen-antibody reactions. Either one of the pair is first bound into the biosensor on which it confers particular optical reflectance or absorbance characteristics: light shone on the system will be reflected in a certain way or some of it will be absorbed by the bound material. When the second member of the interacting pair is also present, and the full antigen-antibody complex is able to form, the degree to which the optical properties are changed can be used as a measure of the quantity of that second component.

Technical developments are proceeding apace, as some of the designs just described illustrate very well. The whole field of biosensors is fairly new; major advances in sophistication and range are certain to emerge in the coming years, enabling biosensors to become both more robust and more versatile. A particularly interesting possibility is to use complete genetically-engineered cells as sensors; these might be designed and constructed to be receptive to a more complex system of chemical signals than can be detected by a simple enzyme or antigen-antibody sensor.

What Can Biosensors Be Used For?

With a field so new one can do little more than guess at what the future applications might be, deriving as much guidance as possible from the few contemporary uses.

One of the most important present needs is for small, hand-held devices for medical diagnostic purposes. These have immediate value not only for rapid diagnosis in the doctor's office or in the hospital ward but also as instruments for the patient's own use, for instance in the routine monitoring by diabetics of blood glucose levels. A drop of blood is all that is needed; the patient pricks his own finger, deposits the blood on the sensor tip and within seconds will have a direct reading of his blood sugar. Devices for such purposes need to be small, robust, reliable and easy for medically unskilled people to use.

The next level of complexity for medical analyses is in hospital laboratories where large numbers of measurements are routinely made. Already automatic analysers are widely used in which samples are put through a predetermined series of operations to yield specific items of information: better and faster results might be obtained by using biosensors as the detecting and measuring elements in place of the existing "wet" chemical and biochemical analytical methods. Further developments might enable the more rapid measurement of circulating levels of specific drugs and other materials during surgery, when speed is of the essence. The results would be available in the operating theatre within seconds rather than having to send samples to the laboratory and await a response.

Patients might benefit enormously from implanted biosensors controlling the release of drugs into the bloodstream, as in the case of insulin administration to diabetics. In a normal individual, the amount of insulin released by the pancreas into the bloodstream is governed by natural sensors which respond to variations in the concentration of circulating glucose after eating and during periods of fasting (overnight, for example). It is impossible for the diabetic patient to receive injections of insulin which respond to blood glucose levels in that way. Rather, the injections have to take place at comparatively long intervals and patients have to be careful about when and what they eat. Clearly they would benefit if an implanted biosensor could measure blood glucose continuously and release insulin from a reservoir when needed. The reservoir and pump might be worn by the patient, or even be miniaturized to such a degree that the whole system could be implanted and the reservoir topped up at infrequent intervals by injection through a special septum; in that way the patient would become independent of external insulin administration for long periods of time.

Some ideas go further yet, away from any sort of pump in the conventional sense. Perhaps very tiny chemical "bags" of insulin, so small that they could circulate in the blood, could be constructed in such a way that their surfaces carried sensors which, when triggered by glucose, allowed some or all of the insulin to escape from the bag. Occasional injection into the bloodstream of large numbers of the insulin bags might be enough to allow patients even more independence than they would have with an implanted pump and reservoir. Although the insulin/diabetes example has been used here because it is familiar to many people, the concept is in no sense confined just to that hormone and that disease. There are many medical conditions that would benefit from the internal release of a hormone or other material exactly when the need for it was triggered by the chemical demand signal.

In a totally different application, biosensors have an existing and growing value for the control of industrial fermentations and other types of biochemical processing. At present there are still so many technical problems associated with actually inserting the biosensor element into the reaction mixture in the processing vat for "online" control that often the sensor is used "offline": a portion of the reaction mixture is automatically passed over the sensor located outside the vessel and is then discharged to waste rather than being recirculated. The data obtained are essentially the same as if the sensor were actually inside.

All of these uses require reliability and accuracy; the instruments must be robust enough for their purpose and calibrated to give the right answer within acceptable limits. Obviously, such requirements are most critical in medical contexts: if the sensor is to be implanted it cannot readily be removed for periodic checks and recalibration.

Bioelectronics: a courtship between technologies

The Future for Biosensors

Because biochemical systems have such a spectacular capacity for combining complexity and specificity, biosensors are likely to become very important for the rapid measurement of biological parameters more complicated than single chemicals, for example, the problem of microbial contamination. Microbes are ubiquitous and, given the chance, will colonize any material which provides them with the nutrients they need. But their nutrients are often also our foods; if we fail to protect them against microbial contamination, they will deteriorate in quality and may become positively harmful.

So one important issue in all products for human consumption and external application is to ensure either total sterility or at least the absence of harmful microbes. Government regulations prescribe acceptable practice for the preparation and handling of foods, drugs and toiletries but, in spite of manufacturers' and retailers' best efforts to prevent such events, contamination with pathogenic organisms does occur from time to time. Sampling for the microbes in water, milk and various foods is undertaken routinely but normal methods may take days to yield results; by that time the product will often already have been sold and consumed. With specific biosensors, sensitive to very low levels of contamination, pathogens or spoilage organisms could be detected much more rapidly, possibly within minutes. Similar methods might also be used to ensure that blood taken from donors is free from hepatitis, AIDS or other viruses.

Environmental protection offers yet another important set of opportunities for biosensors. The control or elimination of pollution is dependent on identifying and measuring the pollutants — continuous measurement with almost immediate results would clearly facilitate proper control measures. Immediate detection of a pollutant inadvertently released to the environment would enable appropriate steps to be taken not only to eradicate materials already released but also to make absolutely sure that no more escapes. And if noxious compounds do find their way into the environment, into water courses, soil or the air, their removal is enormously aided by being able to determine rapidly and accurately the location of the pollutants and the effectiveness of their clean-up.

The evaluation of market opportunities for biosensors is difficult and perhaps premature. Estimates of market size in recent years have been of the order of $50 million annually, with predictions in the multi-billion dollar range for annual sales by the end of the century. As with many other branches of biotechnology, the future for biosensors appears bright if not yet fully understood.

BIOCHIPS: BIOLOGICAL COMPUTERS

The courtship between biotechnology and computing is very much more speculative than the current reality of biosensors; in many ways it is also more challenging and intriguing.

Computers are devices composed essentially of switches and connecting wires; each switch can exist in either an "on" or an "off" position so that current either does or does not flow through it. Altering the switch position in response to a signal is the physical basis of computer

Exploiting biotechnology

memory storage. As electronic computers have undergone development during the past four or five decades two characteristics have marked their progress: progressive miniaturization and an ever larger capacity for processing information.

Enlarged capacity has conferred an ability for dealing with increasing complexity, a tendency that shows no signs of abating. The actual or projected uses of very elaborate computers in weather forecasting, economic modelling, traffic control and for the "star wars" defence proposal have been well publicized. New needs will engender computers with still more capacity: yet more complex models are required for research in science, engineering, economics and other fields, to cope both with the growing sophistication of experimental design and the accumulation of vast amounts of information to be integrated and analysed.

Simultaneously, the computers themselves become physically smaller and smaller as the years pass. The capacity of today's familiar micro- and mini-computers to process information far outweighs that of earlier machines which filled whole rooms. Even simple, wafer-thin, hand-held electronic calculators are faster and can do more than the typewriter-sized apparatus which was the last word in desktop calculating 30 years ago.

In spite of these spectacular developments, serious impediments to future progress exist. Increased information-handling capacity means more switches, and more wires to connect them. The total cumulative length of wire through which electric current flows becomes greater and greater. Unbelievable though it may seem with electricity flowing at 186,000 miles (300,000 kilometres) a second, the length of wiring to be traversed in very complex operations seriously prolongs the processing time; this might have profound consequences not only when tracking an incoming hostile missile but also for the rapid analysis of weather patterns or sorting out an urban traffic jam in as short a time as possible. As manufacturing technology advanced, *silicon chip* switches became smaller which enabled them to be packed closer together, so reducing the total distance which the current had to travel. But current flowing in wires and through switches generates heat, so that the more compact the layout the greater the problems of keeping the system cool. Thus, growing operating complexity would be well served by miniaturizing the switches, leading to less current to operate them and hence generating less heat. It would also offer the prospect of denser packing and shorter current transit distances, thus simultaneously increasing operating speed and reducing heat output.

Molecular Electronics

The requirement for this increased operating speed has stimulated research and development into the possible manufacture of very tiny chips, essentially of molecular size. Individual molecules are very small indeed. A trillion trillion (1 with 24 zeros) molecules of a small compound like common salt weigh no more than about 100 grams (3.5 ounces) while a large molecule like a protein (some 1,000–10,000 × heavier than a molecule of common salt) might be a sphere of such a size that 25 million of them in a row would stretch only about 2.5 centimetres (1 inch). If chips really could be brought down to molecular size, even if each chip comprised dozens of molecules, the aggregate size of the computer, the total distance travelled by the electric currents and the heat output would all be dramatically reduced. If molecular switches were as fast as the present silicon variety (in practice they might well be much faster), the small-sized instruments based upon them would have fantastic powers of computation compared with present-day machines.

Although research into molecular electronics has already been going on for more than 20 years, no marketable products are yet in sight. The obstacles to development include:

- designing molecules with the right switching properties;
- assembling them in the correct relationships to one another;
- inventing ways of allowing them to communicate in a controlled manner — in effect, connecting them with *molecular wires* or designing new systems for transmitting information.

These are all major problems, the last two being perhaps particularly difficult to resolve, but a number of molecular designs have already been developed experimentally which, in principle at least, offer models for switching mechanisms. It is the intercommunication between the molecules, as well as the problems of system assembly, which are particularly intractable and might offer the greatest opportunities for the intervention of biotechnology in computing.

Biomolecular Electronics; the Significance of Self-Assembly

Biological molecules score once again because of their specific recognition properties. In nature, the *organelles* which exist inside individual cells, bodies ultimately responsible for the architecture of the cell itself, assemble themselves spontaneously in their correct formats. In the higher plants and animals the very same factors are responsible for the location of the cells in organs and, beyond that, for the part those organs play in the structure of the whole organism. All this depends on the ability of molecules, mostly proteins, to recognize one another in specific ways and so form the multimolecular conglomerates which constitute the physical structure of the whole plant or animal.

In Chapter 2 we saw how the structure of proteins is encoded in the DNA which specifies the order of amino acids in their chains. That is all the information needed for a protein both to fold up into a specific three-dimensional structure and to carry on its surface one or more binding sites which enable it to join up, again in a specific way, with other proteins having reciprocal and matching binding sites. Thus, genetic information resident in DNA can specify elaborate, self-assembling structures. We know that such structures exist and are determined genetically: we, the authors and you, the reader, are examples of self-assembled systems based on genetic information.

For the moment, the exact mechanisms of protein-protein interactions are largely obscure, but our understanding of protein molecular structure is advancing very rapidly. Before too long precise details will be worked out. Knowledge of the factors in the amino acid sequences of protein chains which make the proteins fold in a particular way is also under intensive study, so one may look forward to the time, probably not more than a decade or two into the future, when it will become possible to design proteins with defined three-dimensional

Exploiting biotechnology

structures, able both to perform specified chemical tasks and to join up with others into self-assembling protein macrostructures. Some people think such structures could form the basis of biological computer chips, or "biochips".

It might be done something like this — a protein (which we will call an "addressable protein") would be designed with three major properties:

- an ability to be addressed, perhaps by an optical system as we shall explore below;
- an ability to link with a number of other addressable protein molecules, each identical with itself;
- a tendency simultaneously also to join up with one or more molecules of another type of protein (a "spacer protein") able to act as an electrical insulator.

A solution of chemicals might be made containing many more addressable than spacer proteins, and a flat surface provided onto which the various proteins were encouraged to attach themselves in the form of a sheet. Because they would be present in the greatest numbers, most addressable proteins would organize themselves into arrays or groups with other identical addressable protein molecules. But some interaction between addressable and spacer proteins would also take place, resulting in each group of addressables being separated from neighbouring groupings by a band of spacers. Beyond each band of spacers would lie another array of addressables, a pattern repeated in all directions over the flat sheet. The precise sizes of the addressable groups and the widths of the spacer bands would depend on:

- the relative numbers of the different types of proteins present in the system;
- the number and types of binding sites with which each category of protein had been provided;
- the comparative eagerness of each type of protein for attaching to identical copies of itself rather than to other types.

All these factors would be built into the design to produce the desired configuration, one which also envisages direct intermolecular electronic communication between the addressable protein molecules within each spatial group, but not between groups because of insulation by the spacers. Figure 12.2 shows the type of molecular pattern which might result.

Some people have conceived of the possibility of making these self-assembling systems by genetic engineering. When the right protein structures can be worked out it will be possible to design and construct genes to specify each of the proteins involved, since the relationship between genetic information and the amino acid sequences of proteins is well understood (Chapter 2). The resulting artificial genes for these proteins would be linked up to appropriate genetic control mechanisms (for "punctuation", etc.) and inserted into a suitable microorganism which could be manipulated at will as a living factory, just as in the case of the therapeutic proteins reviewed in Chapter 7.

Bioelectronics: a courtship between technologies

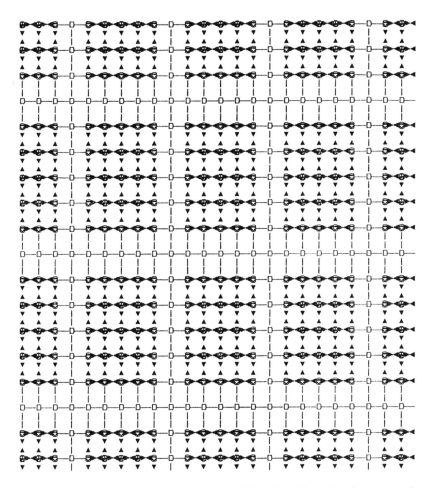

Figure 12.2 Model for a molecular memory store.

Exploiting biotechnology

Problems of Addressing Biochips

Although there are ideas for making very fine "wires" out of conducting molecules, nobody seems yet to have thought of how to use them to join up individual "switching" molecules into chips.

The present ways of manufacturing silicon chips may not be very suitable for molecular electronics, biologically based or not. Factory-made chips are first designed and drawn with a circuitry appropriate for their task. The drawings are reduced in size photographically and projected onto the flakes of silicon, with the conducting and insulating regions constructed on the flakes by the automatic deposition of chemicals in the positions defined by the circuit diagram. Both switches and connecting wires are made in this way and their minimum sizes are limited partly by the technical problems of making very small structures and partly by having to locate the wires and switches far enough apart to ensure that they are properly insulated from one another and devoid of electronic "cross-talk". But the dimensions of those deposited wires are far too large to offer a means of wiring up the very small molecular groups of addressable proteins we have just discussed.

Perhaps, in time, scientists will find ways of connecting proteins with wires also determined genetically and assembling spontaneously with the correct connections. The idea is not ridiculous. The nervous systems of animals are in effect exactly phenomena of this sort, with a very complex ramification of nerve connections throughout the brain, spinal cord and peripheral organs based on inherited genetic information: those nerves are very rarely wrongly connected.

Other ways of addressing the protein groups may be more immediately feasible than wires; optical methods appear to be favoured, although there is much work still to be done to perfect them to the point of their being able to interact with the small individual molecular groupings implied in this model.

Many chemical and biological molecules respond to light, their chemical states thereby being in some way energized and altered. After illumination, the new *excited state* of some molecules is so unstable that the molecule flips back to its original condition in a tiny fraction of a second; the light energy which was absorbed to effect the original excitation (called the *exciting light*) is re-emitted as a light flash, usually of a colour different from the original illuminating beam. This is the nature of fluorescence and, because the exciting and fluorescing lights are of different colours, they can easily be distinguished from one another. In other cases, the excited state of the molecule is more stable and may be able to survive for a long time, especially if the material is kept very cold.

Suppose that addressable proteins were designed in such a way that, when a local group of them was illuminated by a light beam of the right colour, the proteins were energized into a stable excited state. Suppose further that, in the excited state the addressable molecules could be even more excited by light of another colour, after which they would rapidly emit a flash of fluorescent light and return to the first excited state. This might form the (admittedly theoretical) basis of an optically addressable system.

Directing a light beam precisely at particular groups of addressable proteins whose positions on the flat sheet were determined and identified by their spatial coordinates, some groups of

Bioelectronics: a courtship between technologies

proteins are put in the first (i.e. lower) excited state while others are not. The excited groups are equivalent to "on" switches while the quiescent ones are equivalent to "off". Next, all groups are interrogated by light beams of the second colour: those already excited ("on") respond with a fluorescent flash while those quiescent ("off") do not. Hence, on and off are distinguished via an optical address system without the need for actual wires. The following diagram illustrates the model:

(a) without exciting illumination:

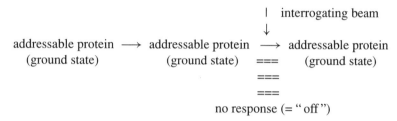

(b) with exciting illumination:

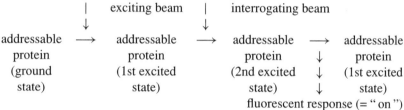

Dreams or Reality?

Nobody has yet produced such a system, nor are they likely to do so for years to come. One major problem is that nobody knows how to make light beams fine enough to target such small molecular groups; indeed, it might turn out to be impossible. However, recent developments with microlasers may offer hope of finding ways of optically addressing very small entities. Scientists have also reported early experiments in the use of light rather than electric impulses to transmit signals in a rapid type of computing called *parallel processing*; this approach simultaneously addresses many different aspects of the overall problem instead of working on them sequentially. Those scientists did not use biological models and their work is a long way from biocomputers (and, indeed, may never lead there).

Nevertheless, since the 1970s large sums of money have been invested in research and development by a number of countries on the grounds that there is nothing in the fundamental science which suggests that biomolecular computer chips are inherently absurd. At a fundamental level, studies on the structure/function relationships of proteins, on

Exploiting biotechnology

protein-protein interactions and on the formation of multimolecular protein complexes are certainly active areas of investigation in many laboratories around the world. Throughout history, experience of technological advances has usually shown that, if a development is both possible and potentially profitable sooner or later it will happen. The science in this case is certainly in good hands and the market will in time decide if biocomputers are likely to be commercially viable.

13 The future: biotechnological bonanzas?

GENERALITIES

While the activity grows ever more important, the word "biotechnology" is rather less glamorous than once it was. Some people feel it is too strongly associated with genetic engineering, too close to academic science and too slow in producing commercial results. The term is beginning to be avoided in favour of such alternatives as *biological engineering* and *integrated pharmaceutical company*. Fashions can be expected to fluctuate and individual firms will try to present their most effective image to those they wish to attract and impress. Should a major disaster be widely attributed to biotechnology, watch the word disappear. But if the general story is one of success, and the word is believed to help, it will certainly be used. Quite a few people see biotechnology as an economic "salvation industry" like computing and aerospace; it is certainly not going to go away.

There is, of course, no magic about spotting the winners in biotechnology. The rules are the same as in other businesses: get the products right (effective, reliable, price-competitive), identify correctly the market niches and the strategy to develop them, and go out there and sell.

The difficulties of meeting these objectives vary from one industry to another. In some cases, perhaps most obviously in the development of new drugs and medical preparations, the technology may involve advances at the very frontiers of science: novel concepts, elaborate experiments and manipulations, and elusive new discoveries. The risks of venturing into the unknown must be taken with caution; nobody can be absolutely sure of facts before they have been proved but good scientists, experienced in their subjects, usually acquire a pretty keen sense of what is likely to emerge from their work. The "facts" when confirmed are rarely totally unexpected. But many hurdles lie ahead in the shape of clinical trials, regulatory approval and successful market launch, perhaps in the teeth of powerful competition.

For some products the main difficulties will arise in the manufacturing and production technologies: the product itself might perform well enough but making it on a cost-effective scale may require a considerable investment of capital both for the development of appropriate production techniques and for the construction of the actual manufacturing facilities.

With other products and services, the obstacles may lie largely in introducing a novel technology to a mature industry with its own pre-existing "culture", a culture possibly far removed from the prevailing concepts in biotechnology: microbial mining and enhanced oil recovery come to mind here. The industries recognize their problems clearly enough and acknowledge that potentially there may be several ways of solving them. Managers will make decisions based on the cost-effectiveness of each proffered solution and its compatibility with existing practice. Inevitably they are likely to feel most comfortable with technologies closest to their own experience.

Exploiting biotechnology

As far as the future is concerned, the authors have no special crystal ball. In the six sector-based chapters of this report we have reviewed the present state of biotechnology, hazarded guesses about possible trends for the near term, and very occasionally have tried to glimpse the more distant future. In this closing chapter we will attempt to collate those guesses and glimpses, and interpret them in the light of some general considerations.

HEALTHCARE

From the beginning of the present biotechnological era, investments in therapeutic drugs, prophylactic vaccines and diagnostic reagents have usually been regarded as those most likely to yield substantial returns in the short and medium term. The reasons are clear: in wealthy economies, the considerable investment needs of biotechnological development mesh readily with the pattern of rapidly rising expenditure on healthcare as well as the substantial prices paid for medically effective drugs. In those countries with national health services, the public continually insist that the government fund them more lavishly; at the same time, individuals almost everywhere are increasingly willing to devote more and more of their personal resources to their own private medical care.

The Prospects for Therapeutics

One may reasonably look forward to a rising level of spend on therapeutic drugs even in cases in which their efficacy has not yet been proven to normally accepted standards — and even when they are very expensive. A case in point is the current pressure on governments to make available to patients treatments for AIDS which are possibly no more than palliative. At the same time there is felt to be an urgent need for funding major research programmes to discover new drugs for that disease.

While infectious diseases are well controlled in the rich industrial countries (although uncomfortable echoes of the past are reappearing in the form of tuberculosis and malaria), they remain major issues in poorer, less well developed parts of the world. In the latter, there are enormous demands both for drugs to treat diseases which have never properly been conquered and for public health programmes together with the prophylactic vaccines and drugs necessary to reduce the incidence of infections. Each pound or dollar spent on health in a poor country will probably yield more in social and economic benefit within that country that it would in a rich one: many of the diseases of the poor are well recognized and easier to treat than those of the rich who no longer suffer to any marked degree from malnutrition, typhoid or cholera. Unfortunately, biotechnology as a source of new therapeutics and prophyalctics tends to be expensive, and commonly beyond the resources of the poorer regions which may manage to improve public health more cheaply and more effectively through improved hygiene, the availability of antibiotics and other tried medication, and better medical understanding in the population.

The future: biotechnological bonanzas?

By contrast, the medical problems of the developed countries are increasingly those of old age, above all vascular disorders and cancer. Since the sufferers will often have retired from active work, the benefits of prevention and therapy are appreciated most by the individual patients and their families, not by the national economy at large. Nevertheless, it is often those very ageing and retired people who have either the personal resources or the insurance cover to gain most from medical and biotechnological advances and there can be no doubt that their diseases and disabilities will remain prime targets for new development.

And for Diagnostics

Diagnostics is equally a major component of biotechnological involvement in the healthcare industry. Perhaps even more importantly for the complex syndromes of an ageing population, accurate diagnosis is likely to be a *sine qua non* for successful treatment. The tendency for the patient to demand the latest in technology, coupled with what appears to be a growing frequency in some countries of legal actions for negligence, is already stimulating the medical profession to play safe and not to skimp on the most up-to-date diagnostic methods available.

The message for biotechnology managers and investors is probably fairly clear: unless there is adequate public funding for development, and subsidy for use, investment in healthcare biotechnology for the foreseeable future will probably generate the best returns in diagnostics, prophylactic vaccines, therapeutic drugs for degenerative diseases like arthritis, cancer, vascular disorders and Alzheimers, and techniques for correcting inherited (genetic) deficiencies. Hardly surprisingly, most investment in healthcare biotechnology will take place in highly developed industrial countries. The ultimate prize is presumably a blockbuster "wonder drug", exactly right for dealing with an important disease, and carrying all before it in the market place. Such drugs do happen — but rarely, and the rewards they bring have to be set against not only their own very high development costs but also against the costs of all the other attempts which failed. A jackpot is nevertheless a real possibility and fortunately some managers and some investors cannot resist aiming for it.

Healthcare is not only about Preventing and Curing Diseases

An important aspect of healthcare in the wider sense is the maintenance of personal well-being, for which individuals are about as eager to spend money as they are for medicines. "Well-being", with its strong emotional and psychological components, leads to a desire and therefore to a need for cosmetics applied directly to the skin or hair. Whether or not skin creams and lotions do what they claim is not for us to judge; the fact of the matter is that people buy them in very large quantities because they believe they are or may be effective. Fashion in cosmetics is an important aspect of marketing, leaving plenty of possibilities for new preparations based on a variety of biotechnological considerations. "Cosmetic" is perhaps also the right word to use in a wider sense for treatment with certain hormones, particularly

Exploiting biotechnology

those which might be able to prevent, or even to some degree reverse, the visible ravages of age. Many people have speculated at the fortunes to be made by really reliable and harmless ways of overcoming the sagging and wrinkling of skin — and the prevention and reversal of male pattern baldness!

Nor is Healthcare Important just for People

Drugs for human use must be properly tested and licensed, usually an extremely expensive undertaking open essentially only to major corporations. There is, however, an economically very important aspect of healthcare which is not quite so expensive as medicines for human use because of the less stringent requirements for testing and licensing: healthcare for animals, both domestic pets and those on the farm. There are said to be many people who would pay almost as much to cure their pets as they would to cure themselves; indeed, in countries with national medical services, they willingly pay more because they expect their own treatment to be free at the point of delivery. So another aspect of increasing wealth in society is likely to be more expenditure on pets, with clear implications for biotechnology; animals are physiologically and biochemically not so very different from ourselves and the advances being made in human prophylaxis and therapy will certainly bring dividends for the treatment of animals as, conversely, drugs for human use are often first tested on animals. In similar vein, diseases of farm animals can cause economic ruin for agribusinesses and farmers are accordingly very interested in keeping their livestock healthy. The market for animal healthcare products and services will surely grow in parallel with its human counterpart.

BIOTECHNOLOGY IN THE CHEMICAL INDUSTRY

There are several important distinctions between the healthcare and chemical industries, none more so than the unsurprising fact that virtually all of healthcare is based on biology while the greater part of industrial chemistry is not.

Very many pages ago, we observed that "The chemical industry converts feedstocks into valuable products. Each production process is based on a sequence of catalysed chemical reactions engineered into a viable production technology." That sets the scene. While all of healthcare has overt or implicit biological significance, the chemical industry worldwide interacts with biotechnology in a much more restricted sense:

- some of its feedstocks are of biological origin;
- some products (drugs, toiletries and cosmetics, pesticides, fertilizers, foodstuffs, etc.) are directed to human, animal or plant use;
- to a limited degree, biological catalysts (almost always purified enzymes or microorganisms) are even now used to catalyse reactions in chemical manufacture and their importance is likely to increase in the future;

- certain enzymes and other biochemicals are already made in large quantities by the industry.

Limited though industry-wide interaction may be, a significant proportion of important chemical companies are deeply involved with biotechnology and have been for a long time.

In order to contribute to chemical manufacture, most relevant biotechnological innovation must be coordinated closely with existing industrial practice, built as it is on a wealth of experience coupled with very large capital investments in equipment and facilities. Any new process at variance with that experience, and requiring abandonment of existing manufacturing facilities in favour of novel equipment designs and practices, will need to be very much more attractive than existing methods before it is accepted. That does indeed happen from time to time, as the history of single-cell protein shows, but most large chemical manufacturing operations continue to have no biotechnological input.

New opportunities will certainly arise in the future. Biological catalysts are very effective and very specific, but also very demanding of the environmental conditions in which they can operate. That surely will be a major area for development: the acquisition from natural sources, perhaps with some structural modification, of new enzymes able to function at (biological) extremes of temperature, pressure, acidity, etc., will enable the efficiency of some existing manufacturing processes to be improved and the generation of undesirable by-products to be reduced or avoided.

Perhaps a little further in the future, it will be possible to design and construct *ab initio* new sorts of enzymes capable of catalysing reactions between molecules wholly non-biological in origin and structure. One notable difference between biological and manufacturing chemistry is that the former is overwhelmingly water-based, while manufacturing processes frequently take place of necessity in organic solvents in the total absence of water. Enzymes, too, are mostly water-based but already there is progress in the techniques of using them in non-aqueous environments, progress which is likely to accelerate with a growing ability to design and make artificial enzymes for specific jobs.

A point made earlier in this book needs to be stressed again here: much of biotechnology is a high-cost activity. It is therefore likely to be used more in the manufacture and isolation of high-value products than for making bulk commodity chemicals.

PROSPECTS IN AGRICULTURE AND FOOD

As we acknowledged at the end of Chapter 9, this is a difficult area for prediction. It is easy to see what could be done technically but political and social factors make it very much more difficult to foretell what actually will happen.

It seems to us that control of diseases in plants and animals will continue to be major areas of progress. Whatever the arguments about excess production, most people will presumably want healthy products and minimal losses in the fields. Continuing progress in the development of novelty products like fruits, and decorative plants and flowers, will influence niche markets without much effect on the economics of major food products. Because of their relatively small impact on global food production compared with their very great local commercial

Exploiting biotechnology

benefits for the less developed countries, it appears highly likely that these activities will grow apace.

So will improving the quality storage properties of a range of foods. Long-life and and firmer tomatoes are already with us; similar improvements for other fruits and vegetables seem certain in the course of time. Companies large and small will undoubtedly continue to look for better manufacturing processes — the example we quoted of upgrading palm oil to the more desirable qualities of cocoa butter is just one of many improvements which will come about with time.

The really big uncertainties relate to the overall quantities of food produced in the world as a whole and its distribution between various countries and regions within those countries.

MINERAL RECOVERIES

There are a number of analogies between the use of biological technologies in chemical manufacturing and their deployment for aiding the recovery of minerals: both industries reflect mature and major activities in which most of the operators are large corporations. Biotechnology, with much to offer for the future of both, has hitherto played no more than a minor role in either.

Modern industrial civilization is totally dependent on raw materials extracted from the earth: materials for building, metals and other substances for manufacture, and coal, oil and gas for use as fuels and chemical feedstocks. Sometimes, as with building materials, peat, some hard and soft coal, bitumen and tar sands, they lie on or close to the surface and can be removed at low cost, an essential factor in their economic value. Such minerals have another advantage — they occur in large deposits and, with the exception of bitumen and tar sands, require little processing before use.

Other raw materials are more difficult and hence more expensive to acquire. They lie below the surface and much capital must be invested to sink shafts, open mines, drill production wells and provide the surface operations that go with them. Minerals located at depth fall into three main categories, each presenting rather different technical and economic problems which can limit their value as raw materials:

- solids (such as metal ores) in which the substance(s) of interest may be present in low concentration, distributed in unwanted rocky mineral and must be extracted and processed;
- solids (such as coal) in which the seam itself is relatively uncontaminated with rock and little subsequent processing is necessary, but which have to be recovered by deep mining;
- fluids (oil and gas) which move under the influence of *in situ* pressure gradients and can readily be brought to the surface once suitable wells have been drilled; a downhole lift pump may be needed to raise oil to the surface if the reservoir pressure is too low.

The economics of recovery are clearly disadvantageous if a high cost of extraction is coupled with a comparatively low market price for the product. The balance changes to the producer's benefit when the product price rises or the extraction costs fall. While the market price of most mineral products is determined by complex global factors beyond the control of a single producer, he may be in a position to invest in new technology to reduce costs, thus changing the local balance to his advantage. It is by such a reduction of extraction costs that biotechnology is currently placed to contribute to the more cost-effective exploitation of poor quality ores and coal reserves, or partially depleted oil reservoirs.

Metal Leaching

The leaching of copper has been going on for so long, and is still being used on so huge a scale, that the advantage of bioleaching as a recovery method for tailings and low grade ores is beyond doubt. Technical advances may be expected to yield both improved strains of microorganisms, which would replace the natural populations in leaching operations, and effective techniques for *in situ* rock fragmentation, leading to greater opportunities for solution mining.

Both of these developments will be driven by the market. Existing sources of high grade metal ores will progressively be depleted — costs will rise as lower quality materials have to be worked, prompting a demand for cheaper extraction procedures which will favour biotechnological developments. To a degree, the effects of declining extraction productivity will no doubt be offset by the discovery of new deposits, although some will be in geographical areas remote, inhospitable and expensive to work, or within the jurisdiction of potentially unfriendly or unstable governments. Furthermore, high prices for some minerals will prompt substitution by new materials, just as plastics have already replaced metals for many purposes in manufacture and construction. The balance of these factors will certainly be different for different metals; a requirement for bioleaching will grow more rapidly for some than for others.

In several countries in which mining for metals is an important factor in the national economy, mining corporations and academic institutions undertake R & D to search for microbial leaching systems which work better and can be applied to a wider range of ores. The future is thus likely to witness the development of the underlying biological technology in parallel with the mining and processing engineering needed to give it practical expression.

Crude Oil Recovery

The overall picture is rather similar for the involvement of biotechnology in crude oil recovery. Although the relevant technologies have been under development for nearly half a century, there have been comparatively few properly designed field tests and the use of these methods in commercial production has tended to be restrained by the state of the oil markets.

When, in the late 1940s, microbial enhanced oil recovery first began to be considered seriously as a production technology, the price of crude was very low and supplies were abundant. Some development work did take place in the succeeding decades, mainly in Eastern European countries chronically short of hard currency for buying crude on the world market. In western industrialized countries concern with microbiology subsided as they were able to benefit from the enormous expansion of low cost primary oil production in the Middle East and elsewhere. By the mid-1970s, however, the political and economic consequences of a growing dependence on those sources began generally to be perceived as a serious drawback. The oil price shocks of that decade served to stimulate efforts to improve the recovery of oil from western domestic sources and they included a renewed interest in microbiological techniques.

In the last 10 to 15 years the pattern of activity has fluctuated with the price of crude. When it approached $40 a barrel in the early 1980s, a good deal of government funding and company interest was directed to exploring microbial methods for enhancing the recovery of oil; significant progress was made on both sides of the Atlantic. In 1986 the price of a barrel of crude collapsed to around $10 and the mood changed again: much of the company interest evaporated and government funding in the area generally declined. Since that time the price appears to have stabilized near $20 and enthusiasm has once more reawakened, particularly with the realization that the economic welfare of important mature oil provinces, including those in the southwestern part of the US, would be boosted substantially by a low-cost technology to recover some of the two-thirds of the original oil-in-place which remains in their reservoirs. A new generation of field tests, on firmer financial and technical bases than has characterized most of the earlier trials, shows signs this time round of leading the breakthrough into a proven production technology. A growing body of opinion anticipates that assisting oil production with microbiology and biotechnology will increase steadily in importance during the coming decade.

THE ENVIRONMENT

The motivation for spending money on the remediation and control of pollution is very different from investment in production. Aside from a wholly genuine desire on the part of many industrialists to play their part in retaining or restoring a pleasant and healthy environment, much of the impetus for instituting environmental protection measures derives from an obligation to meet or anticipate legal requirements, and to project a caring public image.

There is an interesting interplay between technology and both government and public pressures with respect to the environment. Technical improvements in the ability to measure environmental contaminants, coupled with a greater understanding of their potential harm, leads in time to regulatory requirements for their control down to more stringent limits. Further advances then result in better techniques for measurement followed by additional legislation to meet those better detection limits. This pattern of ever-tightening regulations generally (if erratically) appears likely to continue and will, in turn, stimulate yet more development in biotechnology and other fields.

The future: biotechnological bonanzas?

It is the chemical and biological aspects of waste management and pollution control which are those most readily susceptible to biotechnological intervention. It may not be able to help much with rusting bedsteads but real and very extensive opportunities do exist for cleaning up chemical pollution in rivers, lakes and seas and they may exist for degrading old rubber tyres and discarded East German *Trabant* cars (someone apparently came up with a microbial system for digesting their plastic bodies if not their steel gearboxes). The openings for biotechnology will multiply as the complexities of industrial and agricultural life continue to grow, as more and more waste discharges are perceived to be harmful and the public, directly and through its organs of government, demands better prevention controls and insists ever more vigorously that "polluters" should be required to cleanup.

The diversity of actual and potential forms of pollution is so great that it is impossible to predict what will become the most pressing issue next week nor how, and for whom, it will provide a commercial opportunity: a new oil spill in a sensitive area, another discharge of chemical effluent into a waterway or an industrial accident spewing toxic material — all will happen sooner or later. Additional methods for the detection, clean-up and containment of pollution are going to be needed for as far ahead as we can envisage. The problems and the need for vigilance are with us for ever; accidents by their very nature can never entirely be prevented and substances that should not be released into the environment will nevertheless inevitably continue to be generated in manufacturing processes. There will be no shortage of work in pollution control.

ELECTRONICS

Biosensors

These instruments already have the ability to provide, virtually instantly, information on the presence and concentration of certain chemical species. Advances in enzyme technology will extend their sensitivity and improve their reliability as well as the range of chemicals to which they can be made to respond. Miniaturization and remote reporting are two of the attributes which will greatly enhance their value for a whole variety of functions.

There exist at least four significant areas of application. One is clearly in medicine. For a number of medical conditions, continuous *in vivo* monitoring of various chemicals will be a highly desirable feature of diagnosis; coupled with reporting by radio, a patient will be able to be monitored while he goes about his normal business at a distance from his medical attendants. By responding to the appropriate parameters of a patient's internal chemical environment, biosensors might serve to control the continuous administration of therapeutic drugs without the direct intervention of either the patient or his doctor.

A second use category for biosensors is in processing. People managing chemical and biochemical manufacturing processes need to know exactly when certain stages have been completed and that the overall reaction is going according to plan. The more readily such determinations can be made by placing a sensing device within the reaction mixture and having it report remotely to a control centre, the less manpower and effort will be required

Exploiting biotechnology

to take samples manually or note the readings on a dial or register. Thus, incorporating biosensors specific for different parameters in a beer fermentation vat would allow the brew master sitting in his office to maintain a constant check on developments, perhaps while simultaneously partaking of last week's product!

Biosensors have value for quality control. Products must meet defined standards before despatch to the client or customer; it is important to ensure that manufacturing standards have been met and that there has been no deterioration during storage. Rapid determination of key product characteristics with appropriately designed biosensors will enable more of this information to be obtained rapidly and, perhaps once more with remote reporting, recorded electronically for future reference.

Their chemical specificity and low power requirements make them ideal instruments for unattended monitoring in the environment. Made to respond to a variety of potential pollutants, biosensors will detect low concentrations which can be reported or recorded on a continuous basis. A warning can be sounded in the event of a discharge and a record provided of the magnitude of the event and the effectiveness of remedial measures. There is almost no limit to the potential uses of environmental sensors.

Biocomputers

This is the most difficult area of all in which to look ahead with any confidence. The potential technology appears so exciting that, if such computers based on biochemical molecules can actually be made, there seems little doubt of their value. But can they? How? When? The answers are obscured in the mists of the future. We can do no more than repeat what we implied a chapter or so ago: the future is of very long duration and both computing and biotechnology have hardly started. The way science and technology are developing suggests very strongly that sooner or later anything and everything which does not actually contravene the laws of nature will become possible. But just when that will come true in the case of biological computers is another matter.

THE LIMITS TO PREDICTION

Foretelling the future is always risky, the more so when, as in the case of biotechnology, the future depends on so many interacting factors. One can certainly make reasonable guesses about some of the advances in the relevant biological sciences which can be expected during the next few years:

- the Human Genome Project will undoubtedly make available enormous amounts of information of profound significance for understanding the human condition and some of its defects;

The future: biotechnological bonanzas?

- with this and other new knowledge, progress in the techniques of genetic manipulation will permit new forms of therapeutic treatment;
- greater insight into molecular biology and developmental genetics will greatly add to an understanding of degenerative and neoplastic diseases, hopefully leading to more effective therapies;
- parallel progress in plant genetics and the manipulation of plant material will enable new forms of crop plants to be developed, perhaps allowing field crops to be used as more convenient and lower cost factories for animal proteins than microbial systems in fermenters;
- advances in the biochemistry and genetics of nitrogen fixation should reduce the demand for nitrogen fertilizer while biological pest control might in time offer a variety of effective and environmentally acceptable means of such control. Together they will improve the efficiency of agricultural production (though what to do with more produce or unemployed agriculturalists is an unresolved question);
- developments in protein engineering are likely to offer not only better but also new enzyme catalysts: "better" in the sense that they are more robust, survive higher temperatures and harsher conditions, and last longer in service; "new" by being able to catalyse reactions unknown in biochemistry and hitherto inaccessible to biological catalysis;
- the recent discovery of extremely thermotolerant and other previously unknown bacteria offers the prospect of using microbiological procedures in environments too hot, too acid or too salty for the more familiar species.

These are all generalized scientific predictions, not too difficult to make for rapidly advancing, well-established sciences. They simply project existing trends and none is original or innovative. However, the difficulty with predicting biotechnology is that it is not simply a science but a marriage of science, engineering and business. Over and over again people have observed that good experiments do not necessarily lead to successful marketable products and services. Transferring the technology from the laboratory to the marketplace depends on commercial skills capable of recognizing a market opportunity (or creating one where it does not already exist) integrated into a management structure able to coordinate the necessary effort — and, of course, the financial resources to fund the initiative until the revenue comes in. There have been so many examples in industrial and commercial history of industrialists failing to exploit a new scientific advance because they did not perceive sufficient commercial benefit (though competitors may then or later have taken a different view), that it would be unwise to prophesy that even the best and most apposite technical discovery will actually see the light of day as a commercial product. It all depends on the favourable conjunction of too many factors for accurate prediction. It is much safer to stick to the general and say without a shadow of doubt that many new biotechnological products and services will find their way to market in the coming decade but just what they will be, who will undertake their development and in what form they will reach the consumer demands greater foresight than we can justifiably claim.

Appendix 1
Help! Where do I go from here?

BOOKS

There is not the space here to list books at each level of complexity in the basic sciences that individual readers might require. The following are all modern undergraduate texts; they may be too advanced and detailed for readers with no prior science who might therefore like to ask their bookshops for something simpler, or start with:

Biology by Sylvia S. Mader (1990: 3rd edition). Published by Wm. C. Brown Publishers. ISBN 0-697-05638-4.

Biochemistry

Biochemistry by L. Stryer (1988: 3rd edition). Published by W.H. Freeman. ISBN 0-7167-1843 X and 0-7167-1920-7 (international student edition).

Biochemistry by C.K. Mathews and K.E. van Holde (1990). Published by Benjamin/Cummings. ISBN 0-8053-5015-2.

Biochemistry by G. Zubay (1988). Published by Macmillan. ISBN 0-02-432080-3.

Genetics

An Introduction to Genetic Analysis by D.T. Suzuki, A.J.F. Griffiths, J.H. Miller and R.C. Lewontin (1989: 4th edition). Published by W.H. Freeman. ISBN 0-7167-1956-8 and 0-7167-1996-7 (paperback).

Genes & Genomes, a changing perspective by Maxine Singer and P. Berg (1991). Published by University Science Books (US) and Blackwell Scientific Publications (UK). ISBN 0-632-03052-6 and 0-632-02879-3 (paperback).

Exploiting biotechnology

Microbiology

General Microbiology by H.G. Schlegel (translated by M. Kogut) (1993: 7th edition). Published by Cambridge University Press. ISBN 0-521-43372-X (hardback) and 0-521-439809 (paperback).

General Microbiology by R.F. Boyd (1988: 2nd edition). Published by Times Mirror/Mosby College Publishing. ISBN 0-8016-1291-8.

Biotechnology

Most biotechnology books focus primarily on technical matters, with little attention paid to business and commerce, but a new arrival is rather different:

Biotechnology: the science and the business edited by V. Moses and R.E. Cape (1991). Published by Harwood Academic Publishers. ISBN 3-7186-5094-0 (hardback) and 3-7186-5111-4 (soft cover).

A wide-ranging review of the technologies (although not quite so all-embracing as the title implies) is provided by:

Comprehensive Biotechnology edited by M. Moo-Young *et al.* (1985). ISBN 0-08-026204-X. Volumes 1 and 2 (ISBN 0-08-032509-2 and 0-08-032510-6) — *The Principles of Biotechnology*: (1) *Scientific Fundamentals*; (2) *Engineering Considerations*. Volumes 3 and 4 (ISBN 0-08-032511-4 and 0-08-032512-2) — *The Practice of Biotechnology*: (3) *Bulk Commodity* Products; (4) *Speciality Products and Service Activities*. First Supplement: *Animal Biotechnology* edited by L.A. Babiuk *et al.* (1989) (ISBN 0-08-034730-4). Published by Pergamon Press.

Another marathon is currently being published in 12 volumes (1-4 already out, the rest to follow):

Biotechnology — a multi-volume comprehensive treatise edited by H.-J. Rehm and G. Reed. Published by VCH.

The following multi-author publication (intended for a readership familiar with science) is a largely technical summary of biotechnology and associated molecular biology activities in the main areas of human healthcare, plant and animal biotechnology/food, industrial processes and process biotechnology. There is also a section on "boardroom issues":

Industrial Biotechnology International 1993 edited by Rod Greenshields (1993). ISBN 0967-2044. Published by Sterling Publications Ltd.

A good technical presentation (which also includes a brief treatment of legal, social and ethical issues), not too long and with the benefit of the coherent style of a single author, is:

Molecular Biotechnology by S.B. Primrose (1991). Published by Blackwell Scientific Publications. ISBN 0-632-03233-2 (hardback) and 0-632-03053-4 (paperback)

Appendix 1

An interesting book dealing with commercial matters is:

The Business of Biotechnology edited by R. Dana Ono (1991). ISBN 0-7506-9119-0. Published by Butterworth-Heinemann.

PERIODICALS

Advances in the basic sciences of biochemistry, genetics, microbiology, etc. are reported as technical research papers and subject or topic reviews in thousands of more-or-less specialist journals published by learned societies, scientific institutes, national organizations and commercial publishers. There are dozens and dozens of regular publications dealing with biotechnology, some published as often as daily, others no more than once or twice a year. General science journals such as *Nature, Science, New Scientist* and *Scientific American* regularly include material of relevance to biotechnology and also carry many of the job advertisements. However, the main dedicated journals providing a variable mix of technical, industrial and business articles, news items and more jobs vacancies are *Bio/Technology* and *Trends in Biotechnology*. A partial listing of other magazines and newsletters, usually available on a subscription basis, which address various aspects of biotechnology is: *Applied Genetics News, BBI (Biomedical Business International) Newsletter, BioEngineering News, Biotech News, Biotechnology Bulletin, Biotechnology in Japan Newsservice, Biotechnology Insight, Biotechnology Law Report, Biotechnology Progress, Biotechnology News, Biotechnology Newswatch, BioVenture View, Changing Medical Markets, Genetic Engineer and Biotechnologist, Genetic Engineering Letter, Genetic Engineering News, Genetic Technology News, New Biotech Business* and *Scrip*.

DATABASES

In addition to general databases carrying biotechnological information, the following specialize in this field:

Biobusiness (produced by BIOSIS and hosted by Data-Star, Dialog);

BioCommerce Abstracts and Directory (BioCommerce Data; Data-Star, Dialog);

Current Biotechnology Abstracts (Royal Society of Chemistry; Data-Star, ESA-IRS, Maxwell Online);

Derwent Biotechnology Abstracts (Derwent Publications; Dialog, Maxwell Online).

DIRECTORIES AND GUIDES

Many directories are published on an occasional or regular basis; a selection follows, their areas of specialization being fairly obvious from their titles:

Annuaire des Biotechnologies et des Bioindustries (ADEBIO/Biofutur);

Exploiting biotechnology

Australian and New Zealand Biotechnology Directory (ABA/Australian Industrial Publishers);
Bio 1000 (D.J. Mycsiewicz);
BioScan: The Biotechnology Corporate Directory Service (Oryx Press);
Biotech Products for Therapeutic Use (PJB Publications);
Biotechnologie, Das Jahr- und Adressbuch (Polycom);
Biotechnology Directory 1989: Products, Companies, Research and Organization (Stockton Press);
Biotechnology Guide Japan (Macmillan);
Biotechnology Guide USA (Stockton Press);
Biotechnology Marketing Sourcebook (British Library);
Canadian Biotechnology Directory (Winter House Publications);
Directory of British Biotechnology (Longman);
Directory of EEC Information Sources (Euroconfidentiel);
Genetic Engineering and Biotechnology Firms Worldwide Directory: Technical Highlights and Funding Sources (Sittig & Noyes);
Genetic Engineering and Biotechnology Yearbook (Elsevier);
Guide to European Collaboration in Science and Technology (SEPSU, Royal Society);
Healthcare Biotechnology: Company Profiles (PJB Publications);
Industry File for Chemistry and Biosciences (Oakland Consultancy);
Information Sources in Biotechnology (Stockton Press);
International Biotechnology Directory 1992 (Macmillan);
International Biotechnology Industrial Directory (Biotechnology News);
UK Biotechnology Handbook '90 (BioCommerce Data).

MARKET SURVEYS

A number of commercial firms undertake market surveys in various sectors of biotechnology; prices are often in the range of $2,000 and more, with the surveys usually sold on the basis of a prospectus. The firms include:

Business Communications Co. Inc (BCC), 25 Van Zant Street, Suite 13, Norwalk, CT 06855-1781.

Frost & Sullivan, Inc., 106 Fulton Street, New York, NY 10038-2786; UK office: Frost & Sullivan, Ltd., Sullivan House, 4 Grosvenor Gardens, London SW1W 0DN.

Financial Times Management Reports, Customer Services, P.O. Box 6, Cambourne TR14 9EQ.

Innomed International, 4 Embarcadero Center, Suite 5083, San Francisco, CA 94111.

POV Reports, PO Box 238, Cedar Grove, NJ 07009.

Appendix 1

Technology Management Group (TMG) Inc, 25 Science Park, New Haven, CT 06511.
Technical Insights, Inc., Box 1304, Fort Lee, NJ 07024-9967.
Theta Corporation, Theta Building, Middlefield, CT 06455.

CONFERENCES

There is considerable activity worldwide in biotech. conferences. Emphases differ widely: some are mainly technical, others concentrate on business or related activities, and some cover both. None reveals commercial secrets! Advance notices usually appear in the main biotechnology journals and in some of the newsletters.

INDUSTRY ASSOCIATIONS

Their role generally is to
- offer informational and other support to member companies;
- identify key issues affecting the development of biotechnology;
- participate as appropriate in the formulation of local, national and international policies;
- keep a watch on impending regulatory, patent, safety and other potentially relevant legislation;
- promote biotechnological education and training;
- encourage positive public attitudes towards biotechnology.

Belgium

Belgian Biotechnology Coordinating Group (BBCG),
Square Marie-Louise 49,
1040 Brussels.
Phone: +32 +2 238 98 23; Fax: +32 +2 231 13 01

Canada

Industrial Biotechnology Association of Canada (IBAC),
237 Argyle Avenue,
Ottawa, Ontario K2P 1B8
Phone: +1 +613 233 4559; Fax: +1 +613 233 3882

Denmark

Association of Biotechnology Industries in Denmark,
Novo Alle,
DK-2880 Bagsvaerd.
Phone: +45-44448888; Fax: +45-42984627

Exploiting biotechnology

Finland

Kemianteollisuus Ry (Chemical Industry Federation of Finland),
Eteleranta 10,
P.O. Box 4,
SF-00131 Helsinki,
Phone: +358-0-172-841; Fax: +358-0-630-225.

France

Organbio — Organisation Nationale Interprofessionelle des Bioindustries,
28 Rue Saint-Dominique,
F-75007 Paris.
Phone: +33 +1 47 530912; Fax: +33 +1 47 559862.

Ireland

BioResearch Ireland,
EOLAS,
Glasnevin,
Dublin 9.
Phone: +353 +1 370177/370101; Fax: +353 +1 370176.

Italy

Assobiotec — Associazione Nazionale per lo Sviluppo delle Biotecnologie,
Via Academia 33,
1-20131 Milan.
Phone: +39 +2 26810306; Fax: +39 +2 26810-284-310

Japan

Japan Bioindustry Association (JBA),
Dowa Building,
10-5, Shimbashi 5-chome,
Minato-Ku,
Tokyo 105.
Phone: +81 +3 3433 3545; Fax: +81 +3 3459 1440

Netherlands

Nederlandse Industriele en Agrarische Biotechnologie Associatie (NIABA),
P.O. Box 443,
2260 AK Leidschendam.
Phone: +31 +70 327 04 64; Fax: +31 +70 320 57 65

Appendix 1

Spain

Asociacion de Bioindustria,
C/Brue 72-74,
6A Planta,
08009 Barcelona.
Phone: +34 +3 3183383; Fax: +34 +3 3023568.

United Kingdom

BioIndustry Association (BIA),
1 Queen Anne's Gate,
London, SW1H 9BT.
Phone: +44 +171 957-4600; Fax: +44 +171 957-4644

United States

Biotechnology Industry Organization (BIO),
1625 K Street, N.W., Suite 1100,
Washington, D.C. 20006.
Phone: +1 +202 857-0244; Fax: +1 +202 857-0237

International

International Council of Scientific Unions (ICSU),
51 Boulevard de Montmorency,
75016 Paris.
Phone: +33 +1 4525 0329; Fax: +33 +1 4288 9431

The ICSU steering Committee for Biotechnology (COBIOTECH) publishes a journal, runs training courses, conferences and workshops and organizes an International Biosciences Network with specialist activities based on Africa, the Arab countries, Asia and Latin America. The Secretary General of COBIOTECH is Professor K.G. Skryabin, Centre of Bioengineering, Institute of Molecular Biology, Academy of Sciences of the USSR, Ulitsa Vavilova 32, Moscow V-312 (Phone: +7 +095 135 73 19; Telex: 411982 BTREM SU).

Senior Advisory Group Biotechnology (SAGB),
Conseil Europeen des Federations de L'Industrie Chimique (CEFIC),
Avenue E. Van Nieuwenhuyse, 4 - Box 1,
B-1160 Brussels,
Belgium.
Phone: +32 +2 676 72 86; Fax:+32 +2 676 72 88

The purpose of SAGB is to promote a supportive climate for biotechnology in Europe.

Exploiting biotechnology

NATIONAL AND SUPRANATIONAL GOVERNMENTAL AGENCIES

United Kingdom

The Department of Trade and Industry has the lead within central government for fostering biotechnology and an Interdepartmental Committee on Biotechnology was set up in 1982 under the chairmanship of the Government Chemist. A useful publication is *BIOTECHNOLOGY— A Plain Man's Guide to the Support and Regulations in the UK* (January 1991, 2nd edition). ISBN 1 85324 510 0. The main contact address is:

The Biotechnology Unit,
Chemicals and Biotechnology Division,
Laboratory of the Government Chemist,
Queen's Road,
Teddington,
Middlesex TW11 0LY.
Phone: +44 +181 943-7354; Fax; +44 +181 943-7304.

UK readers in particular may be interested in *New Life for Industry: Biotechnology, industry and the community in the 1990s and beyond* by the NEDC Biotechnology Working Party chaired by David Barnes (1991) and published by the National Economic Development Office. The report does not explore the underlying sciences and technologies but reviews the industrial opportunities in biotechnology, discusses social and environmental issues and considers at some length the situation in Britain with respect to the national science base, manpower, the quality of training, UK sourcing of equipment and materials, technology transfer, regulation, finance and national policy.

European Community

Probably the single most useful source of information about Community policy with respect to biotechnology is:

Directorate-General for Science, Research and Development (DG-XII)/E-1,
Commission of the European Communities,
Rue de la Loi 200,
B-1049 Brussels.
Phone: +32 +2 296 56 19; Fax: +32 +2 295 53 65/296 43 22.

The office publishes *EBIS — European Biotechnology Information Service Newsletter* which contains technical, organizational and financial news items about European biotechnology originating from the Commission and elsewhere.
But note that decisions of importance to biotechnology are also made by:

Directorate General for the Internal Market and Industrial Affairs (DG-III),
Commission of the European Communities,
Rue de la Loi 200,
B-1049 Brussels.
Phone: +32 +2 235 96 68; Fax: +32 +2 236 30 28.

Appendix 1

New initiatives are continually emerging from the Community. In addition to obtaining information direct from the Commission, the following publication may prove helpful:

Biotechnology. EEC Policy on the Eve of 1993, available from European Study Service, Avenue Paola 43, B-1330 Rixensart, Belgium (Phone; +32 +2 653 90 19; Fax: +32 +2 652 03 02).

United Nations

The United Nations Industrial Development Organization (UNIDO) has been a sponsor of biotechnology around the world, with most emphasis on developing countries. Useful addresses are:

UNIDO,
Vienna Office,
P.O. Box 300,
A-1400 Vienna,
Austria.
Phone: +43 +1 21131; Fax: +43 +1 230-7355

and

International Centre for Genetic Engineering and Biotechnology,
United Nations Industrial Development Organization,
Padriciano 99,
I-34012 Trieste,
Italy.
Phone: +39 +40-37571; Fax: +39 +40-226555.

Appendix 2

Glossary of technical terms

ACETIC ACID Organic acid, the product of some types of fermentation; most familiar as the main ingredient of vinegar

ACID FRACTURING Technique for cracking and dissolving reservoir rock in order to facilitate the flow of oil

ACQUIRED IMMUNE DEFICIENCY SYNDROME (AIDS) Virus infection which attacks the immune system

ACQUIRED IMMUNITY Immunity to infection, etc. acquired as the result of contact with the infective agent or a vaccine obtained from it

ACTIVATED SLUDGE PROCESS One of the techniques of dealing with sewage

ACTIVE IMMUNIZATION Conferring immunity to a disease by vaccination with the causative organism or a product obtained from it

ACTIVE SITE That part of an enzyme which binds the substrate and participates in the catalytic reaction

ADJUVANT Substance injected together with a vaccine in order to enhance its effect

ADRENAL Gland producing the hormone adrenalin and others

ADRENALIN Hormone produced by the adrenal glands: speeds up heart rate, stimulates glycogen reserve metabolism, etc.

AEROBIC In the presence of oxygen and air; applied to microorganisms able to use, or requiring, oxygen for growth

AEROBIC DEGRADATION Biological breakdown of organic compounds in the presence of air (oxygen)

AFLATOXIN Toxin made by the fungus *Aspergillus* which may cause liver damage; sometimes formed in certain foods kept under inappropriate storage conditions

AIDS See "Acquired immune deficiency syndrome"

ALBUMIN Category of small proteins found in blood

ALCOHOL Class of chemical compound of interest as industrial solvents and fuels; the one in beverages is "ethyl alcohol"/"ethanol"

ALGA (pl. ALGAE) One of the lower orders of plants, mostly microscopic in size but also including seaweeds

ALGINIC ACID Polymer used for thickening

ALLERGEN Antigenic material evoking an allergic reaction

Exploiting biotechnology

ALLERGY Hypersensitive immune response evoked by contact with specific substances in food, the air or on the skin

ALUMINA Oxide of aluminium

AMBERGRIS Waxy substance used in perfumery and made in the stomach of certain whales

AMBIENT PRESSURE/TEMPERATURE Generally, the pressure (one atmosphere) and temperature (roughly 20-30°C) prevailing at the earth's surface

AMINO ACID Type of chemical compound, of which twenty different versions form the building blocks for making proteins

AMYLASE Type of enzyme catalysing the breakdown of starch and glycogen

ANAEROBIC In the absence of oxygen and air; applied to microorganisms able or compelled to grow without oxygen

ANAEROBIC DIGESTION Biological breakdown of organic compounds in the absence of air (oxygen)

ANTIBIOTIC Chemical produced by one microorganism which is capable of killing or preventing the growth of another

ANTIBODY Protein produced in the body of an animal in response to a foreign chemical substance or organism; part of the defence against foreign invaders

ANTICOAGULANT Substance preventing the clotting of blood

ANTIGEN Chemical substance which stimulates animals or humans to produce a specific antibody that can combine with it

ANTIHISTAMINE Pharmacologically active agent countering the effects of histamine, a powerful vasodilator released during allergic reactions, etc.

ANTITHROMBIN Agent used to prevent blood clotting

ANTITOXIN Substance or preparation countering the effects of a toxin

ANTITRYPSIN (1-α-ANTITRYPSIN) Anti-cancer agent

AQUIFER Stratum of water-filled rock which may lie below an oil reservoir

ASPARTAME® Peptide-based artificial sweetener

ASPARTIC ACID One of the twenty amino acids found in proteins

ASPERGILLUS NIGER Fungus used industrially to produce certain organic acids

ATOM Smallest particle of a chemical element

ATP Chemical which plays a central role in energy metabolism in all living systems. Its full name is "adenosine triphosphate"

AUTOCLAVE Device resembling a large pressure cooker used to generate high-temperature steam for sterilization purposes

AUTOIMMUNE DISEASE Condition in which immune responses develop towards a person's own proteins

Appendix 2

BACTERIAL IMMUNOMODULATOR Bacterium or one of its components which enhances the immune response

BACTERIUM (pl. BACTERIA) A category of microorganism, usually comprising one cell only

BARREL Traditional volume measure for crude oil; equal to 159 litres, 35 Imperial gallons and 42 US gallons

BASE PAIR Interacting pair of nucleotides, the units used to build nucleic acid molecules

BATCH CULTURE Technique of growing microbes in a vessel containing a certain quantity of growth medium

BENEFICIATION Removal of some of the rocky substance of an ore to enhance its content of a metal or other valuable material

BIALAPHOS A herbicide

BINDING SITE Specific place on the surface of a protein receptor molecule to which another chemical attaches itself

BIOCATALYSIS See "Biological catalysis"

BIOCHEMISTRY Study of all aspects of the chemistry of living things

BIOCHIP Hypothetical computer chip made with biological molecules instead of being based on silicon

BIOCIDE Substance which kills living cells

BIODEGRADABLE Capable of being broken down by biological catalysis

BIODETERGENT Biological substance with the properties of a detergent. See also "biosurfactant"

BIOGAS Gas containing methane and carbon dioxide; produced in landfills

BIOLEACHING See "Microbial mining"

BIOLOGICAL CATALYSIS Catalysis of biochemical reactions using enzymes

BIOLOGICAL ENGINEERING Use of biological methods as direct aids to engineering

BIOLOGICAL FILTRATION Type of filter system containing microorganisms and used in sewage treatment

BIOLOGICAL OXYGEN DEMAND (BOD) Indicator of pollution by biodegradable materials: a high BOD indicates heavy pollution

BIOLOGICS Biological products

BIOMASS Mass of living (or once living) material

BIOMOLECULAR ELECTRONICS Devices using minute switching and conducting mechanisms based on molecules of biological origin

BIOPLASTIC Biotechnological product with properties resembling those of artificial plastics

BIOREACTOR Vessel for carrying out microbiological or biochemical reactions

BIOREMEDIATION Use of biological methods to clean up polluted soils, etc.

Exploiting biotechnology

BIOSENSOR Device based on immobilized microorganisms or enzymes, used for detecting and measuring specific chemicals

BIOSEQUESTRATION Use of biological compounds for chemical sequestration

BIOSORBENT Material of biological origin with general or specific ability to absorb chemicals

BIOSURFACTANT Compound of biological origin with surfactant properties

BIOTRANSFORMATION Use of microbes or enzymes to effect specific chemical changes in manufacture

BOTULISM Very serious disease of the neuromuscular system caused by eating food containing the toxin made by botulinum bacteria

CALCIUM CARBONATE Chemical compound containing calcium, carbon and oxygen, the basis of chalk and limestone; vulnerable to attack by acids

CARBOHYDRATE Class of chemical compounds made of the elements carbon, hydrogen and oxygen; commonly called "sugars"

CARBON Chemical element, one of the most important in biochemistry and a component of virtually all biochemical compounds

CARBORUNDUM Very hard mineral, used industrially as an abrasive and in oil production as a fracture proppant

CARCINOMA Type of cancer

CARDIATONIC Stimulating heart pumping action

CARNIVORE Animal (or one of a few plants) whose diet consists of other animals

CARRAGEENAN A polysaccharide obtained from seaweeds

CATALYST Substance assisting a chemical reaction without itself being used up in the process

CELL HYBRIDIZATION/FUSION Joining the boundary membranes of two cells together to produce a single hybrid cell

CELLULOSE Long-chain polymer of glucose units; main component of plant cell walls and of cotton

CENTRIFUGATION Sedimentation of solids from their suspension in a liquid by using a centrifuge

CENTRIFUGE Instrument for increasing the gravitational force on a sample; commonly a vertical axis bearing a ring of sample tubes in holders is rotated at high speed causing the tubes to swivel to a horizontal position

CHEMISTRY Study of the elements and their compounds

CHEMOSTAT Device for growing microorganisms on a continuous basis, with a constant inflow of fresh nutrients balanced by a constant outflow of mature culture; also called a "continuous culture"

CHEMOTHERAPY Treatment of diseases with chemical agents

Appendix 2

CHITIN Long chain (mainly carbohydrate) polymer present in the cells walls and exoskeletons of fungi, insects and shellfish

CHITINASE Enzyme which catalyses the breakdown of chitin

CHLAMIDIA Group of bacteria, some of which cause diseases in animals

CHLOROPLAST An organelle found in most plant cells which contains the green pigment chlorophyll; the site of photosynthesis. Also contains DNA

CHROMATIDS the two (normally) identical substructures of eukaryotic chromosomes

CHROMOSOME(S) Elongated structure(s), composed largely of DNA, and the the physical site(s) of the genes; one or more are present in the nuclei of the cells of higher organisms, or lying free in the cells of bacteria

CITRIC ACID Organic acid of industrial value, also used as a flavouring agent; produced by fermentation with certain fungi

CLONAL SELECTION THEORY Hypothesis to account for the observed patterns of antibody formation

CLONE Population of cells or genes derived from an individual cell or gene

CLONING Process of making a clone by allowing a population of identical cells to arise from a single individual by growth and division, or by multiplying the number of cells into which a particular gene has been introduced artificially

COCCIDIOSIS Intestinal disease caused by a protozoon

CO-DISPOSAL (OF WASTES) Practice of using microbes to attack wastes which are difficult to degrade, while simultaneously providing them with more accessible nutrients

CODON Triplet of nucleic acid units (nucleotides) coding for a single amino acid of a protein chain

COENZYME One of several small non-protein molecules required for the activity of many enzymes

COLONY Dense mass of cells produced from a single individual when organisms are cultured on solid jelly media

COMPOUND In chemistry, a substance comprising only one type of molecule

CONCENTRATION In a chemical sense, the amount of a particular compound in a defined volume

CONTINUOUS CULTURE See "Chemostat"

COPPER SULPHATE Compound containing copper, sulphur and oxygen

COPPER SULPHIDE Compound containing copper and sulphur only

CORTICOID Sometimes an abbreviation for "corticosteroid", any of a group of hormones produced by the cortex of the adrenal gland and elsewhere

COWPOX Mild viral infection, immunity to which also confers immunity to smallpox

CULTURE Collection of microbial cells, usually grown in the laboratory

CUTICLE Outer skin or covering

Exploiting biotechnology

CYSTIC FIBROSIS Inherited disease (i.e. a genetic defect) affecting the pancreas, sweat glands and respiratory system

CYTOKINE Substance affecting the growth and division of cells

DDT a persistent insecticide containing chlorine atoms. Its full name is dichlorodiphenyl-trichloroethane (which is why people use the abbreviation!)

DECAY PRODUCT Atom or subatomic particle remaining after a radioactive atom has disintegrated

DEGRADE In biochemical terms, to break down a large compound or structure into smaller units

DEOXYRIBOSENUCLEIC ACID (DNA) Polymer which carries the genes and is the bearer of inherited information; the sugar it contains is deoxyribose, hence its name (sometimes abbreviated to "deoxyribonucleic acid")

"DESIGNER DRUG" One chemically designed to achieve a specific objective

DESULPHURIZATION Removal of sulphur, e.g. from coal

DEXTRAN Polymer, shorter than starch or glycogen, made of glucose units

DIABETES Disease of sugar metabolism, one of the causes being a deficiency of the hormone insulin

DIMER Compound consisting of two monomeric units

DIODE Electrical device permitting current to flow in one direction only

DIPLOID Cells possessing a full set of chromosomes (i.e. both members of each pair). See also "haploid".

DISACCHARIDE Sugar (e.g. sucrose) made of two monomeric units

DNA See "Deoxyribosenucleic acid"

DNA POLYMERASE Enzyme which catalyses the linking into a long polymeric chain of the individual nucleotide units which make up DNA

DOUBLE HELIX Name given to the two strands of nucleic acid in DNA which lie alongside one another in a special relationship and are twisted together into a helix

DOWNHOLE At the bottom of an oil well

DOWNHOLE PACKER Device inserted into the bottom of an oil well to control the flow of oil and/or water

DRY ICE Frozen carbon dioxide: a solid which vaporizes at -78°C and is a convenient means of providing a very cold local environment

DUMP LEACHING Use of microbes to dissolve metals from crushed ore piled in dumps or heaps in the open air; see also "Microbial mining"

ELECTROLYSIS Decomposition of chemical compounds by passage of an electric current through a solution

ELECTRON Subatomic particle carrying a negative electric charge; normal constituent of atoms

Appendix 2

ELECTROPORATION Use of a strong electric field to allow the entry of DNA into cells

ELEMENT Substance comprising one type of atom only

EMULSIFY Form an emulsion

EMULSION Fine droplets of a liquid dispersed in another with which it does not mix (e.g. oil in water)

ENDOCRINE SYSTEM All of the glands in an animal which produce, or are capable of producing, hormones

ENERGETICS In a biochemical context, the study of energy provision and use in living organisms

ENHANCED OIL RECOVERY See "Tertiary (oil) production"

ENSILAGE Process of making silage

ENZYME Biological catalyst, always a protein

ENZYME-LINKED IMMUNOASSAY Technique for detecting antigen-antibody interactions

ENZYME-LINKED IMMUNOSORBENT ASSAY Technique for detecting antigen-antibody interactions

ERYTHROMYCIN Antibiotic active against certain bacteria

ETHANOL Alcohol present in intoxicating beverages

ETHYL ALCOHOL See "Ethanol"

ETHYLENE Gas, one of the simplest hydrocarbon chemicals

ETHYLENE OXIDE Chemical compound, the product of adding an atom of oxygen to ethylene

EUKARYOTE Organism whose cells possess defined nuclei bounded by membranes and with chromosomes consisting of DNA plus certain proteins

EXCITED STATE In chemistry, an unstable configuration of a molecule from which it tends to return to a stable condition (the ground state) with the emission of energy (e.g. in the form of light)

EXCITING LIGHT Light of a wavelength suitable for evoking a fluorescent response

EXOSKELETON Hard exterior supporting structure of insects, crabs, etc., containing chitin

EXPONENTIAL PHASE That period of microbial growth when nutrients are available in excess and the growth rate is limited only by the organism's own biochemical capabilities

EXPRESSION (OF GENES) Process of actually synthesizing specific proteins on the basis of genetic information

EXTRACELLULAR Outside the cell

FACTORS VIII AND IX Blood proteins important for clotting, one of which may be absent in hæmophiliacs

FEEDSTOCK(S) Starting material(s) for a chemical process; often with a manufacturing connotation

Exploiting biotechnology

FERMENTATION Microbial or enzymic breakdown of complex chemicals to yield simpler products, usually in the absence of air

FERMENTER Vessel used for carrying out fermentations

FERRIC/FERROUS SULPHATES Compounds containing iron, sulphur and oxygen

FIBRINOLYSIN Enzyme which breaks down fibrin, a blood clotting protein

FILTER STERILIZATION Removal of all microorganisms from a fluid by passage through a filter fine enough to prevent their passage

FILTRATION Removal of solids from a liquid by passage through material containing fine holes

FINES Smallest particles of a ground mineral such as coal

FLOC Clump

FLOCCULATE Forming flocs or clumps

FLUIDIZED BED PROCESS Technique for effecting chemical reactions. It uses a catalyst immobilized onto solid particles light enough to be dispersed throughout the vessel but large enough to tend to settle to the bottom

FLUORESCENCE Phenomenon in which a compound in an excited state emits light when returning to the ground state

FLUORESCENCE ASSAY Technique for sensitive measurement based on fluorescence

FOLIC ACID Chemical essential in metabolism which bacteria can make for themselves but which humans need to ingest fully-formed

FRACTIONATE To separate into fractions

FRACTURING Procedure in oil production of injecting water into the reservoir under high pressure in order to crack the rock and allow oil to flow more easily to the well

FREEZE-DRYING See "Lyophilization"

FRUCTOSE Type of monosaccharide sugar

FUNGUS (pl. FUNGI) Type of microorganism, usually filamentous in form, though some, like yeast, are unicellular; also called "mould"

FURUNCULOSIS Skin condition characterized by multiple boils

GAMETE Haploid reproductive cell (egg or sperm)

GASOHOL Mixture of petrol (gasoline) and (ethyl) alcohol used to power motor vehicles

GEL A jelly-like preparation behaving essentially like immobilized water

GENE Unit of inheritance; in molecular terms, a gene is a specific section of DNA carrying the information for either (i) the structure of a certain protein or (ii) controlling the synthesis of certain proteins

GENE THERAPEUTICS A possible simple way of treating patients suffering from genetic disorders

GENE THERAPY Procedures for curing inherited diseases by correcting genetic defects

Appendix 2

GENETIC ENGINEERING Collection of techniques enabling genes from one species to be transferred, and possibly expressed, in another; also called "recombinant DNA technology"

GENETIC INFORMATION Instructions for making proteins encoded in the genome

GENETICS Study of genes and inheritance

GENOME Collective noun for all the genes of an organism

GENOTYPE Genetic informational complement of an organism

GERM Popular word for microorganism, with a hint of the disease-causing varieties

GLAND Organized group of cells exporting a specific product or products (hormones, for example) to other parts of the organism

GLUCONIC ACID Organic acid produced by a fungus and used industrially

GLUCOSE One of the most common monosaccharide sugars, a building block for polysaccharides like starch, glycogen and cellulose

GLUCOSE ISOMERASE Industrial enzyme used to make high fructose syrup

GLUTAMIC ACID One of the twenty amino acids found in protein; see also "Monosodium glutamate"

GLYCOGEN Sugar food reserve material stored by animals

GLYPHOSATE A herbicide

"GREENHOUSE" GAS One, like carbon dioxide and methane, which absorbs heat radiated from the earth just as glass retains heat in a greenhouse

GRISEOFULVIN Antibiotic active against certain fungi

GROWTH HORMONE Protein hormone produced by the pituitary gland at the base of the brain, a lack of which results in stunted growth of an animal

GUAR GUM A polysaccharide obtained from the guar plant

GYPSUM Form of the compound calcium sulphate; used in plaster

HAEMOPHILIA Genetically-determined disorder in which blood clotting is impaired due to a deficiency of a blood protein (often Factor VIII)

HAIRY CELL LEUKAEMIA Variety of leukaemia, cancer of the blood-forming tissues

HALOGEN "Salt-forming" elements which include chlorine and iodine

HAPLOID Cells (such as gametes) having only one member of each pair of the normal complement of chromosomes. See also "diploid".

HEAVY HYDROCARBON FRACTIONS Non-volatile compounds in crude oil remaining when the lighter components evaporate; also called "heavy ends"

HEAVY METAL Category of generally toxic metals with large-sized atoms, including cadmium, lead, mercury and many others

HELPER CELL Type of T-lymphocyte undertaking a coordinating role

HERBIVORE Animal which consumes only plants

Exploiting biotechnology

HIGH FRUCTOSE SYRUP Syrup containing sugars, a significant proportion of which is fructose; used as a low calorie sweetener

HORIZONTAL DRILLING Technique used in the oil industry to steer the drill bit from the vertical to the horizontal direction, so producing a long and effective horizontal drainage well lying within the oil-bearing zone of a reservoir

HORMONE Chemical substance released into the blood by a special group of cells in a gland in order to evoke a response in other distant cells of the body

HORMONE DEFICIENCY DISEASE Disorder resulting from the lack of a particular hormone

HOST Cell acting as the recipient in the transfer of genetic information; or an organism or cell invaded by an infectious microbe

HYBRIDOMA Cell produced by the fusion of an antibody-producing cell with a cancer cell

HYDROCARBON Category of chemical compounds containing only carbon and hydrogen

HYDROGEN Lightest and simplest of all the elements; found in almost all biochemical compounds

HYDROXYAPATITE Form of calcium phosphate, the major inorganic constituent of bone

HYPERSENSITIVITY In immunology, the phenomenon of an excessive reaction to the presence of an antigen

HYPOTENSION Low blood pressure

ICE-MINUS BACTERIA Genetically modified bacteria unable to act as nucleation centres for the formation of ice

IMMISCIBILITY Inability to mix, e.g. oil and water

IMMOBILIZATION Techniques for attaching enzymes and microorganisms to insoluble support materials

IMMUNE COMPLEX Combination of antigen with antibody

IMMUNE RESPONSE Series of responses exhibited by the immune system when challenged by an antigen

IMMUNE SYSTEM Entire system of cells and proteins comprising the defence of an animal against foreign organisms, viruses and some chemical substances

IMMUNOGENIC Evoking an immune response (i.e. as by an antigen)

IMMUNOGLOBULIN Type of protein which includes antibodies

IMMUNOSENSOR Substance of immunological origin used as a chemical detector

IMMUNOSTIMULANT Substance which stimulates the immune system

IMMUNOTHERAPY Use of the immune processes for therapeutic purposes

INDIGO Blue dye of plant origin

INFECTION Invasion of an organism, or part of an organism, by pathogenic bacteria, viruses or other microorganisms

Appendix 2

INFORMATIONAL MACROMOLECULES Proteins and nucleic acids, in which the precise sequence of the individual units in the polymeric chains is of profound significance (in polysaccharides, by contrast, the unit sequence is uniformly repetitive)

INOCULATION Introduction of a small number of individual microorganisms (inoculum) into a culture medium with the expectation of a larger number ultimately developing by growth and reproduction

INOCULUM See under "Inoculation"

INORGANIC (COMPOUNDS) Apart from carbon monoxide and dioxide, those chemical compounds, mostly of mineral origin, which do not contain the element carbon

INSULIN Protein hormone produced by part of the pancreas gland; one of its main effects is to lower blood glucose and its deficiency is one of the causes of diabetes

INTELLECTUAL PROPERTY Patents, copyrights, trade marks and plant breeders' and designers' rights, as well as confidential information

INTERFERON Protein produced by a mammalian cell in response to viral infection: renders other cells resistant to infection by viruses

INTERLEUKIN Protein produced by helper T-cells of the immune system; encourages the growth and development of antibody-producing B-cells

INTRACELLULAR Contained within the cell

INTRONS Sections of "nonsense" DNA interspersed between "sense" sections in the genes of organisms other than bacteria and viruses; also called "intervening sequences"

"IRON SULPHATE" General term to include ferrous and ferric sulphates

ISOMERIZATION Chemical/biochemical process in which the atoms of a compound are rearranged into a different structure

ITACONIC ACID Organic acid produced by a fungus and used industrially

KETONE Type of chemical compound, the most familiar being acetone

KILLER CELL See "T-lymphocyte"

LACTIC ACID Organic acid produced in milk souring and cheese making

LACTOFERRIN Iron-containing protein found in milk; used for treating certain diseases

LAG PHASE Acclimatization period of a non-growing microbial inoculum when supplied with fresh nutrients

LANDFILL GAS See "Biogas"

LANOLIN Greasy substance present in wool

LARVA Stage in animal development very different from the adult, e.g. tadpole vs. frog, caterpillar vs. butterfly

LEACHATE Liquid which flows from leaching activities or from landfills

LEACHING Subjecting ores and other materials to the action of percolating water in order to remove soluble chemicals

LEGUME (adj. LEGUMINOUS) One of the plant family which includes peas and beans

Exploiting biotechnology

LEUCOSIS Abnormal whitening of some parts of the body

LIGHT HYDROCARBON FRACTIONS Those small-sized chemical compounds in crude oil which evaporate easily; also called "light ends"

LIGNIN Major component of wood; a complex three-dimensional polymer of variable composition which it is very difficult for microbes to degrade

LIGNOCELLULOSE Bulk material of most plants, containing lignin and cellulose

LIPASE Enzyme catalysing breakdown of fats and oils

LIPID Category of biochemical compounds which include fats and oils

LIPOSOME Artificially constructed sphere bounded by fat-containing membrane

LOCUST BEAN GUM Polysaccharide obtained from the fruit of the carob tree

LONGEVIN A protein made by bacteria in the intestine of carp and other animals which appears to combine with tithonin (q.v.) made by the host to influence the phenomena of ageing

LYMPH Fluid bathing the tissues in an animal body

LYMPHOCYTE Category of white blood and lymph cells: B-lymphocytes produce antibodies while T-lymphocytes coordinate the immune response and kill infected cells

LYMPHOKINE Soluble protein produced by some cells of the immune system which stimulates other cells to kill invading organisms; there are several kinds of lymphokines

LYMPHOMA Cancer of the lymphatic system

LYOPHILIZATION (FREEZE-DRYING) Procedure for removing water from a frozen sample by placing it in a high vacuum; the water vaporizes without the sample melting

LYSINE One of the twenty amino acids found in proteins

MACROMOLECULE Loose term distinguishing large molecules like proteins, nucleic acids and polysaccharides from the many small chemicals in living cells

MACROSTRUCTURE In biochemistry, an organized and defined structure containing many molecules, often of protein

MATRIX (pl. MATRICES) In an oil reservoir, the porous rock containing the oil

MATRIX ACIDIZING Using acid introduced into or generated within the porous rock of an oil reservoir to enlarge the drainage channels

MELANOMA Type of skin cancer, often pigmented

MEMORY CELL One of the types of immune system cells, usually producing antibody, which survives for a long period of time following exposure of animals or people to an antigen

MERISTEM Most active of the plant tissues, giving rise to new cells and tissues. Located in root and shoot tips, leaves and fruits

MEROZOITE The form of the malarial parasite resident in human liver cells

METABOLISM Total system of biochemical reactions and interactions of the molecules which make up living cells

METALLOID Certain elements with metal-like properties but which do not behave chemically as metals: arsenic and antimony are examples

METALLOTHIONEIN Type of protein able to bind tightly and specifically to certain metals

METHANE Simplest of all hydrocarbons, the main component of natural gas, used as a fuel and a common hazard in coal mines; also called "marsh gas" and "fire damp"

METHANOL Alcohol resembling ethanol but chemically simpler; also called "methyl alcohol", "wood alcohol" or "wood spirit"

METHYL ALCOHOL See "Methanol"

MICROBE Synonym for "microorganism"

MICROBIAL MINING Low-cost use of microorganisms to dissolve metals from ores

MICROBIOLOGY Study of microorganisms

MICROFLORA Population of microorganisms

MICROINJECTION Technique for introducing chemicals into cells with very fine syringes handled in a micromanipulator

MICROMANIPULATOR Instrument for surgical manipulation of cells under a microscope using tiny forceps, scalpels, syringes, etc.

MICROORGANISM Organism belonging to the categories of viruses, bacteria, fungi, algae and protozoa; most of them are small and properly seen only under a microscope

MICROPROJECTILE TECHNIQUES See "Particle acceleration"

MITOCHONDRION (pl. MITOCHONDRIA) Organelles which are the sites of most energy metabolism and found in most cells other than bacteria; they contain DNA distinct from that in the nuclear chromosomes

MINIMATA DISEASE Poisoning incident in Japan caused by the consumption of fish contaminated by mercury pollution

MOBILE OIL That portion of the crude oil in a reservoir which can in principle be swept to a producing well by a waterflood

MOLASSES Viscous sugar syrup produced during refining

MOLECULAR BIOLOGY Study of protein structure and its genetic specification and control; virtually indistinguishable from biochemical or molecular genetics

MOLECULAR ELECTRONICS Electronic systems using minute switching and conducting devices based either on individual molecules or on comparatively small numbers of them

MOLECULAR GENETICS Study of genetics at a biochemical level

MOLECULE Grouping of certain specified atoms, joined together in a particular way; the smallest particle of a chemical compound

MONOCLONAL ANTIBODY Specific antibody of unique structure produced by a clone of cells obtained by fusing an antibody-producing cell with a cancer cell

MONOCYTE Type of white blood cell

Exploiting biotechnology

MONOMER Single unit; a building block for constructing the chain of a polymer such as a polysaccharide, protein or nucleic acid

MONOSACCHARIDE Single sugar unit (e.g. glucose or fructose)

MONOSODIUM GLUTAMATE Flavouring agent containing the amino acid glutamic acid

MOULD See "fungus"

MULTICELLULAR Common types of living organisms in which each individual is composed of many cells

MUSCULAR DYSTROPHY Inherited disorder with degeneration of muscle fibres

MUTAGEN Chemical substance, or physical agency, which causes mutations

MUTAGENESIS Phenomenon of causing mutations

MUTANT Organism or gene carrying a mutation

MUTATION Inheritable change in the structure (sequence) of a gene

MYELOMA Type of bone marrow or plasma cell cancer

NATURAL SELECTION Darwinian concept of competition resulting in the propagation primarily of those individuals and species best able in their natural environment to produce healthy and viable offspring

NEOMYCIN Antibiotic active against certain bacteria

NEUROTOXIN Poison affecting the nervous system

NITROGEN Chemical element, present in amino acids, proteins, nucleic acids and other biochemicals, and hence essential to the nutrition of all living organisms; constitutes about 80% of the earth's atmosphere

NITROGEN FIXATION Biochemical process for incorporating the element nitrogen, one of the gases of the atmosphere, into the biochemical compounds of metabolism

NUCLEATION CENTRE Particle acting as the focus around which certain physical or chemical events may occur

NUCLEIC ACID Either of two complex organic acid polymers (DNA and RNA) which carry the inherited genetic information used to make proteins

NUCLEOTIDES Chemical units from which nucleic acids are built

NUCLEUS Membrane-bounded body in the cells of all organisms; contains the chromosomes and hence is the primary site of inherited genetic information. In bacteria and viruses the chromosome is free in the cell and not enclosed in a nucleus

NYSTATIN Antibiotic active against certain fungi

OIL SHALE Type of rock bearing solid hydrocarbon ("kerogen") which therefore does not flow; the rock has to be mined, crushed and heated to obtain the hydrocarbon

OLEOPHILIC FERTILIZERS Fertilizers for use in oil spill clean-up. Nutrient preparations containing nitrogen and phosphorus are compounded together with substances which ensure that the fertilizer sticks to the spilled oil and is not washed away into the surrounding water

Appendix 2

ORGANELLES Structures within most cells in which certain specific functions are localized

ORGANIC (COMPOUND) Chemical compound containing the element carbon (see "inorganic" for certain exclusions); all such natural compounds are of biological origin and many artificial ones can also be made

ORGANIC ACID Category of chemical compounds, many of them biological in origin, which have acidic properties generally weakly resembling those of mineral acids like sulphuric and hydrochloric

ORGANIC SOLVENT Liquid capable of dissolving substances (like oils) which are insoluble in water

OSMOTIC PUMP One which depends on the flow of water through certain types of membrane

OXIDATION Chemical process in which oxygen is added to a compound or hydrogen is removed; reactions of this type are important in converting the chemical energy in food to other forms

OXYGEN Gaseous chemical element forming 20% of the earth's atmosphere; important in many biochemical compounds

PALINDROMIC SEQUENCE In molecular genetics, a section of double-stranded DNA conveying identical information when "read" in either direction

PANCREAS Vital organ which produces digestive enzymes as well as the hormone insulin

PARKINSON'S DISEASE A chronic disorder of the central nervous system

PARTICLE ACCELERATION Technique for injecting DNA-coated gold particles into the target cells by accelerating them with an electric discharge

PASSIVE IMMUNIZATION Administration to man or animals of a preparation containing an antibody which neutralizes a toxin or combines with an invading microorganism or virus

PATENT Form of legal protection enabling an inventor to stop others using his invention

PATHOGENIC Capable of causing disease

PECTINS Plant polymers which often form gels when isolated; they help the setting of jam and cause cloudiness in fruit juices

PECTINASE Enzyme catalysing the breakdown of pectins

PENICILLIN Antibiotic produced by certain fungi, used in medicine to treat some bacterial infections

PENICILLIN ACYLASE Enzyme involved in the manufacture of derivatives of natural penicillins

PENICILLIUM Fungus which produces the antibiotic penicillin

PEPTIDE Chain of two or more amino acids joined in a certain way; when such chains are long enough the are usually called "proteins"

PERCOLATING FILTRATION One of the techniques used in sewage treatment

Exploiting biotechnology

PHENOTYPE Collective physical and chemical properties of an organism determined by the actual proteins it contains at any one time

PHENYLALANINE One of the twenty amino acids found in proteins

PHENYLKETONURIA Inherited disease characterized by the appearance of phenylpyruvic acid in the urine

PHENYLPYRUVIC ACID Breakdown product of the amino acid phenylalanine formed in certain diseases of genetic origin

PHEROMONE Chemical released by an animal as a signal to another of the same species, often as a sex attractant

PHOTOBIOREACTOR Vessel supplied with illumination for carrying out photosynthetic microbiological reactions; see also "Bioreactor"

PHOTOCELL Device which generates a voltage when illuminated; abbreviation for "photo-electric cell"

PHOTOSYNTHESIS Process in green plants for making sugars and other chemicals from carbon dioxide in the atmosphere; needs light, usually from the sun

PHYSIOLOGY Study of the functioning of organisms and of their constituent parts, usually at a level of lesser chemical detail than biochemistry

PHYTOALEXINS Plant proteins with antibacterial and antifungal activity produced as a consequence of wounding

PIEZOELECTRIC CRYSTAL Materials whose electrical properties are sensitive to surface distortion

PITUITARY Gland situated at the base of the brain which produces a number of different hormones, including growth hormone

PLASMA CELL Antibody-secreting cell, part of the immune system

POLLEN A powder produced by seed plants; composed of fine granules which contain the male gametes

POLYCHLORINATED BIPHENYLS Class of toxic industrial chemicals

POLYETHYLENE/POLYTHENE Non-biodegradable plastic produced from petrochemicals

POLYHAEMOGLOBIN Preparation used in blood replacement

POLYHYDROXYBUTYRATE Biodegradable bacterial product with many useful plastic-like properties

POLYMER Long chain of individual chemical units all of the same general type (sugars, amino acids, etc.), the precise number not being specified

POLYMER FLOODING Addition of polymer to a waterflood in order to achieve a uniform pattern of water flow and thereby improve oil recovery

POLYMERASE Enzyme which catalyses the linking into a long chain (polymer) of the individual units of which it is composed

Appendix 2

POLYMERASE CHAIN REACTION Enzymatic process for greatly increasing the quantity of a particular section of DNA

POLYSACCHARIDE Long-chain polymer composed of monomeric sugar units

POLYSTYRENE Non-biodegradable plastic produced from petrochemicals

PORE THROAT Smallest connecting space between the pores of rock

PRECIPITATE In chemistry, the deposition of a solid from a solution

PRIMARY IMMUNE RESPONSE Response evoked by first contact with an antigen

PRIMARY (OIL) PRODUCTION Early stage of working an oil well in which oil reaches the surface under the original pressure in the reservoir or is brought up by pumping

PROCESS PATENT Form of legal protection for a new production method

PROFILE IMPROVEMENT Use of plugging agents to control the direction of water flow in the porous rock of an oil reservoir

PROKARYOTES Organisms (e.g. bacteria) composed of simple cells lacking both a membrane-bounded nucleus as well as some of the other subcellular structures found in higher plants and animals

PRONUCLEUS Haploid nucleus of the egg or sperm after fertilization but before fusion to form the diploid nucleus of the zygote

PROPHYLAXIS Preventive medicine

PROPPANT Hard particles injected into artificial fractures in oil reservoirs in order to prevent their resealing themselves under the weight of the overlying rock

PROPYLENE Gaseous hydrocarbon produced from petrochemicals

PROTEASE See "Proteinase"

PROTEIN Molecule comprising one or more precisely ordered chains of amino acids

PROTEINASE Type of enzyme catalysing the breakdown of proteins; also called "protease"

PROTOPLAST Plant cell with the wall removed

PROTOZOON (pl. PROTOZOA) Smallest types of microscopic animals

PROVESTEEN Single-cell protein product based on yeast

PRUTEEN Single-cell protein feed supplement based on bacteria

PSEUDOPLASTICITY Dependence of the viscosity of a liquid on shear rate

PULLULAN A biological polymer

PURIFICATION In chemistry, the isolation of a compound in a state free from all other materials

PYRITIC SULPHUR That form of mineral sulphur which is contained in crystals of iron pyrite and similar materials

QUORN Microbial protein product made from a filamentous fungus

RADIOACTIVITY Phenomenon of atomic instability resulting in the disintegration of certain atoms with the emission of high energy rays

Exploiting biotechnology

RADIOIMMUNE ASSAY Type of immunological measurement using radioactivity as a sensitive detector

RADIOISOTOPES For any particular element, those versions of its atoms which decay by radioactivity

RAFFINOSE A trisaccharide sugar (i.e. one comprising three monosaccharide units)

RECEPTOR/RECOGNITION SITE Highly selective location on a protein or nucleic acid molecule to which certain other molecules can bind specifically

RECOMBINANT Cell whose DNA comes from more than one source

RECOMBINANT DNA TECHNOLOGY See "Genetic engineering"

REFRACTORY CHEMICAL/SLUDGE One not susceptible to further breakdown by microbial action

RENNET Enzyme used in cheese making; obtained from the lining of calves' stomachs

REPLICATION Synthesis of multiple identical copies

RESIDUAL OIL Generally, oil remaining in a reservoir after any particular recovery process is complete; often used for the oil remaining after waterflooding

RESTRICTION ENZYMES Enzymes which break foreign DNA in a manner characteristic for the particular enzyme; properly called "restriction endonucleases" because they break the DNA internally

RETROVIRUSES See "RNA viruses"

RETTING Microbiological process for treating flax

RHEOLOGY Study of flow and change of shape, e.g. of viscosity

RIBONUCLEASE Enzyme catalysing the breakdown of RNA

RIBOSENUCLEIC ACID (RNA) Type of polymer, varieties of which play important roles in using the information of DNA to make specific proteins; the sugar it contains is ribose, hence its name (sometimes abbreviated to "ribonucleic acid"). In some viruses RNA, not DNA, is the bearer of inherited information

RNA See "Ribosenucleic acid"

RNA VIRUSES Those whose genetic information is encoded in single-stranded RNA rather than the more usual double-stranded DNA; also called "retroviruses"

ROTARY BIOLOGICAL CONTACTOR Device for accelerating certain biological waste degradation procedures

SACCHARIDES General chemical name for sugars

SALVARSAN Drug specially designed for the treatment of syphilis which, while effective against the disease, showed undesirable side effects

SARCOMA Type of cancer

SCAVENGER CELL Type of blood cell responsible for ingesting bacteria and debris from dead tissue cells

SCLEROGLUCAN Type of biological polymer

SCREENING Technique of isolating a microorganism from a mixed population by obtaining a clone from each individual and testing it separately for the desired property; also applied to the examination of individual natural or artificial chemicals for specific properties

SECONDARY (OIL) PRODUCTION Means of obtaining more oil from a reservoir after the end of the primary production stage. Water or gas are injected to increase the underground pressure and drive oil to the producing wells

SEED CULTURE Culture of microorganisms used to start (inoculate) a fermentation

SEMI-SYNTHETIC PENICILLINS Antibiotics produced from natural penicillins by making certain chemical changes to their structures

SEQUENCE Order of the building blocks ("monomers") along the chains of polymers such as proteins and nucleic acids

SEQUESTRATION The phenomenon of binding a chemical, thereby preventing it from participating in reactions

SERUM PROTEIN One of the many types of protein dissolved in the clear liquid (plasma) of blood

SHEAR A type of physical deformation produced by two planes sliding over one another or by a liquid being forced through a small orifice

SHIKONIN Red plant pigment used both as a dye and as an anti-inflammatory for hæmorrhoids

SILAGE Crop harvested when green and kept succulent for fodder by partial fermentation

SILICON CHIP Electrical circuit manufactured on a "chip" (flake) of silicon

SINGLE-CELL PROTEIN Product made of specially cultured microbial cells and intended as a human or animal feed supplement

SLUDGE FARMING Practice of disposing of sewage and other sludges by digging them into soil

SOLUBILIZE Render capable of being dissolved

SPECIFICITY Characteristic ability of certain biochemicals (particularly proteins and nucleic acids) to recognize and interact, each with a very limited range of other chemical species

SPORES Small, resistant bodies formed by microorganisms when their living conditions become difficult

SPOROZOITE The form of the malarial parasite injected into the human blood stream by mosquitoes

STARCH Polymer made of glucose units; the common reserve material in plants

STATIONARY PHASE End of microbial growth in a culture, when all nutrients have been exhausted

STERILIZATION In microbiology, the procedure of eradicating all forms of living matter. More generally, the word also implies removal of the power of reproduction; the difference between these usages is not real because reproduction is one of the innate properties of

Exploiting biotechnology

living matter. Thus an organism unable to reproduce is, in a sense, biologically dead even if it continues to exhibit biochemical activity

STEROID Type of chemical compound of plant or animal origin; many have pharmacological activity

STICKY ENDS Sections of single-stranded DNA extending from the double stranded portions where they have been cut by a restriction enzyme

STOCK CULTURE Sample of a particular microbial population used as a reference source and maintained under conditions minimizing any risk of change

"STOPCOCKING" The practice of controlling water coning in oil production by intermittently stopping the flow of oil and water

STRAIN Population of identical microbes with some characteristic property or properties distinguishing them from related strains of the same species

STRAND In nucleic acid terms, one polymeric chain of DNA or RNA

STREPTOKINASE Bacterial enzyme useful for helping to disperse blood clots

STREPTOMYCIN Antibiotic active against certain bacteria

SUBSTRATE Either the substance(s) on which an enzyme acts or the food(s) for a microorganism

SUCRALOSE® Artificial sweetener based on a modification of sucrose

SUCROSE Table sugar; obtained from sugar cane or beet, it is a dimeric compound comprising one monomeric unit each of glucose and fructose

SUGAR Class of chemical compounds; also called carbohydrates

SULPHATE Chemical compound containing sulphur and oxygen together with a metal or other entity

SULPHIDE Chemical compound composed of sulphur together with one or more metals

SULPHONAMIDES Category of artificial anti-bacterial drugs

SULPHUR DIOXIDE/TRIOXIDE Gaseous compounds containing sulphur and oxygen

SULPHURIC ACID A strong mineral acid obtained when sulphur trioxide dissolves in water

SURFACTANT Chemical substance capable of dissolving both in water and in fats and oils, thus helping them to mix; virtually synonymous with "detergent"

SURFACTANT FLOODING Addition of surfactants to a waterflood in order to mobilize otherwise non-mobile oil

SYNERGY Working together of two or more components to produce a greater effect than would be expected from the simple addition of their separate actions

SYNTHESIS In chemistry, the manufacture of compounds

TAR SAND Mineral deposit in which a tarry form of hydrocarbon is mixed with sand

TARTARIC ACID Organic acid found in grapes

TERTIARY (OIL) PRODUCTION Also called "enhanced oil recovery": that phase of working an oil reservoir after production by secondary recovery (pressure maintenance) is complete. Tertiary recovery involves special techniques including the injection of polymers and/or surfactants

TETRACYCLINE Antibiotic active against certain bacteria

TETRAMER Short chain consisting of four monomeric units

THERMISTOR Electrical semiconductor sensitive to temperature

THERMOSTABLE Does not break down at higher than normal temperature

THIEF ZONE High permeability rock structure through which water tends to flow preferentially

THIOCYANATE Toxic chemical produced during the processing of certain gold ores

THIXOTROPY The property shown by a fluid which becomes less viscous when stirred

THYROID Gland producing the hormone thyroxine

THYROXINE Hormone essential for normal growth and development; produced by the thyroid gland

TISSUE PLASMINOGEN ACTIVATOR Enzyme involved in blood clotting

TITHONIN A protein found in animals which plays a part in ageing see also "longevin"

TOXIN Biological poison

TOXOPLASMOSIS Disease caused by a protozoon; shows enlarged liver and spleen

TRANSCRIPTION Process of constructing RNA molecules on the basis of DNA templates

TRANSDUCER A device for changing one form of energy into another

TRANSGENIC Animals and plants containing foreign genes introduced by genetic engineering

TRANSLATION Process of constructing protein molecules on the basis of RNA templates

TRIMER Short chain consisting of three monomeric units

TRIPLET CODE Way in which information for making proteins is encoded in DNA. Each triplet (group of three consecutive units) signifies a particular amino acid; the linear sequence of the triplets in the DNA chain is the same as that of the amino acids in the protein

TRUNNION In a centrifuge, the pair of opposing pins suspending the tube holder and allowing it to swivel from a vertical to a horizontal position

TRYPANOSOMIASIS Disease (called "sleeping sickness" in man) caused by the trypanosome protozoon

TRYPTOPHAN One of the twenty amino acids found in proteins

TUMOUR NECROSIS FACTOR Cytokine causing disintegration of tumours

TYROSINE One of the twenty amino acids found in proteins

Exploiting biotechnology

ULTRAVIOLET LIGHT Highly energetic light, invisible to humans, which can cause chemical damage to DNA, with possible genetic consequences

UNICELLULAR Organism composed of a single cell

VACCINATION Use of a vaccine to evoke an immunological response

VACCINE Preparation containing certain antigens used to evoke immunological resistance to a disease as a precautionary measure before possible actual exposure to a particular pathogen

VACCINIA Name of the virus causing cowpox

"VANGUARD" Type of microbe which helps to start the breakdown of wastes

VANILLIN Flavouring compound obtained from vanilla and now also available synthetically

VAT LEACHING Use of microbes to dissolve metals from crushed ore contained in a vessel; see also "Microbial mining"

VENTURE CAPITALIST Financier specializing in providing relative high-risk funds for start-up companies or for expansion, etc.

VERTEBRATE Animal possessing a backbone as distinct from worms, insects and shellfish which do not

VIRUS Minute infectious agent composed of nucleic acid (DNA or RNA) and protein, which is totally inert outside the host cell

VITAMIN Essential chemical which an organism requires in small quantities but is unable to make for itself

WATERFLOODING Injection of water into an oil reservoir in order to sweep residual oil to a producing well

XANTHAN A bacterial polymer

YEAST A fungus, usually growing as single cells, or in clumps of cells, but not in filaments

ZYGOTE Diploid cell formed by the union of gametes (egg and sperm) from which a whole plant or animal will subsequently grow.

Index

Abattoirs (*see* slaughterhouses)
Absorbent wound dressings (*see* wound dressings)
Absorbents, problems with, 239
Abstracting services, 72
Accountants (accounting), 65, 80, 93
Acetic acid, 148
Acetone, 8, 141
Acid(s) (acidity), 53-54, 63, 152, 156, 179, 189, 198, 200-202, 215, 227, 238, 246, 267
    leachates (*see* leachates, acid)
    organic, 147-148, 206
    rain (*see* rain, acid)
Acidulants, 148
Acquired immune deficiency syndrome (virus of), 47-48, 108, 111-113, 136, 255, 264
Adenosine triphosphate (ATP), 122
Adhesives (glues), 156, 159-160, 163
    from mussels, 160
Adjuvants, 125, 127
Administration, 57
Administrators, 109
Adrenal glands, 126
Adrenalin, 126
Advertising, 65
Aeration, 52-53
Aerosols, 180
Affinity, 55
Aflatoxin, 187
Africa, 236
Agar, 45
Age deterioration (ageing), 134, 184, 265
Agitation, 52-53
Agribusiness (agricultural industry), 73, 163-184
Agriculture, 2, 4, 72, 145, 157, 163
    commercial prospects for biotechnology in, 267-268
    economic and social effects of, 196
Agrobacterium, 177
AIDS (*see* acquired immune deficiency syndrome)

Air, 52, 60, 200, 219, 234-235, 255
Alanine, 14
Alaska, 228-229
Alcohols, 4, 8, 12, 15, 19, 47, 51-52, 59, 141, 157-158, 206
    as microbial feedstocks, 63
    beverages containing, 145, 185-186
Alfalfa (lucerne grass), 171, 175, 180
Algae, 41, 164, 178, 239
Alginic acid, 148, 189
Alkali, 53, 227, 238
Allergens, 115
Allergic reactions (allergies), 66, 115, 125
    desensitization to, 115
Alloys, 8, 223
Alumina, 161
Aluminium hydroxide and phosphate, 127
Alzheimer's disease, 265
Ambergris, 191
Amino acids, 10-14, 19-20, 22, 24, 27, 32, 36, 43-44, 115, 118, 132, 151, 155, 171, 192-193, 257-258
    essential for human nutrition, 193
    production of, 147
Ammonia, 176-177, 193, 233, 235
Amylases, 152
Anabastine, 174
Anaerobic,
    conditions (*see* conditions, anaerobic)
    digestion (*see* digestion, anaerobic)
    microbes (*see* microbes, anaerobic)
Analgesic alkaloids, 174
Analyses, 5, 54
    automatic, 253
    chemical, 249
Anaplasmosis, 181
Animal(s), 4, 9, 12, 32, 40-41, 46, 48, 61, 105, 108, 117-118, 139-140, 150, 155, 158, 164, 223, 257
    cells of, 38, 165
    diseases of, 164, 179, 267
    farm, 179, 187, 266
    feed supplements for, 63, 153, 193-194

genetically modified, 4, 38, 68, 164, 181-182
healthcare of, 4, 179-184
    commercial prospects for biotechnology in, 266
herds, 163
marine, 239
proteins of, 43
transgenic, 164, 181-182
ways of vaccinating, 180
Annual general meetings (*see* companies, annual general meetings of)
Anti-cancer,
    drugs (*see* drugs, anti-cancer)
    therapies, chemical, 128
    therapies, immunologic, 114, 129-130
Anti-coagulant drugs (*see* drugs, anti-coagulant)
Anti-fertility drugs (*see* drugs, anti-fertility)
Anti-foaming chemicals (*see* chemicals, anti-foaming)
Anti-inflammatory agents (*see* drugs, anti-inflammatory)
Anti-viral,
    drugs (*see* drugs, anti-viral)
    proteins (*see* proteins, anti-viral)
Antibiotics, 3, 12, 57, 104, 121, 123-124, 127, 145-147, 154, 186, 236
    potency of, 147
    resistance to, 121, 124, 126, 128
Antibodies, 105-116, 125-127, 131, 174, 187, 250, 264
    as antigens, 113-114
    conventional production of, 109
    heavy and light chains of, 106
    monoclonal, 109-110, 115, 117, 122, 135, 181, 186
Antibody-producing cells (*see* cells, antibody-producing)
Antidepressants, 147
Antigenicity, 105
Antigens, 105, 107-108, 110-111, 113-116, 121, 125, 250
    interactions with antibodies, 116, 253
Antihistamine drugs (*see* drugs, antihistamines)
Antileukaemic drugs (*see* drugs, antileukaemic)
Antimalarial alkaloids, 174
Antimony, bioleaching of, 201
Antisera, 109, 117

Antithrombins, 184
Antitoxins, 107-109
Antitrypsin, 183
Apples, 191
Aquifers, 204-205, 207-208, 226, 230
Architecture, biochemical and cellular, 140-141, 250, 257
    implications for metabolic efficiency, 140
Arizona, 178, 198
Arsenic, 225, 238
    bioleaching of, 201
Arthritis, 146, 150, 161, 265
Artificial,
    enzymes (*see* protein engineering)
    genes (*see* genes, artificial)
    insemination (*see* insemination, artificial)
    skin grafts (*see* skin grafts, artificial)
    vaccines (*see* vaccines, artificial)
Asbestos, 238
Asian rice gall midge, 172
Aspartame®, 192-193
Aspartic acid, 192
Aspergillus niger, 148
Asphalt extenders, 156
Aspirin, 146
Associations, industry, 93, 279-281
Atmosphere, 9, 175, 220
Atoms, 7, 12, 16, 117
Auditors, 93
Australia, 181, 202
Autoclaves, 46
Autoimmune diseases (*see* diseases, autoimmune)

B-cells (B-lymphocytes), 105-109, 125-126
Bacille Calmette-Guerin (BCG), 129
Bacillus thurigensis, 172, 180
Bacteria, 12, 18, 20, 28-30, 33, 37-50, 54-55, 61, 104, 106-107, 111, 126, 128, 140, 148, 150, 160, 165, 171-172, 175, 177, 184, 187, 189, 192, 198, 206-207, 211, 234, 238-239, 246
    ice-minus, 71
    photosynthetic, 178
    "rock-eating", 61
Bacterial
    immunomodulators (*see* immunomodulators, bacterial)

Index

infant enteritis, 183
Baking, 54, 153, 189
Baldness, 115, 266
Bananas, 167
Banks, 90, 93
   loans from, 95, 96
   at low-interest, 99
Barley, 153, 189
Basic research (*see* research, basic)
Batch cultures (*see* microbes, grown in batch culture)
Batteries, 223
Beaches, contaminated, 59, 230
   clean-up of, 228-229
Beans, 175, 191
Bearings, 159
Beer, 18, 49, 59, 142, 153, 185, 271
   low alcohol, 186
Beet sugar (*see* sugar, beet)
Belgium, 279
Bialaphos, 171
Binding sites, specific, 105, 113, 115, 125, 257-258
Bioabsorbents (biosorbents), 239-240
Bioaffinity sensors, 252-253
Biochemical engineering, 194
Biochemicals, 5
Biochemistry, 7-18, 20, 31, 39, 77, 118, 141, 145-146, 275
Biochips, 5, 255-262
   commercial prospects for, 271
   problems of addressing, 260
Biocides, 238
Biodegradability and biodegradation, 146, 152, 161, 171, 219, 221-222, 225, 236
Biogas, 226, 235-237
Bioleaching, 197-203, 246
   economic efficiency of, 202-203
   in situ, 202
   injector wells for, 202
   mechanism of, 199-202
Biological,
   computer chips (*see* biochips)
   engineering, 263
   evolution (*see* evolution, biological)
   filtration (*see* filtration percolating)
   oxygen demand, 233
   sciences (biology), 1, 14, 101
      commercialization of, 7

Biomass, 54, 59-60, 141, 143, 145, 155-158, 193, 239-240
Biomolecular electronics (*see* electronics, biomolecular)
Biopesticides, 164, 180
Biopharming, 4, 182
Bioplastics, 3, 160
Biopolymers, 149
Bioreactors, 232
Bioremediation (*see* environments, contaminated, bioemediation of)
Biosensors, 5, 116, 249-255
   commercial prospects for, 255, 271-272
   design of, 250-251
   environmental uses for, 255
   mechanism of action, 251-253
   uses for, 253-254
Biosequestration, 226
Biosurfactants (biodetergents), 213, 230
Biotechnology,
   importance of healthcare in, 103
   public perception of, 65, 74-75
Biotransformations, 149
Bismuth, bioleaching of, 201
Bitumen, 268
Blackflies, 172
Bladders, 115, 129
Bleaching, 189
Blockbusters (*see* drugs, wonder)
Blood, 32, 106-108, 111-112, 115-117, 120-121, 129-130, 153, 183-184, 187, 250, 253-255
   clotting of, 24, 31, 123
Blue,
   jeans (*see* jeans, blue)
   tongue, 181
Board(s) of directors, 79, 90-92, 101
Boll weevils, 172
Bone, 4, 163
   marrow, 105, 128, 133
   replacement material for, 161-162
   resorption of, 161
Books, 72
Booms for containing oil spills, 228
Botulism, 108, 186
Bovine,
   leucosis, 181
   rhinotracheitis, 181
   somatotrophin (BST) [growth hormone], 182

Brains, 165
Brazil, 157-158, 199
Bread, 153, 163, 185
Breeding procedures, conventional, 65
Brewing, 2, 59, 142, 153, 185, 189, 206, 271
Budget(s), 90-92
    forecasts, 90-91
Builders and maintenance men, 93
Building blocks, 10
Buildings, 57, 74, 90-91, 94, 158, 220, 231
Buses, fuel for, 158
Business, 65, 78, 101
    development, 79
    performance, 91
    plans, 90, 98
    strategies, 91-92, 101
Butane, 204
Butterflies, 172
Buyers, 94
By-products, 41, 151

Cadmium, 223, 225, 238-239
    bioleaching of, 201
Caesium, 241
Calcium,
    carbonate, 206
    phosphate, 161
Calculators, electronic, 256
Calf foetal serum, 184
California, 71, 178, 204
Calves, 181
Canada, 73, 202, 212, 279
    Research Councils of, 73
Cancers, 41, 48, 104, 114-115, 126, 129-130, 134, 150, 191, 265, 273
    growth of, 128
    specific vaccines for, 130
    therapy for, 127-130
Cane sugar (*see* sugar, cane)
Canine heartworm, 181
Capital and capitalization, 94-96, 100, 199, 263, 268
    costs and expenditures, 90, 198, 217, 233, 237, 245
    depreciation, 57
    investment, 3, 198, 233, 237
Carbohydrates, 10, 61-62, 105
Carbon, 8, 16, 60-61, 139, 200, 204, 242
    dioxide, 9, 10, 15-18, 52, 61, 122, 200, 215, 219-220, 225, 232, 235-236, 242, 246
    atmospheric, 169
    inorganic, 61
Carborundum, 149
Carcinogens, 237
Carcinomas, 115
Cardiatonic agents, 174
Carrageenan, 189
Carrier molecules, 17, 125
Carrots, 167
Case,
    histories (*see* histories, case)
    studentships (*see* studentships, CASE)
Cash flow, 88, 90-91, 98
Cassava, 63
Catalysis and catalysts, 11, 13-14, 19-20, 24, 59, 116, 139-140, 145, 155, 162, 266
    efficiency of, 267
    with dead cells, 142
    with immobilized systems, 143, 148, 192, 240, 250-252
    with isolated enzymes, 142
    with living, growing cells, 141-142
    with living non-growing cells, 142
Caterpillars, 172
Cattle, 41, 152, 179, 181-183, 192
Celery, 167
Cells, 12, 14, 24, 28, 30, 37-39, 42-44, 47, 52, 54, 108, 110, 124, 133, 140, 143, 165, 167, 250, 257
    antibody-producing, 110
    blood, 128
    breaking (opening) of, 54, 141, 160
    cancer, 105, 107, 110, 114-115, 126-128
    dead (decomposed), 54, 142
    diploid, 30
    division of, 30, 110, 115, 125
    fusion of, 110, 165-166
    haploid, 30, 183
    helper, 107
    hybrid, 38, 166-167
    immobilization techniques for, 143
    immortal, 110, 167
    naked, 165
    non-growing, 141
    of animals (*see* animals, cells of)
    of plants (*see* plants, cells of)

reproductive (*see* gametes)
scavenger, 106-108
surfaces of, 47, 140-141, 239
    receptors and antigens of, 111, 130
walls of, 127, 164-165
Cellulon, 159
Cellulose, 156, 158, 165
  bacterial, 159
Centrifugation, 54-55, 142
Cereals, 185, 189
Cetus Corporation, 51
Chalk, 204, 206-207
Champions, 92, 95
Cheese, 18, 142, 145, 148, 185, 189-190, 206
  manufacture, clotting in, 190
  whey as a microbial feedstock, 63
Chemical,
  analyses (*see* analyses, chemical)
  change, 8
  energy, 16
  formulae, 15
  industry, 2, 73, 124, 162, 196, 224, 226
    biotechnology in, 139-155
    commercial prospects for biotechnology in, 266-267
  products, 50, 54, 59, 139, 266
  reactions, 8-9, 252
Chemicals, 3, 5, 198
  antifoaming, 52
  agricultural, 174
  biotechnological production of, 145-151
  commodity, 57, 162
  complex, oxidation of, 60, 61
  de-icing, 171
  from plants, 174-175
  industrial, 222
  inorganic, 139, 220, 225
    disposal of, 238
  made in tissue culture, 174-175
  naturally-occurring, 65-66, 144, 146
  novel, 65
  oilfield, 211
  organic, 61, 139, 156, 225
  wood preserving, 226
Chemistry, 7-9, 15, 101, 117
Chemotherapy, 115, 128
Chicken feed, 192
Chief executives (Chief Executive Officers [CEOs]) (*see* managing directors)
Chile, 198-199
China, 199, 236
Chips,
  biological (*see* biochips)
  silicon, 256
Chitin and chitinase, 181
Chlamydial abortion, 181
Chlorine, 8, 225, 238
Chloroplasts, 38
Chocolate, 188-189
  mousse, 228
Cholera, 264
Cholesterol, 134
Chromatids, 28-30
Chromium, 226
Chromosomes, 28-31, 118
Citric acid, 145, 147
Civil engineering, 2
Clay, 215
Clients, 79, 86, 92-94
Clinical trials, 93-94, 129, 134, 136, 195
  costs and time of, 70, 136
Clonal,
  propagation, 167
  selection theory, 106
Clones and cloning, 40, 106, 109-111, 126, 167
Clostridium botulinum, 186
Clothes washing, 153
Cloth-making, 158
Cloudiness in drinks, removal of, 153, 189
Clover, 175-176
Cloves, 191
Coal, 49, 139, 197, 225, 237, 268
  as a fuel, 241-242
  chemistry of, 241-242
  desulphurization of, 5, 49, 243-245
    microbial methods for, 220, 226, 245-247
  hard (bituminous), 242
  high-sulphur, 242-243, 246
  low-sulphur, 242, 246
  organic sulphur in, 245-246
  pyritic sulphur in, 246
  soft (lignite), 241
  washing of, 244-245
Cobalt, bioleaching of, 201
Cobalt-chromium alloy, 161
Coccidiosis, 181
Cocoa butter, 188, 268

Coconut, 163
Codeine, 174
Codes and coding of nucleic acids, 20-22, 33, 37, 39, 43, 47, 117-118, 131, 171
Codons of nucleic acids, 21-22, 32, 36, 43-44
Coenzymes, 140, 142
Coking ovens, 226
Colchicine, 174
Colibacillosis (*see* scours)
Collaborative programmes (*see* programmes, collaborative)
Collaborators, 92
Colon, 115
Colorado beetle, 172
Combustion products, 252
Commercialization, 7
Commercial,
    confidentiality, 65
    exploitation, 2, 109
    management, 80
    opportunities, 72, 98, 109
    use of inventions, 67
Common cold, 48
Communication about biotechnology, 65, 71-72, 134
Companies, 2, 77-80, 86, 90-95, 97-101
    annual general meetings of, 91
    cars, provided by, 80
    chairmen of, 79, 91
    expansion of, 100
    first-round funding for, 99
    founders of, 98-99
    funding of, 93-94, 102
    large, 77-78, 135
    listings of, 72
    owners of, 79
    physical facilities of, 90
    policies of, 79-80
    private, 72, 96, 134
    public, 96
    reports to shareholders of, 91
    resources available to, 81, 82, 99
    second-round funding for, 100
    seedcorn funding for, 99
    small, 77, 79, 92
    start-up, 72, 78, 80-81, 95-101
        support grants for, 99
    subsidiary, 92
Competition and competitiveness, 72, 83-84, 90, 94-95, 98, 134
Competitive advantage, 94
Competitors, 78, 86-87, 92, 94, 134
Complement, host-specific, 131
Compost heaps, 63
Compounds, chemical, 8
    pure, 8, 49
Computers and computing, 1, 72, 256
    biological, 5, 255-262
Confectionery, 148
Conferences, 72, 80, 90, 279
    organizers of, 93
Confidentiality, 135
    protection of intellectual property by, 67-68
Consultants and consultancies, 72, 79, 90
Contaminants, microbiological, 46, 53, 59, 141, 144, 178, 185-187, 255
Contaminated sites, clean-up of, 4, 49, 232, 270
Continuous cultures (*see* microbes, grown in continuous culture)
Contraceptives, 150
Contracts, 72, 90, 95
Cooking, 158, 186, 189, 191, 236
Copper, 8, 238
    mining and bioleaching of, 198-201, 203, 269
    sulphate and sulphide, 200
Coproduced water, disposal of in oil production, 231
Copies and copying (*see* replication)
Coronary vasodilatory substances, 147
Corporate partnerships, 73
Corrosion, 52, 157, 206, 230
Corticoids, 150
Cosmetics, 70, 159, 174, 265-266
Cost-effectiveness, 50, 142, 145, 157, 159, 193, 205, 213, 217, 219, 243, 246-47, 263
Costs, 57, 68, 78, 87, 89-90, 143-144, 151-153, 159, 167, 179, 191, 205-206, 210, 215, 219, 234, 236, 239, 267
    operating, 90, 198, 201-203, 212
    overhead, 57, 90, 216
Costs and benefits, social and economic, 75, 196, 243, 264
Cotton, 158-159, 163, 165
Cowpox (vaccinia), 108, 113

Index

Cream, 188-189
Crops and crop plants (*see* plants, crop)
Cross-pollination, 38, 171
Crown gall disease, 38, 168
Crude oil (*see* oil, crude)
Cuba, 191
Cucumbers, 191
Culture and culture clashes, 78, 153, 263
Culture vessels (*see* fermentation vats)
Customers, 65, 82, 92, 94
Cutting enzymes (*see* enzymes, restriction)
Cyanide, 201-202, 238
Cyclamates, 191
Cystic fibrosis, 134
Cytokines, 125, 129

Darwin, Charles, 25
Databases, 135
   on-line, 6, 72, 277
Daughters, 28, 30-31, 42, 47
Dead bodies, decay of, 155-156, 176
Death, 42, 237
Decision,
   -maker of last resort, 161
   making, 79, 81-86, 90
Degree courses, 74
Denmark, 69, 75, 279
Deoxyribonucleic acid, 10-11, 20-40, 47, 112, 117-120, 165-167, 173, 183, 257
   double-stranded, 23, 28-29, 36, 120, 127
   exchange between bacteria in nature, 146
   fingerprinting (*see* genetic fingerprinting)
   fragments, electrophoretic separation of, 120
   from different individuals, 119
   polymerase, 121
   probes, 118-119, 121-122, 181
   repeat sequences in, 117-118, 120
   replication of, 22-24, 40, 43, 114, 127-128
      accuracy of/errors in, 20, 24, 43
   sequences, amplification of, 121
Depression, 130
Desensitization (*see* allergies and allergic reactions, desensitization to)
Designer drugs (*see* drugs by design)

Desulphurisation of coal (*see* coal, desulphurization of)
Detergents (*see* surfactants)
   laundry, 152
Development, 65
   costs of, 88, 94
   grants for, 74
Dewatering, 54, 56, 144
Dextran, 148
Diabetes, 32, 66, 116, 126, 133, 151, 251, 253-254
Diagnosis, 104, 116-122, 179, 253, 271
   of microbial infections, 121
   with antibodies, 116
   with monoclonals antibodies, 109, 117
Diagnostics, 3, 134, 137, 265
   commercial prospects for biotechnology in, 265
   for plant diseases (*see* plant diseases, diagnostics of)
   in food industry, 186
Diamond, 60
Diarrhoea, 181
Diesel fuel (*see* fuel, diesel)
Diet, 103-104, 132, 193
Digestion, anaerobic, 226, 235-237
Digestive system, 111, 125
Digoxin, 174
Dimers, 10
Dining facilities, 80
Diosgenin, 174
Diphtheria, 108
Directories and guides, 72, 277
Directors, 90
   executive, 79
   non-executive, 79
Disaccharides, 62, 188
Discharge and effluent process streams (*see* wastes, industrial)
Diseases, 3, 32, 186, 267
   autoimmune, 126
   bacterial, 41, 173
   degenerative, 273
   fungal, 41, 173
   human, 66, 75, 103-104, 108, 116, 122, 146
   inherited, 31, 120, 132
   of the poor, 264
   resistance to, 4, 163-164, 168-169, 172, 184
   symptoms of, 110-112, 115, 121

viral, 41, 47-48, 173
Dispersants, 229-230
Distribution of products (*see* products, packaging and distribution of)
DNA (*see* deoxyribosenucleic acid)
Doctors, 80, 137
Dogs, 191
Domesticated animals and crops, 1
Double helix, 21
Dough, 185, 189
Downstream processing (*see* products, extraction, separation and purification of)
Dragline silk, 160
Drains and drainage channels, 153, 199, 202, 206-207, 209, 215
Drill,
   bits, 149
   strings, 149
Drought, 164, 169, 179
Drugs, 2, 103-104, 122, 128, 132, 139, 252, 255, 266
   antihistamine, 115
   anti-cancer, 3, 70, 104, 125, 136, 150
   anti-coagulant, 147
   anti-fertility, 174
   anti-inflammatory, 147, 174-175
   anti-leukaemic, 174
   anti-viral, 125
   artificial chemical products as, 124
   by design, 123-124, 146
   delivery methods for, 125
      oral, 125, 134
   efficacy of, 70, 136
   for animal applications, 179, 195
   for human applications, 67, 70
   for specific diseases, 146
   immunosuppressive, 126, 131
   injection of, 125, 132
   natural products as, 123
   sulphonamide, 127
   testing and evaluation of, 70
   wonder, 88, 265
Due diligence, 98
Dwarfism, 150

Education, 74, 135
Eggs, 19, 29-30, 183, 187, 189
   fertilized, 14, 31
   yolks of, 188
Ehrlich, Paul, 146

Electrical devices, instruments and procedures, 38, 249, 252-253, 256
Electricity generation, 236-237, 241, 243
Electronics, 3, 5, 74
   biomolecular, 257-262
   commercial prospects for biotechnology in, 271-272
   molecular, 256-262
Electrons, 117
Electroporation, 38, 165
Elements (chemical), 8, 117
Embryos, 180
   cloning of, 183
Employees, 78, 101
Employment, 74, 79, 158, 196, 277
Emulsification and emulsifiers, 188, 230
Encapsulation, 125
Encoding (*see* codes and coding of nucleic acids)
Energy and energetics, 8-9, 14-18, 28, 52, 60-61, 156, 176-177, 198, 200, 221, 251
   transduction of, 252
Engineering, 84, 101, 193
Engineers, 72, 80, 82, 92, 205
England, 193, 195, 207, 237
Enhanced oil recovery (EOR) (*see* oil, tertiary production of)
Ensilage (*see* silage)
Environmental,
   benefits, 199
   biotechnology, market prospects for, 247-248, 270-271
   damage, 203
Environments,
   anaerobic, 141, 186, 206-207, 215, 234
   biochemical, chemical and physical, 18, 32, 53, 142, 165
   changes, 164
   cellular, 38, 141
   contaminated, bioremediation of, 4, 144
      injection wells for, 227
   friendly and harsh, 163, 165, 176, 215
   hot, 140
   industrial, 50
   intracellular, 38, 104, 141
   microbial, 33, 41-43, 54
   natural, 15, 50, 59, 68, 197, 224, 232-233
      biosensors, uses in, 255

monitoring of, 271
non-aqueous, 155, 267
oil-soaked, 50
Enzyme-linked,
  immunoassay (EIA), 116, 186
  immunosorbent assay (ELISA), 116, 186
Enzymes, 11-15, 19-20, 24-25, 32, 43, 59, 63, 116, 123, 127, 130, 140, 142, 145, 149, 157, 160, 165, 172, 177, 188, 190-192, 197, 222, 229, 239, 245, 249-250, 267, 271
  commercial uses for, 152-153
  defective, 132
  heat-resistant, 140, 152, 154, 192
  immobilized, 143, 148, 152
  in detergents, 153
  in food industry, 189
  in non-aqueous environments, 155
  industrial, 140, 151-155
  isolated (purified), 142
  naturally-occurring, 140
  purification of, 140
  restriction, 33, 37, 120-121
  sources of, 152
  specificity of, 11, 139-140, 155
  stability of, 151
  techniques for immobilizing, 143
Equine,
  herpes, 181
  infectious anaemia, 181
Equipment, 5, 46, 49-51, 57, 71, 90-92, 94, 153, 157-158, 186, 206
Erythromycin, 147
Ethics, 1, 75, 137, 182
Ethyl alcohol, 17, 18, 157 (*see also* alcohols)
Eukaryotes, 28-30, 40
EUREKA, 73
Europe, 136, 229, 235
  Eastern, 215, 270
  Western, 219
European Community, 62, 237
  Commission of, 69, 73, 75, 282
  Free Trade Association (EFTA), 62
Evolution, biological, 14, 20, 25, 104, 223
Excreted products (excretion), 43, 52, 54, 141
Exhibitions, 72
Expenditure, 90-91

Experience, 51, 79, 80-81, 93, 96, 99, 101, 212
Explosions and explosives, 147, 156, 202-203, 220, 236
Exxon Valdez, 228-229

Fabrics, 150, 153
  dyeing of, 150
Factors viii and ix (*see* haemophilia, factors viii and ix)
Faeces (dung), 236
Failure(s), 49, 82, 96-99
Far East, 236
Farmers, 177, 196
Farming, organic, 176
Fashion, 150
Fathers, 28-30, 118
Fats, 10, 153, 188, 213
Fatty acids, 178
Feedstocks, 9-10, 19, 44, 46, 50, 53, 57, 59-64, 139-145, 151, 156-157, 185, 191, 197, 206, 213
  agricultural, 144
  conversion to products, 59
  economics of, 64
  microbial, low-cost, 62
  options for, 61-63
  petrochemical, 61, 144, 157, 160, 193, 203, 212
  water-insoluble, 144
Fees for services (*see* services, fee-generating)
Feline,
  infectious peritonitis, 181
  leukaemia, 181
Fermentation, 2, 15, 18, 47, 52-53, 57, 59, 61, 74, 122, 145, 147-148, 150, 154, 157, 185, 192, 206, 215, 254
  broths, 55, 142
  technology of, 193-194
  vats (fermenters), 15, 51-54, 141-143, 150, 160, 164, 172, 193, 195, 236, 250, 271
Ferric sulphate, 200
Ferrous sulphate, 200
Fertilizers, 164, 175, 196, 224, 266, 273
  artificial, 176
  oleophilic, 229
Fever, 130
Fibres, 145, 148, 152, 158-160, 163, 222
  high tensile, 159-160

Field trials, 94, 196, 213, 216-217, 246, 269
Filaments, 46-47, 54
Filter beds, 234
Filtration, 54-55, 57, 142, 240
   percolating, 234
Finance, 79, 95
   for biotechnology, 94-102
Financial,
   bottom line, 81
   community, 72, 80
   inducements and assistance (*see* taxes, benefits and inducements)
   management, 90-91
   plans, 98
   resources and obligations, 90, 95
Finland, 280
Fireflies, 122
First round funding (*see* companies, first round funding for)
Fish, 224, 229, 231, 233
   and chips, 149
   canned, 153, 186
   farming, 228
Flak jackets, 160
Flavour(s) and flavouring agents, 4, 59, 145, 148, 153, 164, 174, 189-191
   artificial, 191
   enhancers, 149, 192
Flax, 145
Fleming, Alexander, 146
Flocs and flocculation, 142, 234
Floods, 164, 169
Flowers, decorative, 163-164, 167, 173-174, 196, 267
Flue gases, 242
   desulphurization of, 245
Fluidized bed combustion, 244
Fluorescence, 116, 121, 252-253, 260-261
   immunoassay (FIA), 116
Foaming, problems of, 52
Foetus, 109
Folic acid, 127
Folk remedies (folklore), 123, 146
Food and foodstuff, 4, 10, 12, 14-15, 18-19, 39, 42-43, 59-61, 139, 143-144, 148-149, 153-154, 159, 176, 185-195, 196, 222, 225, 255, 266
   additives in, 174, 196
   fermented, 185
   fresh, handling of, 187
   grains, 51, 157
   industry, 73
      commercial prospects for biotechnology in, 162, 267-268
      traditional technologies in, 185-186
   plants, 68
   poisoning, 111
   processing, 187
   products, 68, 70
   storage properties of, 164
   world shortages of, 61, 193
Foot-and-mouth disease, 41, 181
Forecasting (*see* future of biotechnology, difficulties of forecasting)
Foreign,
   currency, 158, 194, 217, 270
   products and proteins, 39, 51, 105, 109-110, 131, 134, 150-151, 182-184
Forensics, 120
Forests, 220-221
Founders (of companies) (*see* companies, founders of)
Fowl plague, 181
Foxes, 181
Fragrances (*see* flavours and flavouring agents)
France, 69, 280
Free-market economy, 77
Freeze-drying, 56
Fringe benefits, 80, 90, 101
Frontier research, 2
Frost damage to plants, control of, 71
Fructose, 52, 62, 188, 191-192 (*see* high fructose syrup [sweetener])
Fruits, 152, 185, 189, 196, 267-268
   juices of, 153, 189
Fuels, 9, 60-61, 139, 144-145, 152, 155-158, 163, 176, 236, 241, 268
   diesel, 158, 230
   fossil, 220-221
   high-sulphur, 220
   nuclear, 224, 241
      reprocessing of, 241
   storage tanks, leaks from, 230
Funding, 92
   for research and development, 65, 73, 93
   from in-house resources, 73

from private sources (*see* venture
    capital funds)
from public sources, 80, 90, 95
matching, 95
rounds (*see* investments, stages of)
Fungi (moulds), 30, 40, 41, 46-47, 54,
    104, 146-148, 154, 171-172, 179,
    187, 194, 234
Fungicides, 171
Fur, 158
Future of biotechnology, difficulties of
    forecasting, 263, 272-273

Gallium, bioleaching of, 201
Gametes, 29-30
Garbage (*see* wastes, domestic)
Garlic, 191
Gases, 54, 204-205, 207, 209, 215, 220,
    250
    as microbial feedstocks, 62
    nerve, 252
    toxic, 53
Gasohol, 158
Gasoline (*see* petrol)
Gear wheels, 159
Gelatin, 153
Gels, 189, 209
Genentech, 2
Genes, 20, 24, 29-33, 37-40, 111, 118,
    126, 130, 134, 165, 190, 258
    artificial, 258
    expression of, 31, 40
    probes (*see* deoxyribonucleic acid
        probes)
    therapeutics, 134
    therapy, 3, 24, 104, 132-134
Genetic,
    diseases, inheritance of, 31, 122, 132
    engineering, 2, 19-20, 31-41, 51, 65,
        68-70, 74, 110-113, 121, 123,
        126, 131, 133, 150-152, 159,
        162-164, 180, 182, 185-186,
        191, 203, 223, 240, 246, 258,
        263, 272
    faults and defects, 24, 104, 132, 134,
        265
    fingerprinting, 3, 117-121
    information and its transfer, 10, 14,
        19-21, 26, 28, 31, 33, 39-40,
        43, 47, 110, 177, 257
    inheritance, 14-15, 25

instructions, 19
manipulations, 39
    guidelines and rules for, 69-71
material, 22, 28, 30, 33, 39
messages, 25, 39-40
methods, 43-44, 154
organization, 38, 177
relationships, 120
Genetically modified organisms, release of
    into the environment, 68, 70-71, 99,
    196
Genetics, 19-38, 46, 49, 77, 132, 141,
    145, 275
    of development, 273
Genomes, 38, 40, 117, 166, 171-172, 183
Genotypes, 40
Geography, 169, 269
Geology, 205, 220
Germany, 60, 75
Glands, 124, 126
Global warming, 169, 220
Gluconic acid, 148
Glucose, 16-18, 62-63, 116, 148,
    156-157, 191-192, 251, 253
    isomerase, 154, 191-192
Glutamic acid, 192
Glycogen, 10
Glyphosate, 169, 171
Goats, 183
"Going public", 91, 97
Gold, 239
    bioleaching of, 201
Governments, 74, 86, 95, 99, 134, 137
    agencies of, national and
        international, 73, 93, 95,
        134-135, 281-283
    departments of, 93
    research institutes funded by, 77
Granting agencies and bodies, 93 (*see also*
    companies, funding of *and* research,
    grants and contracts for)
Grants, 72, 95
Graphite, 60
Gravity, force of, 55
Grease, 153, 213
Greenhouse gases, 219
Greenhouses, 196
Griseofulvin, 147
Growth, 61
    hormone, 150, 184

bovine (*see* bovine somatotrophin (BST) [growth hormone])
human, 132
microbial, 19, 42-46, 50-51, 59, 141, 143, 209, 228, 233-234
  exponential phase of, 42, 45, 47
  lag phase of, 42
  stationary phase of, 42, 43
of cells, 110, 115, 125
Guar, 189
Gulf War, 228
Gums, 148, 189
Gypsum, 245
Gypsy moth, 172

Haber process, 176-177
Haemophilia, 20, 24, 31
  factors viii and ix in, 24, 183
Haemorrhoids, 175
Hafnium, 240
Hair, 115, 128, 153, 265
Hairy cell leukaemia, 129
Halogens, 238
Healthcare, 3
  animal (*see* animals, healthcare of)
  commercial prospects for biotechnology in, 264-267
  marketing of, 93-94, 135-136
  paying for developments in, 137
  products, 68, 70, 93
Heart attacks, 123-124
Heat (*see* temperatures, mainly high)
  –sensitive materials, 57
Heating, 236
Helium, 47
Helix, 21
Helper cells (*see* cells, helper)
Hemp, 145, 163
Hepatitis, 255
Herbicides, 169, 187
  resistance to, 4, 168, 171
Heroin, 117
High fructose syrup (sweetener), 145, 154, 191-192
Hip joint replacements, 161
Histamine, 115-116
Histories, case, 72
HIV (human immunodeficiency virus) (*see* acquired immune deficiency syndrome)
Holidays, 80

Hormones, 4, 31-32, 104, 117, 124, 130, 150, 164, 186, 186-187, 249, 265
  deficiency diseases, 123
  endocrine, 126
  of animals, 164
Horses, 108-109, 117
  proteins of, 108-110
Hospitals, 93, 253
Hosts, 47, 107, 111-112, 146
Hot springs, 192
Human,
  beings, 28, 30, 32, 41, 48, 61, 105, 109, 126
    material from, 32
  cells, 118
  chromosomes, 28, 134
  diseases, 179, 233
  genes, 68, 117
  genetic manipulations, 132-133
  genome project organization (HUGO), 134-135, 272
  monoclonals, 110
  nutrition, 61
  organs, 131
  proteins, 43, 105, 108, 123, 131, 150
    production of in bacteria, 43, 50, 115, 144, 150
  resources, 80, 90
  vaccines, 127
Hunting and gathering, 163
Hybridomas, 110
Hydrocarbons, 193, 204, 221, 230, 232
Hydrochloric acid, 206-207, 210
Hydrogen, 16-18, 117, 176-177, 204, 241
Hydroxyapatite, 161
Hygiene, 264
Hypersensitivity, 115
Hypotension, 130
Hypotensive substances, 147

Ice, 56
  cream, 189
  crystals, 71
Immigration authorities, 120
Immobilized catalysts (*see* catalysis with immobilized systems)
Immune,
  complexes, 106-107
  response, 104, 107-109, 112
    modifications to, 125-127
    primary, 107

secondary, 107, 111-112, 115
system, 104-109, 111-112, 125-126, 131, 179
Immunity,
    acquired, 107-109
    active, 107-108, 111
    passive, 108-109
Immunoassays, 122, 174, 186
Immunoglobulin, 115-116
Immunological,
    incompatibilities, 131
    responses, 133
Immunomodulators, bacterial, 129
Immunosensors, 253
Immunostimulants, 126
Immunotherapy, 125, 128-129
Immunotoxins, 130
Imperial Chemical Industries (ICI), 193, 195
Incineration, 233, 237
Income, 90-91, 136
Indexing services, 72
India, 199, 236
Indigo, 150, 174
Indigofera, 150
Indium, bioleaching of, 201
Individuals, 74-75
    initiatives by, 86
Industrial,
    community, 80
    development, 2
    laboratories, 77
    products, 3, 186
    wastes (*see* wastes, industrial, disposal of)
Industry,
    nuclear, 240
    "post-industrial", 74
Infections and their agents, 47-48, 53, 104-105, 107-109, 111-112, 124, 128, 131, 146, 168, 187, 264
    viral, 126, 150
Inflammations, 125, 129
Influenza, 41, 108, 111
Information, 89, 134
    exchange of, 72
    sources of, 6, 32
Informational macromolecules, 11
Inheritance, 1, 10, 27, 29, 31
Inks, 171
Innovation, 65, 85

Inoculants, microbial, 185, 230
Insecticides, 4, 112, 171, 174, 187, 221
    resistance to, 114
Insects, 172, 223
    cuticles of, 181
    mating of, 172
Insemination, artificial, 183
Insulin, 32, 43, 66, 126, 130, 133, 151, 175, 254
    human, 66, 151
    porcine, 66, 151
Insurance, 90, 93
    medical, 137, 265
Integrated pharmaceutical companies, 263
Integration,
    of staff into a team, 93
    of technologies and industries, 49
    within biological systems, 47, 141
Intellectual property rights, 65-69, 95, 98, 134 (*see also* patents)
    protection of,
        by confidentiality (*see* confidentiality, protection of intellectual property by)
        by patenting (*see* patents, protection with)
Interest payments and receipts, 90, 95, 97
Interferons, 126-127, 129, 150
Interleukin-2, 126, 129, 136, 150
Interleukins, 125, 175
International agencies (*see* governments, agencies of, national and international)
International Centre for Genetic Engineering and Biotechnology (ICGEB), 73, 283
Intestines and their linings, 111, 128, 184
Intramolecular rearrangements, 191
Introns, 39-40
    editing to remove, 39-40
Inventors, 67
Investments and invested assets, 72, 85, 95-99, 158, 270
    encouragement and inducements for, 99
    returns on, 84, 94, 97, 134, 217
    seedcorn, 99
    stages, 100
    strategies, 96
Investors, 78-79, 91, 93, 95, 97-100

exit routes for, 97, 100
    lead, 98
Iodine, 8, 238
Ireland, 69, 280
Iron, 201
    pyrite (fools' gold), 200, 242, 246
    scrap, 200-201
    sulphate, 200
Isomerizations, 191-192
Israel, 178
Itaconic acid, 148
Italy, 280

Jams, 153
Japan, 162, 175, 224, 280
Jasmin, 174
Jeans, blue, 150, 174
Jellies, 153
Jobs (*see* employment)
    satisfaction in, 78
Joint projects (*see* programmes, collaborative)
Journals, scientific, 72, 89-90, 135, 277
Jurassic Park, 75

Kentucky, 243
Ketones, 132
Kevlar, 160
Kidneys, 136
Killer cells, 107

Laboratories, 139
    activities and operations in, 49, 146
    dating records of, 66
Lactic acid, 18, 148, 179
Lactoferrin, 183
Landfill, 221-223, 226, 232, 235-238, 248
    gas (*see* biogas)
    sites and management of, 4, 62, 145, 238
Landlords, 93
Lakes, 236
Lawyers, 80, 93
Leachates, acid, 236
Lead, 8, 239
Leather, 60, 153, 158, 225
Leaves, 152, 165, 172
Lectures, 80
Leeches, 184
Legislation, 72, 99, 242, 270
Legumes, 164, 175-179, 187

Lemons, 147-148, 191
Lentils, 175
Leprosy, 108, 126
Lettuce, 224
Libraries, 6, 80
Licences and licensing, 65, 67, 90, 95, 109, 134, 136
Light-emitting diodes, 252
Light, 260
    coloured, 252, 260
Lignin, 156-158
Lignite, 241
Lignocellulose, 63
Lime, 147, 153
Limestone, 204, 206-207, 244
Limited partnerships, 73, 95
Linen, 158
Lipases, 153
Lipids, 153
Liposomes, 125
Liquefied petroleum gas, 158
Liver, 112, 125, 132-133
Local authorities, 93
Lock-and-key interactions, 32, 55, 115
Locust bean (carob), 189
Longevin, 184
Long-term considerations, 3-5, 92, 95
Losses, 90-91
Lubricants, 149
Lymph, 165
    nodes and vessels, 105
Lymphocytes, 105, 110, 112
Lymphokines, 107, 125
Lymphomas, 115
Lyophilization (*see* freeze-drying)
Lysine, 193

Macromolecules, 124
Magazines, 75, 277
Magic bullets, 146
Maize, 63, 157, 167-168, 171, 176-177, 179, 191
Malaria, 103, 264
    parasites of, 108, 112
Malignant cells (*see* cells, cancer)
Malnutrition, 264
Malt, 189
Management, 57, 65, 82, 84-85, 90-91, 93, 99, 101, 158
    culture of, 78, 81
    in-house structures for, 79, 91

of basic research, 77
of biotech companies, 77-103
structures, 79
Managers, 80-81, 90, 93
   managing, 79, 81-82, 90-93, 99, 101
      remuneration of, 101
      relationships with boards of directors, 101
   scientific (see science, managers of)
   senior, 92
Manganese, bioleaching of, 201
Manufacturing, 18, 43, 59, 77, 79, 82, 87, 90-91, 94
Marine thermal vents, 192
Market(s), 72, 82-83, 90, 92, 94, 186, 191
   exclusivity, 136
   niches, 82, 194, 267
   opportunities, 3, 212, 255
   position, maintenance of, 89
   potential, 135
   pull, 94
   research and surveys, 72, 82, 92, 98, 278
   share, 150
   sizes, 145, 255
Marketing, 65, 79, 92
   of biotechnology, 93-94
   of healthcare (see healthcare, marketing of)
   staff, 80
Marketplace, 1, 3, 79, 87, 92, 139
Measles, 108
Meander streams, 240
Meat, 184, 189, 196
   canned, 153, 186
   extenders and replacers, 194
   extracts, 61
Medicine(s), 72, 80, 145-146, 150, 237, 271
Medium-term considerations, 91-92
Melanomas, 129-130
Membranes and membrane barriers, 28, 38, 112, 125, 159, 164, 250-252
Memory cells, 107
Mental retardation, 132
Mercury, 8, 223-225, 238
Meristems, 167
Merozoites, 112
Metabolic,
   balance, 18-19
   by-products, 15

   organization, 13-15
   degradation of, 142
Metabolism, 12, 14-18, 32-33, 50, 54, 59, 61, 151, 198, 223
Metallothioneins, 240
Metals, 9, 139, 197, 225-226, 235, 268-269
   absorbents for (see bioabsorbents)
   bioleaching of, 5, 198-203
      commercial prospects for, 269
   heavy, toxic, 164, 223-224, 233
      disposal of, 226, 237-241
   ores, beneficiation of, 198
   processing of, 4, 148, 223
   plants for, 199
   refining of, 198, 203
   separation of, 49, 144, 240
   smelting of, 199, 203, 224
   strategic, 201
Methane, 4, 62, 193, 204, 220, 235-236, 238
Methanol (methyl alcohol), 63, 158, 193
Methylophilus methylotrophus, 193
Mice, 110, 133-134, 183-184
Microbes (microorganisms), 1, 9, 12, 14-15, 18-19, 24, 32-33, 40, 46-47, 50-52, 54, 59, 61, 104-105, 107-108, 111, 117, 139, 141-142, 145-146, 149-150, 152, 155, 157, 180, 191, 206, 220-223, 226, 230, 232, 266
   anaerobic, 15-18, 186, 211-213, 215
   cells of,
      counting individual, 122
      surfaces of, 105, 111
   collections of, 5, 50
   colonies of on nutrient jelly, 44-47
   cultures of, 141
   grown in batch culture, 44-45, 53, 143, 186
   grown in continuous culture, 45-47, 53, 143, 186
   heat-tolerant, 154, 273
   marine, 224
   mining with, 5, 15, 144
   naturally-occurring, 65-66
   pathogenic, 70, 107, 121, 167, 233, 255
   populations of (see populations, microbial)
   products manufactured by, 50-57

pure strains of, 44
technologies for handling, 49, 163
vanguard, 227
Microbial enhancement of oil recovery, 5, 62, 203-217, 227
commercial prospects for, 216-217, 269-270
Microbiology, 38-44, 49, 146, 276
Microinjection, 165-166, 183
Microlasers, 261
Micro-manipulators and micro-manipulation, 38, 183
Microprojectiles, 165
Micropropagation, 167
Microscopy, 43, 121-122
Middle East, 270
Milk, 18, 182-184, 188-189, 194, 225
digestability of, 189
sour, 163, 185, 190, 206
Mineral salts, 60, 193, 246
Minerals, 225, 242
microbial oxidation of, 200
ores of, 5, 8, 197-198, 268
recovery and production from deposits of, 2, 5, 197-218
commercial prospects for biotechnology in, 268-270
Mines (and mining), 3, 197-198, 223, 268
tunnels and voids in, 202, 220
Miniaturization, 256, 271
Minimata disease, 244
Mitochondria, 28-29, 38, 183
Mixing,
of oils and water, 193
of substances, 49
Molasses, 62, 197, 206
Molecular,
biology, 31, 170, 273
electronics (*see* electronics, molecular)
memory stores, 259
Molecules, 7-8, 12, 19, 24, 28, 36, 42, 118, 189, 223, 256
excited states of, 260
light-sensitive, 260
shapes of, 155
specific binding between (*see* binding sites, specific; enzymes, specificity of; specificity, biochemical)
Molybdenum, bioleaching of, 201
Money, 2, 79, 81, 87, 90-91, 99

Monitoring,
of company progress, 72
of production processes, 53-54, 254
Monoclonal antibodies (*see* antibodies, monoclonal)
Monoculture, 172
Monocytes, 105
Monomers, 10, 21
Monosaccharides, 62-63, 188
Monosodium glutamate, 192
Monsters, bug-eyed, 75
Morphine, 117, 174
Mosquitoes, 112, 172
Mothers, 28-30, 118, 183
Moths, 172, 191
Motives and motivations, 81
Motor vehicles, 61, 145, 158, 220
Moulding (of plastics), 160
Moulds (*see* fungi)
Mousses, 189
Mouth feel, 189
Multi-disciplinary teams, 81
Multi-enzyme systems, 140, 250
Multi-molecular assemblies, 257, 262
Mumps, 41
Municipal solid wastes (*see* wastes, municipal)
Muscles, 111, 134, 165, 184
relaxants of, 174
Muscular dystrophy, 134
Mushrooms, 46
Musk, 191
Mutagens and mutagenesis, 25, 44, 171
Mutation and variation, 25-26
Mutants and mutations, 31, 43-44, 111-112, 118, 141, 143-144, 167, 232
deletion, 26, 36, 43
insertion, 26, 36, 43
lethal, 43
replacement, 25-26
selection of, 167, 231
Mycorrhiza, 161, 179
Myelomas, 110

Nagana (*see* trypanosomiasis)
Nails, 128
Natural,
gas, 61-62, 193, 220, 268
selection (*see* evolution, biological)
Nematocides, 171

Index

Nematodes, 171
Neomycin, 147
Nerve gases (*see* gases, nerve)
Nerves, 260
Netherlands, 69, 75, 280
Neurotoxins, 186
New,
   activities, 94, 96
   drugs, 104, 134, 136-137
   inventions, 65-66
   materials, 2, 3, 158-162
   Mexico, 214
   organisms, 65
   plant strains, 4
   products, 3, 72, 86-88
      ideas for, 85-86, 89
      testing of for safety and efficacy, 70
   services, 3, 72
   ventures, 92
      proposals for, 98
News, company and financial, 72, 277
Newsletters, 72, 277
Newspapers, 75
Nickel, 226
   bioleaching of, 201
Nicotine, 174
Nitrate, 175-176, 221, 237
Nitrogen, 8, 41, 52, 60, 175-177, 193, 204, 229, 236, 242, 273
   fixation of, 175-180, 196, 273
   fixing bacteria, 175-176, 179
   oxides of, 220
Nodding donkeys (*see* oil pumps)
Nodules, nitrogen-fixing, 175, 177
North Sea, 231
Nuclei of cells, 28, 38, 110, 166, 183
Nucleic acids, 10, 20
   complementary copies of, 23-24, 26-27
   nucleotide bases of, 43-44, 118-119, 134
   pairing of bases in, 21, 23, 25, 29, 34-35, 43, 118-119
   palindromic sequences in, 33, 36
   strands of, 23, 27, 29, 33, 119, 121
Nucleotides, 10, 20, 121
Nutrients and nutrient media, 42, 46, 50, 52-54, 59, 122, 141, 143, 160, 187, 200, 206-207, 211, 223, 225-226, 228-230, 234, 245
   as jellies (for microbial and single-cell growth), 44, 167
   high-cost, 61
   value of, 193
Nylon, 159
Nystatin, 147

Objectives, commercial, 78, 81, 86
Odours, 189, 191
Offspring, 10, 22, 28, 31, 42-43, 118
Ohio, 243
Oil, 60, 139, 144, 187-188, 193, 197, 225, 268
   companies, 203, 216
   crude, 61, 149, 158, 193, 203-204, 222-223, 228, 232, 242
      prices of, 61, 157-158, 160, 193, 212-214, 270
   degradation by microbes, 229-230
   droplets, entrapment of, 213-214
   industry, 73, 149
   primary production of, 206-209, 211, 217, 270
   production, 5, 12, 49, 144, 158, 221, 226
      coning in, 207-209, 211
      coproduced water in, disposal of, 231
      single well stimulation for improved, 215
      use of separators in, 231
   prospecting for, 205
   pumps, 208, 210, 215
   recoverable, 217
   recovery, commercial prospects for biotechnology in, 216-217
   refining, 204, 226
   reserves, 203
   reservoirs, 5, 59, 149, 203-204, 207 (*see also* rocks)
      acid fracturing of, 206
      horizontal drilling in, 207
      in carbonate rocks, 206-207, 210, 215
      in sandstone rocks, 204, 206, 215
      indigenous microbes in, 216
      matrix acidizing of, 206-207, 209-210, 216
      microbial survival in, 215-216
      polymer flooding of, 212-213, 227
      pressure in, 209
      profile improvement of, 211
      residual oil in, 211, 214, 270

    selective plugging of, 211
    surfactant flooding of, 213, 227
    sweeping oil through, 149, 211
    thief zones in, 211
    water flows in, 209, 211-212, 216
    waterflooding of, 210, 231
    water injection into, 209-210, 213
residual, in production fluids, 232
secondary production of, 209-211
spills,
    in the sea, 221, 226, 228
    microbial clean-up of, 4, 12, 226, 228-231, 271
    non-biological treatments for, 228-229
    weathering of, 228
tertiary production of, 3, 149, 211-215, 217, 230
wells, 149, 205, 208, 210
    casing and perforations of, 206
    drilling of, 149, 205-206, 268
    injection, 149, 209-210, 213, 215, 231
    production, 209, 268
    scale deposits in, 210, 215
Oilfield operators, 209
Oilfields, offshore and onshore, 209
Oilseed rape, 171
Oklahoma, 213
Olive oil, 191
Optical effects and instruments, 252, 260
Oranges, 191
Organelles, 28, 30, 250, 257
Orkney Islands, 228
Orphan drug status, 136
Osmotic pumps, 125
Oxidations, 9, 15-18, 60, 122, 148, 200, 232, 234, 242
Oxygen, 8, 15-17, 42, 46-47, 52-54, 60, 155, 177, 204, 222, 228-230, 234, 242, 246
Ozone, 169

Paddy fields, 178
Paints, 145, 148, 159
Palm oil, 188, 191, 268
Pancreas, 66, 133, 151, 188
Paper, 60, 152, 163, 165, 220, 222
Parallel processing, 261
Parasites, 47, 108, 111-112
Parents, 10, 22, 24, 29, 118

Parkinson's disease, 134
Particle acceleration, 38, 165
Particulate matter, 54
Partners, 95
Parvovirus, 181
Pasteurellosis, 181
Patents, 50, 65-67, 94, 98, 134
    agents for, 93
    claims in, 67-68
    date of filing, 66
    date of invention, 66
    effect of publication on eligibility of, 66
    European rules for, 66
    exploitation of, 67
    fees, 90
    filing an application for, 66
    grace period for filing, 66
    holders of, 67
    international conventions on, 66
    laws and procedures relating to, 65
    licensing of (see licences and licensing)
    lists of those granted, 72
    prior disclosure of, 66
    priorities for, 66
    process, 65
    protection by, 109, 135
    requirement for novelty in, 66
    rights of sale and licensing of, 67, 95
    risk of therein claims being circumvented, 68
    searches and search period for, 67
    United States rules for, 66
Pathogens (see microbes, pathogenic)
Pears, 191
Peas, 175, 191
Peat, 139, 268
Pectinases, 153
Pectins, 153
Penalties, 82
Penicillin, 123-124, 127, 145-147, 154
    acylase, 154
Penicillium, 146
Pensions and contributions to, 80
Percolating filtration (see filtration, percolating)
Perfumes, 163, 174
Periodicals, 277
Personnel, 92
    news of, 72
Pesticides, 4, 149, 172-173, 196, 222, 266, 273

resistance to, 4, 172
Pests, 172-173, 223
  genes conferring resistance to, 68, 164
Petrochemical(s), 223
  feedstocks, 61, 157
  industry, 61, 145
Petrol (gasoline), 8, 157-158, 222, 226, 230, 232
Petroleum engineering, 72
Pets, 179, 266
Pharmaceutical preparations, 51, 57, 175, 238
Pharmaceutics, 2
Phenotypes, 40
Phenylalanine, 132
Phenylketonuria, 132, 192
Phenylpyruvic acid, 132
Pheromones, 191
Phillips Petroleum, 194
Phoenicians, 198
Phosphorus, 8, 60, 177, 229, 238
Photobioreactors, 178
Photocells, 223, 252
Photographic
  emulsion, 153
  film, 117, 119-120
Photoresponses, 260-261
Photosynthesis, 9, 38, 156, 167, 178-179, 221
Photosynthetic bacteria (see bacteria, photosynthetic)
Phytoalexins, 173
Pickled cucumbers, 185
Piezoelectric crystals, 252
Pigments, 150, 160, 174-175, 178
Pigs, 152, 180-183
  pancreases of, 66, 151
"Pill, The", 150
Plankton, 231
Planning, 91
Plant(s), 12, 40-41, 48, 61, 139-140, 148, 150, 155, 157, 223, 257
  breeding, 167-168, 170
  cells of, 166
  crop, 68, 164, 175, 177, 196, 221, 273
    improvement of, 169-173
    quality of, 163
    rotation of, 176
  diseases, 41, 267
    diagnostics for, 164, 173-174
  frost damage control in (see frost damage to plants, control of)
  genetically modified, 4, 38, 68, 164-168
    making foreign proteins in, 175
  genetics, commercial prospects for, 168, 273
  green, 9, 60-61, 221
  transgenic, 38, 196
  viral diseases of, 173
Plasma cells, 106-107
Plaster, retardants for setting of, 148
Plasticizers, 160
Plastics, 139, 160, 223-224, 235, 269
  biodegradable, 160, 171, 233
  stabilizers for, 223
Plutonium, 241
Plywood, 156
Poisons, 52, 172, 186, 224, 241
Poliomyelitis, 41, 108
Pollen, 68, 115, 196
Pollination, 171
Pollutants,
  biodegradable with difficulty, 222-223
  chemistry of, 222-224
  readily biodegradable, 222
  resistant to biodegradation, 223
Pollution, 75, 176, 219-224
  atmospheric, 220
  chemical, 219, 271
  control of, 4, 224-226, 233, 243, 270-271
  detection of, 255
  from mine drainage, 221
  in the sea, 221
  in rivers and lakes, 221
  on land, 221
Polychlorinated biphenyls, 223
Polyethylene, 161, 223
Polyhaemoglobin, 184
Polyhydroxybutyrate, 160-161
Polymerase chain reaction, 121
Polymers, 3, 10, 12, 125, 149, 152, 159, 209, 211, 239
  vinyl, 148
  water-soluble, 148, 213
Polysaccharides, 10, 105, 153, 189, 212
Polystyrene, 223
Populations,
  biological, 33, 42, 120
  genetic, 40

human, 118, 137, 176
insect, 173
microbial, 42, 44-45, 59, 141, 143, 206, 213, 227
mixed, from natural environments, 50, 147, 227-228
Potassium, 177, 237
Potatoes and potato plants, 63, 71
Poultry litter, 236
Precipitates, 54-55, 239
Predators, insect, defence against, 171-173
Pressure (mainly high), 9, 19, 46, 144, 149, 176-177, 204-209, 215, 267
Prices, 1, 62, 78, 83, 87, 92, 94, 150, 171, 176-177, 186, 198, 212, 236, 269
Pricing, policies for, 94
Private sector, 73, 77, 79, 137
investment by, 95-96
Probiotics, 164
Processing, 158, 271
of products, 56, 144
Products, 8, 19, 43-45, 49-50, 65, 77, 82-83, 85, 90-94, 100, 139, 141-144, 151, 219
agricultural, 61, 158
biotechnological, regulations for, 68
complex, oxidation of, 144
development of, 86-89
stages in, 87
drying of, 56, 178
extraction, separation and purification of, 9, 39, 50-51, 54-56, 77, 115, 139, 141-142, 144, 148, 193
high added value, 140, 144, 162
life cycle of, 89
manufacture and synthesis of, 145
packaging and distribution of, 50, 57, 142
petrochemical, 236-237
sales of, 90, 152
secondary, 59, 62
solubility in water of, 151
storage of, 4, 196, 268, 271
water-insoluble, 144
yield of, 4, 50, 53, 59, 163
Production, 60-61, 65, 84, 92, 191, 270
biological and microbial, 54, 142, 159
choosing the best organism or system for, 50-51
control of, 59

in bacteria, 46, 160
methods, protocols and technologies, 59-60, 92, 139, 141, 145, 263
efficiency of, 139, 143-144, 185-186, 196
optimizing and maximizing, 19, 46, 53-54, 59
monitoring of (*see* monitoring of production processes)
predictability of, 59
Professional societies, 93
Profits and profitability, 78, 84, 86, 90-91, 98, 158, 216
Progeny (*see* offspring)
Programmes, collaborative, 77, 95
Prokaryotes, 28, 40
Promotions, staff, 80, 85
Pronuclei and pronuclear injections, 183
Propane, 204
Propeller blades, 52
Prophylaxis, 103, 111-116, 137, 164, 264
Proppants for rock fracturing, 149
Propylene and propylene oxide, 155
Prostheses, orthopaedic, 161
Proteins, 10-12, 14-15, 19-20, 22, 24-27, 31-33, 36, 39-40, 43-44, 47, 55, 105, 110-118, 122, 124-126, 131, 151, 154, 164, 171, 177, 184, 186, 193, 240, 256-258
addressable, 258, 260-261
anti-viral, 126
antigenic, 108
chromophoric, 259
engineering of, 19, 173
fibrous and sticky, 158-160
from plants, 43
interactions between, 257, 262
spacer, 258
surface, 108
synthesis of, 26-27, 47, 59, 171
three-dimensional structures of, 11-13, 257
Proteinases (proteases), 153, 172
inhibitors of, 172
Protoplasts, 38, 166
fusion of, 38, 165-168
Protozoa, 41, 111-112, 234
Provesteen, 194
Pruteen, 193-195
Pseudoplasticity, 148

Public,
  agencies (*see* governments, agencies of, national and international)
  domain, 66
    information in the, 77
  opinion, 93, 270
  policies, 1, 74, 137
  relations, 80, 84, 93
  sector, 77
    investment by, 95
Publishers, 93
Pullulan, 148
Purees, 189
Pyrethrins, 174
Pyruvic acid, 16-18

Quality and quality standards, 4, 49, 59, 169, 184, 271
Quinine, 174
Quorn, 194

Rabbits, 183
Rabies, 41, 181
Race horses, 184
Radiation, 9, 128, 133
Radio, remote reporting of biosensor information by, 271
Radioactive
  compounds, 5, 116
  elements, 4, 116, 221, 224
Radioactivity, 25, 119-120, 122
Radioimmunoassay (RIA), 116, 186
Radioisotopes, 117
Raffinose, 188
Rain, acid, 220, 241-242
Raw materials, 4, 9, 42, 163, 171, 197, 222, 268
  in the food industry, 187, 189
Reaction vessels (*see also* fermentation vats [fermenters])
Reading frame, 22, 25, 36, 39
Receptor proteins, 226
  interactions of, 105, 124-125
  specific sites on, 32, 124, 126
Recognition, 43, 55
  sites and interactions, 32-34, 47, 155, 257
Recombinant DNA technology (*see* genetic engineering)
Recombination, genetic, 36
Recycling of materials,
  from industrial processing, 157, 202, 224
  in natural environments, 139
Redundancy, genetic, 43
Refrigeration, 186
Regulation, biological, 20
Regulations, official, 65, 68-69, 74, 93-94, 219, 231, 234, 255
  impact of on biotechnology businesses, 99
Regulatory,
  agencies, 49, 67-69
  approval of new products, 67-71, 87, 136, 195-196
  guidelines, 71
  legislation, 67-68
  submissions, 94
Renal cell carcinoma, 129
Rennin and rennet, 190
Reproduction, 19, 27-28, 59, 142, 160, 167
  sexual, 30
Research,
  and development (R and D), 57, 77-78, 90, 92, 94-95, 135
  basic, 77, 93
    funding for, 65
  curiosity, 77-78, 177
  frontier, 2
  groups, 77
  institutes, 65, 77-78, 93, 95, 109
  pre-competitive, 73, 77-78
  pure (*see* science, fundamental)
Reservoir engineering, 212-213
Responsibilities, 79, 92
  delegation of, 93
Retroviruses, 47, 112, 130, 133
Revenue (*see* income)
Rewards, 82, 87, 96
Rheology, 148
Rheumatism and rheumatoid arthritis, 109, 146
Riboflavin, 145
Ribonuclease, 13
Ribonucleic acid (RNA), 13, 26-27, 32, 47, 112
  viruses, 47
Rice, 63, 178
Rights, 95
  issues, 96
Rinderpest, 181

Risks and risk taking, 53, 68, 81-82, 84-85, 95-97, 99-100, 111
Rivers, 165, 219, 226, 233, 271
RNA (*see* ribonucleic acid)
Rocks, 8, 60, 149, 200, 225
    cracks and fractures in, 204
    crushing of, 201, 203, 269
    fracturing of, 149, 204, 206, 210
    impervious, 204-205
    microheterogeneity of, 212
    permeability of, 211
    porous and pores in, 149, 204-205, 210
Romania, 208
Romans, 198
Roots, 165, 167, 175, 177, 179
Rope, 163
Rose oil, 174
Rotary biological contactors, 234
Rotenone, 174
Roundworms (*see* nematodes)
Royalties, 67, 90, 109
Rubber reinforcer, 156

Saccharides, 10
Saccharin, 191
Safety, 70, 93, 136, 185
    review committees, 71
Sage, 191
Salaries, 78, 80-81, 90, 94, 101
Sales and sales staff, 79-80, 82, 84, 87, 90-92, 94, 100
Salinity, 215
Saliva, 183
Salmon, 184, 228
    furunculosis of, 181
Salt(s), 238
    common, 8, 189
Salvarsan, 146
Sand and sandstones, 149, 204, 206, 215
Sarcomas, 115
Sardinia, 246
Sauerkraut, 185
Sausages, 189
Scale-up of manufacturing, 53, 82
Scavenger cells (*see* cells, scavenger)
Science,
    culture of, 78, 81
    fundamental, 2, 72, 80, 135
    literature of, 72

    managers of, 80
    parks, 74
Scientists, 51, 72, 74, 77, 79-82, 92, 96, 98, 109
Scleroglucan, 148
Scours, 181
Screening procedures, 123-124, 147
Sea, 231, 233, 271
    water, 229, 239
    weed, 148, 189
Seasons, 167, 168
Second round funding (*see* companies, second-round funding for)
Secrecy, 68
Seeds, 179
    cultures of, 53
    producers of, 177
Selective,
    binding and attachment, 55
    breeding, 168, 182
Selenium, bioleaching of, 201
Self and non-self (recognition in immunology), 108-109, 126, 131
Self-assembly of biochemical structures, 257-259
Selling (*see* sales)
Semen, 120
Seminars, 72, 80
Sequences of units in macromolecules, 10-12, 19-20, 22, 26-27, 36, 118, 151, 155, 257-258
Sequestering of materials, 225
Serine, 44
Serum and serum proteins, 109, 184
Services, 49, 65, 77, 82, 90, 94, 100, 219
    fee-generating, 77
Sewage, 221
    treatment of, 2, 59, 145, 178, 226, 233-235, 247
        activated sludge process in, 235
        deep shaft process in, 235
Sewers, 219, 224
Sex,
    control of, 181
    hormones, 150
    linkage of genes, 31
Sexual reproduction, 30
Shareholders, 79, 91, 93, 96
Share(s), 79, 91, 96-100
    capital, 90

Index

options (*see* stock options)
prices, 91, 95, 97, 100
Shear and shear rates, 148
Sheep, 152, 181, 183
Shellfish, 228
Shikonin, 175
Short-term considerations, 3-5, 91-92, 95
Shrubs, ornamental, 163-164, 173, 267
Siblings, 118
Sick leave, 80
Silage and silage inoculants, 164, 179
Silicon chips, 256, 260
Silk and silk worms, 158-159, 163
Silver, 8, 153, 239
Single,
   antibodies using monoclonals, 109
   cells, 166-167
      preparations of, 38
Single-cell protein, 4, 63, 193-195, 267
   reasons for economic failure of, 194
Sisal, 163
Skills, 79, 89-90, 92, 98-99, 101
Skimmers for oil clean-up, 228
Skin, 158, 165, 265
   creams, 265
   grafts, artificial, 159
   rashes, 152
Slaughterhouse, material from, 66, 151-153
Sleeping sickness, 108, 111
Sludge(s), 226
   farming, 221, 226, 232
   refinery, disposal of, 232
Smallpox, 108, 111
Smoking, 103
Soap powders, 223
Social/economic effects and benefits (*see* costs and benefits, social and economic)
Soft drinks, 148, 154, 191
Soil, 41, 164, 167, 175, 223, 232
   bioremediation of, 226-227, 230
      use of waterflooding in, 227
   contaminated, 59, 224, 230
   inoculants for, 179
   texture of, 227
Solids, 54-55, 233, 250
   as feedstocks, 63
   as support matrices, 55, 115-116, 143, 152
Solubility of materials, 155, 188

Solvents, 215
   extractions with, 55, 233
   organic, 155, 160, 215
Soma, 167
Somaclonal variation, 167
"Someone skilled in the art", 66
Soot, 60
Sorghum, 63, 179
Soups, 153, 189
Souring agents, 148
South Africa, 202
Soviet Union, 193-194
Soy sauce, 145, 185
Soybean, 171, 175, 178, 194
Spain, 198, 281
Specificity, biochemical, 226, 241, 245, 249, 251, 257
Sperm, 29-31, 183
Spiders, 160
Spirits (beverage), 18
Spleen, 105, 110, 125
Spoilage and spoilage organisms, 185, 225
Spores, 42, 47, 172,
Sporozoites, 112
Staff, 86, 90
   recruitment of, 74, 80, 93
   retirement of, 85
   training of, 89
Stainless steel, 161
Stains, 153
Starch, 10, 61, 63, 152, 157, 191
   degradation of by enzymes, 63, 152
   industry, 152
Starter cultures, 185
Steam, 53, 156, 236
Stems and shoots, 165, 167
Sterilization, 46, 53, 142, 144, 172, 186
Steroids, 3, 117, 150, 187
Sticky ends, 35, 37
Stirring (*see* agitation)
Stock,
   in companies, 95
   cultures, 50
   market(s), 96-97
      analysts, 91
      quotations, 100
      rules, 96
   options, 80, 98, 101
Stopcocking, 209
Straw, 156

329

Strawberry plants, 71
Streptokinase (fibrinolysin), 123, 135, 184
Streptomycin, 147
Strontium, 241
Structures, three-dimensional, 257
Students, 74
Studentships, CASE (Cooperative Awards in Science and Engineering), 73
Sub-atomic particles, 117
Subsidies, agricultural, 158
substrates (see feedstocks)
Success, 49, 82, 96, 186
Sucralose®, 192
Sucrose, 62, 188, 191-192
Sugar(s), 4, 8, 10, 12, 15-17, 32, 43, 61, 131, 148, 151, 157, 192, 209, 239
    as microbial feedstocks, 61-62
    beet, 62, 157, 171, 192
    cane, 62, 157, 167, 191
    crystallization of, 188
    market price of, 158
Sulphides, 200, 246
Sulphates, 246
Sulphur, 49, 200-201, 204, 238, 242
    dioxide, 199, 226, 242
    in coal, problems of dealing with, 242
    oxides of, 220
    trioxide, 242
Sulphuric acid, 200, 242
Sunlight, 60, 178, 223
Suppliers of equipment and chemicals, 72, 86, 93, 150, 177,
Supplies, 5-6, 94
    seasonal fluctuations of, 148
Support staff, 80, 90
Surfactants, 3, 12, 50, 149, 153, 156, 213, 215, 223
Surgery, 122, 124, 132, 254
Sweeteners, 4, 145, 191-192
    high intensity, 192
Swine dysentery, 181
Switches, electrical, 224-225
    model of for use in biological computers, 261
Syphilis, 146
Syrups, 189

T-cells(T-lymphocytes), 105, 107-108, 112, 125-126, 129-130
Tartaric acid, 148
Tars, 222, 238, 268

Tastes, 185, 191
Tax, 137
    allowances and deductions, 95, 158
    benefits and inducements, 74, 99
    holidays, 99
    regimes, 99
Taxpayers, 139
Technical papers and reviews, 72
Technologies,
    compatibilities of, 49
    early stage, 77, 85
Technology,
    base, 98
    push, 94
    transfer, 273
Tellurium, bioleaching of, 201
Tempeh, 185
Temperature (mainly high), 9, 19, 53-54, 144, 152-154, 156-157, 169, 176-177, 179, 189, 192, 200, 204, 215, 252, 267, 273
Templates, 23-24, 26-27, 47
Terminology, 1, 40
Tertiary oil recovery (see enhanced oil recovery)
Testing of drugs (see clinical trials)
Tetanus, 108-109, 111-112
Tetracycline, 147
Tetramers, 10
Texas, 207, 214
Textbooks, 93, 275-276
Textiles, 235
    sizing and desizing of, 152
Texture and texturing agents, 153, 159, 185
Thallium, bioleaching of, 201
Thames, River, 184
Thatching materials, 158
Therapeutics and therapeutic agents, 3, 6, 93, 126, 150, 164, 264, 271, 273
    commercial prospects for in biotechnology, 265
Therapy, 104, 107, 122-135, 137
Thermistors, 252
Thickening (see viscosity)
Thiocyanate, 202
Thixotropy, 149
Threadworms (see nematodes)
Thymus gland, 105
Thyroid gland, 126
Thyroxine, 126

Index

Timing of commercial actions, 83, 94
Tin, 8
　bioleaching of, 201
Tissue(s), 4, 106, 120-121, 152, 165, 172
　cultures, 4, 130, 164-168, 171, 174-175, 184, 191
　plasminogen activator, 123, 135, 184
　transplants and grafts, 110
Titanium alloy, 161
Tithonin, 184
Tobacco, 168-169, 171-172, 175
Toiletries, 139, 255, 266
Tomato(es), 167, 171-172, 191, 268
　ketchup, 148
Tourists, 228
Townsend, Charles, 176
Toxins and toxicity, 104, 125, 129, 172, 186-187, 202, 223, 225, 231-233, 271
Toxoplasmosis, 181
Transcription, 26-27
Translation, 27
Transplants,
　interspecific (e.g. pig to human), 131
　intraspecific (e.g. human to human), 131
　of organs, 131
　　rejection responses to, 126, 131
Travel agents, 93
Trees, 63, 156, 179, 221
Trimers, 10
Trinidad, 204
Triplet groups, 37, 43
Trisaccharides, 188
Trypanosomiasis, 181
Tryptophan, 193
Tuberculosis, 108, 129, 181, 264
Tubocurarine, 174
Tumours (see cancer),
Tumour necrosis factor (TNF), 129, 150
Tungsten, 7
Turnover, 89
Twins, 105, 131
Typhoid, 108, 111, 261
Tyrosine, 44, 132

United Kingdom, 69, 73, 75, 109, 153, 178, 211, 233, 236-237, 243, 245, 281
　Department of the Environment, 237, 282
　Research Councils, 73
Ultraviolet light, 25, 121, 169
United Nations Industrial Development Organization (UNIDO), 73, 283
United States, 69, 73, 75, 136, 153, 157, 191, 194, 199, 210-211, 213, 215, 217, 219, 229, 233, 235, 237, 243, 270, 281
　Food and Drug Administration, 69, 136
　National Institutes of Health, 69, 73
　National Science Foundation, 73
Universities, 65, 72, 74, 78-81, 85, 89, 93, 95
　laboratories in, 77, 92
Uranium, 239, 241
　bioleaching of, 201-202
Urine, 116, 132, 183, 250

Vaccination, 103, 108, 134, 137
　of animals, techniques for, 180
Vaccines, 3, 41, 104, 108, 111-112, 121-122, 164, 195, 264
　against cancer, 114
　against rabies, 187
　artificial, 111
　intra-nasal administration of, 180
　oral administration of, 111, 125, 134
　production of, 61
Vaccinia (see cowpox)
Vacuum and vacuum pumps, 224
Vanadium, 239
Vanilla beans and vanillin, 174, 191
Variants and varieties, 53
Variation and variability, 118, 167,
Vascular disorders, 265
Vat and dump leaching, 201
Veal, 190
Vegetable(s), 189, 191, 268
　matter, 235-236
Vegetarians, 190, 194
Venture,
　capital funds, 97
　capital investment, 97
　capitalists, 96, 98, 100
Vinblastine, 174
Vincrystine, 174
Vinegar, 145, 148
Viral,
　diseases, 126-127
　proteins, 47, 127, 173

331

Virulence, 108
Viruses, 28, 38, 41, 47-48, 107-108, 111-113, 126-127, 167, 171, 183, 233, 255
Viscosity and thickening, 148-149, 159, 188-189, 208-209, 212-213
Vitamins, 44, 61, 145, 149, 178

Wales, 198, 237
Warrants, 95
Wastes and waste products, 219, 237
   agricultural, 219, 236
   as microbial feedstocks, 63
   biological disposal of, 226, 237, 241
   burial and entombment of, 224-225, 238, 241
   chemical treatment of, 237
   co-disposal of, 238
   disposal of at sea, 231, 233, 237
   disposal of in landfills, 220
   disposal of in mineshafts, 237
   domestic, 219, 221-222, 233
   hazardous, 237-239, 245
   hydrocarbon, disposal of, 222, 226, 228, 231-233
   industrial, 4, 49, 220-221, 224-226, 232-233, 237-241, 248, 271
   management of, 2, 4, 15, 145, 219, 224-226, 271
   municipal, 219, 238
   oily, 232
   organic, biological disposal of, 226, 237-241
   radioactive, 224, 241
   refinery, disposal of, 232
   solidification and vitrification of, 237
Water, 8, 16-17, 54, 60-62, 144, 149, 154, 159-160, 187, 202, 206, 208-213, 220, 225, 228, 230, 232, 250, 255
   coproduced in oil production, disposal of, 231
   natural, improvement of, 4
   rain and storm, 200, 233
Watercress, 191
Waxes, 215, 222
Weeds, 168, 171
West Virginia, 243
Whales, 191
Wheat, 63, 167, 177
Whooping cough, 108, 111
Wildlife, 228
Willow trees, 146
Wine, 18, 163, 185-186, 189
Wires, electrical, 255-256
Wood, 60, 156, 158, 163, 222, 225
   industries based on, 63
   pulp, 163
Wool, 60, 158, 163, 184
Working conditions and loads, 80
World food shortages (*see* foods and foodstuffs, world shortages of)
Worms, 149
Wound dressings, 159
Write-offs (*see* failures)

Xanthan, 148-149, 189
Xenografting, clinical, 131
Xenotransplantation, 131
X-rays, 25
X and y chromosomes, 31, 181

Yeasts, 15, 17, 19, 40, 46-47, 52, 54-55, 59, 61, 150, 157, 175, 186, 193, 206
Yellow fever, 108
Yields of crops and products, 163-164, 169, 171, 177, 182, 273
Yoghurt, 18, 142, 145, 148, 185, 206

Zambia, 199
Zinc, 8, 239
Zirconium, 240
Zygotes, 183